# INTRODUCTION TO MODEL THEORY

# ALGEBRA, LOGIC AND APPLICATIONS

A series edited by

R. Göbel
*Universität Gesamthochschule, Essen, Germany*
A. Macintyre
*University of Edinburgh, UK*

**Please see the back of this book for other titles in the Algebra, Logic and Applications series.**

# INTRODUCTION TO MODEL THEORY

**Philipp Rothmaler**

*Institut für Logik*
*Christian-Albrechts-Universität*
*Kiel, Germany*

Published in 2000 by
Taylor & Francis Group
270 Madison Avenue
New York, NY 10016

Published in Great Britain by
Taylor & Francis Group
2 Park Square
Milton Park, Abingdon
Oxon OX14 4RN

© 2000 by Taylor & Francis Group, LLC
Originally published in German in 1995 as Einfuhrung In Die Modelltheorie by Spektrum Akademischer
Verlag, Heidelberg. © 1995 Spektrum Akademischer Verlag, Heidelberg.

No claim to original U.S. Government works
Printed in the United States of America on acid-free paper
10 9 8 7 6 5 4 3

International Standard Book Number-10: 90-5699-313-5 (Softcover)
International Standard Book Number-13: 978-5699-313-9 (Softcover)

Library of Congress Cataloging-in-Publication Data

Taylor & Francis Group
is the Academic Division of Informa plc.

Visit the Taylor & Francis Web site at
http://www.taylorandfrancis.com

# Contents

# Preface

The purpose of this book is to give an insight into the model theory of first-order logic and its potential for algebraic applications. Acquaintance with logic—though useful—is not required. Only an undergraduate preparation in algebra (groups, rings, fields, and vector spaces) is assumed on the part of the reader.

The book grew out of a first course in model theory taught at the Christian-Albrechts-Universität in Kiel, Germany in the fall semester of 1992–93. The manuscript for the original German version (published by Spektrum Akademischer Verlag in 1995) was produced in collaboration with one of the students and Word Perfectionists, Frank Reitmaier. Translating it into English and LaTeX I enjoyed the assistance of my student Karsten Guhl. It is my great pleasure to thank them both for their enormous efforts. A number of students, colleagues, and other friends have contributed to both versions of this book with valuable comments and corrections. I am especially indebted to my students Matthias Clasen and Thomas Rohwer. Further thanks are due to Joel Agee, Andreas Baudisch, Paul Moritz Cohn, Ulrich Felgner, Wilfrid Hodges, Rahim Moosa, Arnold Oberschelp, Anand Pillay, Klaus Potthoff, Hans Röpke, Thomas Wilke and (last, but not least) Martin Ziegler. Preparing the final version of the translation I enjoyed the very pleasant hospitality of the Universitá degli Studi di Trento, Italy, and I am most grateful to Stefano Baratella for this. Finally I would like to thank the editors, Rüdiger Göbel and Angus Macintyre, for inviting me to publish a translation into this series.

The English edition differs from the German original in three ways: naturally, corrections and revisions have been made and the bibliography has been updated, second, more exercises, as well as hints and solutions to a selection of them, have been added, and finally, the dimension theory for strongly minimal theories scattered over text and exercises in the original has been made the topic of a separate (penultimate) chapter, which contains, as a particular case, Steinitz' dimension theory for algebraically closed fields.

Attention!
Kozma Prutkov (1803-1863)
*Fruits of Reflection* (Aphorism 42)

# Introduction

Model theory—like mathematical logic in general—is a relatively young field. It deals with the relationship between sets of formal sentences and their models, hence with the relationship between the syntax and semantics of a formal language. We restrict ourselves to the model theory of *first-order* logic, since this has particularly nice features. In their full generality these depend on the axiom of choice, which we apply without much ado, mostly in the equivalent form of Zorn's lemma.

The development of model theory went along with its applications to other mathematical disciplines, mainly to *algebra*, which we concentrate on. Hints to other fields of application and to the model theory of other logics can be found in the references listed at the end of the text. We try to make immediate use of introduced concepts and methods. Therefore the text does not fall into an abstract (theoretical) and a concrete (applied) part, but rather proceeds by letting these two alternate. Thus we present some non-trivial applications of the finiteness theorem (in Part II) before we turn to the central model-theoretic concepts and methods (in Part III).

The table of contents, fairly detailed as it is, may serve as a guide for the beginner when entering new and possibly unfamiliar territory. Its order reflects to a large extent the history of the subject. There are only two exceptions, Cantor's fundamental order- and set-theoretic instruments (§§7.3-6); and the proof of the finiteness (compactness) theorem in §4.3, which is not derived from Gödel's completeness theorem for first-order logic, as is usually the case, but rather uses the later developed ultraproducts. This allows us to completely avoid the necessary calculus of formal derivations and to argue only semantically.[1] Apart from this we get the following chronological picture.

---

[1] I thank Thomas Wilke for suggesting we banish formal derivations from such a course in model theory.

Part I deals with the basics that were developed in the 20s and 30s. These are the concept of structure (Ch. 1), the corresponding first-order languages (Ch. 2), as well as their connection via Tarski's concept of truth (Ch. 3).

Part II contains the fundamental finiteness theorem (Ch. 4) and first model-theoretic results of the 30s and 40s, which were obtained mainly using this theorem. It is interesting to note, however, that Malcev's deeper group-theoretic results (Ch. 6) (which for linguistic and political reasons were taken note of only later) already anticipated A. Robinson's diagrams and the method of interpretation that was developed in the 50s by Tarski, Mostowski and (R. M) Robinson for decidability theory and that today plays a central role in stability theory. Further, it is explained why the finiteness theorem is also called compactness theorem (§5.7).

Part III is dedicated to the machinery developed in the 50s and to the corresponding results about the relationship between models. One could say that this part deals with the category of models of a theory whose morphisms are the elementary maps—however we will make no further reference to category theory. We then present two important algebraic applications of this material. One is Robinson's proof of Hilbert's Nullstellensatz as a consequence of the model completeness of the theory of algebraically closed fields. The other is a theorem of Chevalley about projections of constructible sets as a consequence of quantifier elimination of the same theory, proved by Tarski and Robinson (§9.5, cf. also §9.6). General model-theoretic applications are e.g. various preservation theorems (§§6.1 and 10.2–3).

Part IV starts with work from the late 50s and early 60s, when the concept of type enriched and refined the theory considerably. This part leads more or less directly to, and culminates in Vaught's theorem saying that a countable and complete theory cannot have exactly two countable models. The theorem in itself may seem quite exotic, its proof, however, is intertwined with fundamental model-theoretic methods like saturated and atomic models, omitting types etc. (As often in mathematics, the proof is more consequential than the result.) §11.3 contains, as part of the exercises, a new definition of stability due to I. Herzog and the author.

Part V deals with two rather different applications of the material presented before. They can be studied independently.

The first of these, Ch. 14,[2] concerns the models of the so-called strongly minimal theories—a natural generalization of that of algebraically closed fields, inasmuch as it admits a similar dimension theory. In fact, Steinitz'

---

[2]This chapter is new. In the German original, some of the material was scattered over text and exercises.

well-known dimension theory for such fields is obtained as a special case of the more general model-theoretic theory here. The most recent result in the book is F. Wagner's confirmation of a conjecture of Podewski about strong minimality of certain fields, which constitutes part of the exercises of this chapter.

The last chapter, Ch. 15, is devoted to the models of a concrete theory, the complete theory of the abelian group of the integers. In passing, some stability-theoretic notions are introduced and their relevance is pointed out by referring to the recent literature.

Exercises are scattered about the text, mostly at the end of the sections. There is an appendix giving hints to some of them, and another one containing selected solutions.

A separate appendix contains the bibliography and some hints to the literature.

While a few historical comments are to be found in the text, for a more complete account the interested reader is referred to the illuminating comments at the ends of chapters in Wilfrid Hodges' *Model Theory*.

The logical structure of the text is largely linear. Only the last three chapters are more or less independent.[3] Items marked by * can be skipped. One could also skip the field-theoretic applications. However, this would be against the author's intentions, for Steinitz' theory of (algebraically closed) fields can be seen as a paradigm for the model-theoretic classification (or stability) theory of Morley and Shelah, which constitutes one of the central parts of contemporary model theory and which the interested reader may want to go on studying next.

—After all, nothing's ever said
that wasn't said before, in olden times.
Surely, therefore, you will forgive and understand
if we, though modern, follow where the ancients trod.
Listen well then, be nice and quiet, pay close attention,
and it will all be clear...

P. Terentius Afer (195–159 B.C.)
*Eunuchus* (Prologue)[4]

---

[3]Note, §11.5 is used only in Ch. 14, and one can pass to Ch. 15 directly after §12.1, cf. the chart on p. xiv.

[4]Translation from the German translation of the Latin original by Joel Agee.

# Interdependence chart

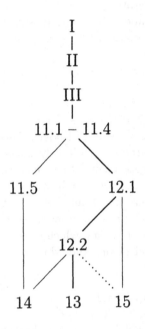

# Notation

This is a list of notation and terminology that will be assumed to be known.

| | |
|---|---|
| $X \subseteq Y, Y \supseteq X$ | $X$ is a subset of $Y$, $Y$ contains $X$ |
| $X \subset Y, Y \supset X$ | $X$ is a proper subset of $Y$, $Y$ contains $X$ properly |
| $\mathfrak{P}(Y)$ | power set of $Y$, i. e. $\{X : X \subseteq Y\}$ |
| $X \Subset Y$ | $X$ is a finite subset of $Y$ |
| $X \cup Y$ | union of $X$ and $Y$ |
| $X \sqcup Y$ | disjoint union of $X$ and $Y$, i. e., formally, $X \sqcup Y = (X \times \{0\}) \cup (Y \times \{1\})$ |
| $X \cap Y$ | intersection of $X$ and $Y$ |
| $X \smallsetminus Y$ | difference of the sets $X$ and $Y$ |
| $X \times Y$ | cartesian product of $X$ and $Y$, i. e. $\{(x,y) : x \in X \text{ and } y \in Y\}$ |
| $\lvert X \rvert$ | power (or cardinality) of $X$ |
| $\emptyset$ | empty set |
| $^Y X$ or $X^Y$ | set of all maps from $Y$ to $X$ |
| $(a_i : i \in I)$ | family indexed by $I$, i. e., formally, a function from $^I\{a_i : i \in I\}$ with $i \mapsto a_i$ (in case $I$ is well-ordered, this is called a sequence) |
| $(a_0, \dots, a_{n-1})$ | $n$-tuple, i. e. a sequence of length $n$ |
| $\bar{a}$ | tuple, i. e. a finite sequence |
| $X^n$ | set of all $n$-tuples with entries from $X$ |
| $l(\bar{a})$ | length of the tuple $\bar{a}$, i. e. $n$ if $\bar{a} \in X^n$ |
| $\bar{a}\,\hat{}\,\bar{b}$ | concatenation of the tuples $\bar{a}$ and $\bar{b}$, i. e. $(a_0, \dots, a_{n-1}, b_0, \dots, b_{m-1})$, if $\bar{a} = (a_0, \dots, a_{n-1})$ and $\bar{b} = (b_0, \dots, b_{m-1})$ |
| $f : x \mapsto y$ | $f$ maps $x$ to $y$ |
| $x \mapsto y$ | $x$ is mapped to $y$ |
| dom $f$ | domain of the map $f$ |
| $f \upharpoonright X$ | restriction of $f$ to $X \subseteq$ dom $f$ |
| $f[X]$ | image of $X \subseteq$ dom $f$ under the map $f$, i. e. $\{f(x) : x \in X\}$ |
| $f[\bar{a}]$ | $(f(a_0), \dots, f(a_{n-1}))$, where $\bar{a} = (a_0, \dots, a_{n-1})$ and $a_i \in$ dom $f$ $(i < n)$ |
| $fg$ | composite of the maps $f$ and $g$, i. e. $(fg)(x) = f(g(x))$ ('first $g$, then $f$') |
| $\mathrm{id}_X$ | identical map on a set $X$ |

| id | identical map if the domain is clear |
| $\mathbb{N}, \mathbb{Z}, \mathbb{Q}, \mathbb{R}, \mathbb{C}$ | sets of all natural numbers (including 0), all integers, all rational numbers, all real numbers, and all complex numbers, respectively |
| $\mathbb{P}$ | set of all prime numbers |
| $\mathcal{R}[x_1, \ldots, x_n]$ | ring of all polynomials in the indeterminates $x_1, \ldots, x_n$ and coefficients from $\mathcal{R}$ |
| $\mathcal{H} \lhd \mathcal{G}$ | $\mathcal{H}$ is a normal subgroup of the group $\mathcal{G}$ |
| iff | if and only if |

# Part I

# Basics

In this first part we fix our terminology concerning structures and see how languages can be used to talk about them. Anybody acquainted with the beginnings of mathematical logic will only have to leaf through this part in order to confirm terminology and notation.

# Chapter 1

# Structures

Let us first look at some examples. By specifying certain neutral elements and operations, we may view the set $\mathbb{Z}$ of integers as an additive group, as a multiplicative semigroup, or as a ring. In the three cases given these would be $(0; +)$, $(1; \cdot)$, or $(0, 1; +, \cdot)$, respectively. We could also add the inverse operation $-$ or the ordering relation $<$. It is this choice of *signature*, as we say, that determines which structure on $\mathbb{Z}$ we are dealing with.

## 1.1 Signatures

A **signature** $\sigma$ is a quadruple $(\mathbf{C}, \mathbf{F}, \mathbf{R}, \sigma')$ consisting of a set $\mathbf{C}$ of **constant symbols**, a set $\mathbf{F}$ of **function symbols**, a set $\mathbf{R}$ of **relation symbols**, and a **signature function** $\sigma' : \mathbf{F} \cup \mathbf{R} \to \mathbb{N} \setminus \{0\}$, where we assume the sets $\mathbf{C}$, $\mathbf{F}$, and $\mathbf{R}$ to be pairwise disjoint. The elements of $\mathbf{C} \cup \mathbf{F} \cup \mathbf{R}$ are also known as the **non-logical symbols**. For simplicity we often identify a signature with its set of non-logical symbols. Accordingly, by the **cardinality** or **power** of $\sigma$, in symbols $|\sigma|$, we simply mean the cardinality of the set $\mathbf{C} \cup \mathbf{F} \cup \mathbf{R}$. Unary relation symbols are also called **predicates**. A signature with $\mathbf{C} = \emptyset$, $\mathbf{F} = \emptyset$, or $\mathbf{R} = \emptyset$ is said to be **without constants**, **without functions**, or **without relations**, respectively. A signature that has neither constants nor functions, is called **(purely) relational**.

The signature function assigns to each symbol from $\mathbf{F} \cup \mathbf{R}$ its arity, i. e. $f \in \mathbf{F}$ is a $\sigma'(f)$-ary function symbol and $R \in \mathbf{R}$ is a $\sigma'(R)$-ary relation symbol. Since a constant symbol $c \in \mathbf{C}$ may be viewed as a (constant) function with unique value $c$, we may think of $c$ as a 0-place function. Accordingly the signature function can be extended to all non-logical symbols by setting $\sigma'(c) = 0$ for all $c \in \mathbf{C}$.

3

When explicitly writing down a signature we separate the sets $\mathbf{C}$, $\mathbf{F}$, and $\mathbf{R}$ by a semicolon. E. g. writing $\sigma = (0, 1; +, \cdot; <)$ and $\sigma'(+) = \sigma'(\cdot) = \sigma'(<) = 2$ fixes a signature with the constant symbols 0 and 1, two binary function symbols $+$ and $\cdot$, and a binary relation symbol $<$. If the arities are understood (e. g. by some suggestive choice of symbols as above), we omit the signature function altogether.

## 1.2   Structures

Let $\sigma = (\mathbf{C}, \mathbf{F}, \mathbf{R}, \sigma')$ be a signature.

Given a set $M$, we may give $\sigma$ a 'meaning' in $M$ by choosing elements from $M$ and functions and relations on $M$ which are to be denoted by the non-logical symbols from $\sigma$. Every such choice determines a so-called *structure* of that signature on $M$, which assigns to each constant, function, or relation symbol a well-defined interpretation in $M$ meeting the constraints given by the signature function.

A $\sigma$-**structure** $\mathcal{M}$ is a quadruple $(M, \mathbf{C}^{\mathcal{M}}, \mathbf{F}^{\mathcal{M}}, \mathbf{R}^{\mathcal{M}})$ consisting of an arbitrary set $M$ (the **underlying set** or **universe** of $\mathcal{M}$), families $\mathbf{C}^{\mathcal{M}} = (c^{\mathcal{M}} : c \in \mathbf{C})$, $\mathbf{F}^{\mathcal{M}} = (f^{\mathcal{M}} : f \in \mathbf{F})$, and $\mathbf{R}^{\mathcal{M}} = (R^{\mathcal{M}} : R \in \mathbf{R})$, where $c^{\mathcal{M}} \in M$ for all $c \in \mathbf{C}$, $f^{\mathcal{M}}$ is a $\sigma'(f)$-ary function from $M$ to $M$ for all $f \in \mathbf{F}$, and $R^{\mathcal{M}}$ is a $\sigma'(R)$-ary relation on $M$ (hence a subset of $M^{\sigma'(R)}$) for all $R \in \mathbf{R}$. The **cardinality** or **power** $|\mathcal{M}|$ of a $\sigma$-structure $\mathcal{M}$ is simply the cardinality $|M|$ of the underlying set $M$. For $P$, a non-logical symbol from $\sigma$, the object $P^{\mathcal{M}}$ is said to be the **interpretation** of $P$ in $\mathcal{M}$. Given a signature $\sigma$ without constants, $\emptyset_{\sigma}$ is used to denote the empty $\sigma$-structure.

Note that empty $\sigma$-structures exist precisely when $\mathbf{C}$ is empty, i. e. when $\sigma$ is without constants.

The notation for structures follows the guidelines fixed for signatures in the previous subsection. Further, if $R \in \mathbf{R}$ is an $n$-ary relation symbol and $(a_0, \ldots, a_{n-1}) \in R^{\mathcal{M}}$, we also write $R^{\mathcal{M}}(a_0, \ldots, a_{n-1})$ or even $\mathcal{M} \models R(a_0, \ldots, a_{n-1})$, this referring to the satisfaction relation to be defined below. In case of an $n$-place function $f$, we write $f^{\mathcal{M}}(a_0, \ldots, a_{n-1}) = b$ or $\mathcal{M} \models f(a_0, \ldots, a_{n-1}) = b$, accordingly. As usual, tuples are denoted e. g. by $\bar{a}$; writing $f(\bar{a})$ or $R(\bar{a})$ then tacitly assumes $f$ and $R$ to have the arity corresponding to the length of $\bar{a}$.

**Example.** Consider $\sigma = (0, 1; +, \cdot; <)$, where $\sigma'(+) = \sigma'(\cdot) = \sigma'(<) = 2$. Every ordered ring $\mathcal{R}$ (see §5.5 below) can be regarded as a $\sigma$-structure $(R; 0^{\mathcal{R}}, 1^{\mathcal{R}}; +^{\mathcal{R}}, \cdot^{\mathcal{R}}; <^{\mathcal{R}})$, where the non-logical symbols from $\sigma$ are interpreted by the corresponding constants, functions, and relations, respectively.

Then, for example, $\mathcal{R} \models 0 < 1$ or, equivalently, $0^{\mathcal{R}} <^{\mathcal{R}} 1^{\mathcal{R}}$.

**Exercise 1.2.1.** Find a signature appropriate for the description of vector spaces over a given field $\mathcal{K}$.

## 1.3 Homomorphisms

In order to compare two $\sigma$-structures $\mathcal{M}$ and $\mathcal{N}$ we need maps between them that preserve certain features of these structures.

A **homomorphism** from $\mathcal{M}$ to $\mathcal{N}$ is a map $h : M \to N$ satisfying

(i) $h(c^{\mathcal{M}}) = c^{\mathcal{N}}$, for all $c \in \mathbf{C}$,

(ii) $f^{\mathcal{N}}(h(a_0), \dots, h(a_{n-1})) = h(f^{\mathcal{M}}(a_0, \dots, a_{n-1}))$, for all $n \in \mathrm{N}$, all $a_0, \dots, a_{n-1} \in M$, and all $f \in \mathbf{F}$ with $\sigma'(f) = n$,

(iii) $R^{\mathcal{M}}(a_0, \dots, a_{n-1}) \Rightarrow R^{\mathcal{N}}(h(a_0), \dots, h(a_{n-1}))$, for all $n \in \mathrm{N}$, all $a_0, \dots, a_{n-1} \in M$, and all $R \in \mathbf{R}$ with $\sigma'(R) = n$.

We write $h : \mathcal{M} \to \mathcal{N}$ for short.

A homomorphism $h : \mathcal{M} \to \mathcal{N}$ is said to be **strong** if for all $n \in \mathrm{N}$, all $R \in \mathbf{R}$ with $\sigma'(R) = n$, and all $b_0, \dots, b_{n-1} \in h[M]$ with $R^{\mathcal{N}}(b_0, \dots, b_{n-1})$, there are $a_0, \dots, a_{n-1} \in M$ such that $R^{\mathcal{M}}(a_0, \dots, a_{n-1})$ and $h(a_i) = b_i$ for all $i < n$.

The structure $\mathcal{N}$ is a **(strong) homomorphic image** of the structure $\mathcal{M}$ if there is a (strong) homomorphism from $\mathcal{M}$ *onto* $\mathcal{N}$.

The difference between the notations $g : M \to N$ and $h : \mathcal{M} \to \mathcal{N}$ thus is that $g$ is merely a map between the universes (as sets), while $h$ is a homomorphism of *structures*.

Using the notation $h[(a_0, \dots, a_{n-1})] = (h(a_0), \dots, h(a_{n-1}))$ from the list before Chapter 1, we can rewrite (ii) and (iii) more concisely as follows.

(ii') $f^{\mathcal{N}}(h[\bar{a}]) = h(f^{\mathcal{M}}(\bar{a}))$ for all $f \in \mathbf{F}$ and $\bar{a} \in M^{\sigma'(f)}$,

(iii') $R^{\mathcal{M}}(\bar{a}) \Rightarrow R^{\mathcal{N}}(h[\bar{a}])$ for all $R \in \mathbf{R}$ and $\bar{a} \in M^{\sigma'(R)}$.

A **monomorphism** from $\mathcal{M}$ to $\mathcal{N}$ is, by definition, an injective and strong homomorphism, i. e. an injective homomorphism that satisfies also the inverse implication of (iii). We use $h : \mathcal{M} \hookrightarrow \mathcal{N}$ to denote that $h$ is a monomorphism from $\mathcal{M}$ to $\mathcal{N}$.

An **isomorphism** from $\mathcal{M}$ to $\mathcal{N}$ (or between $\mathcal{M}$ and $\mathcal{N}$) is, by definition, a surjective monomorphism from $\mathcal{M}$ onto $\mathcal{N}$. We write $h : \mathcal{M} \cong \mathcal{N}$ if $h$ is such an isomorphism. If $h$ fixes a set $X \subseteq M \cap N$ pointwise (i. e. $h$ extends the map $\mathrm{id}_X$), we write $h : \mathcal{M} \cong_X \mathcal{N}$ and speak of an **isomorphism over** $X$ or just an $X$-**isomorphism**. The structures $\mathcal{M}$ and $\mathcal{N}$ are said to be **isomorphic** (over $X$), and $\mathcal{N}$ is called an **isomorphic image** of $\mathcal{M}$ (over

$X$), if there is an isomorphism between $\mathcal{M}$ and $\mathcal{N}$ (over $X$). The notation $\mathcal{M} \cong \mathcal{N}$ (or $\mathcal{M} \cong_X \mathcal{N}$) is used to indicate this. An **isomorphism type** of $\sigma$-structures is an equivalence class of $\sigma$-structures modulo the equivalence relation $\cong$.

**Remark.** That $\cong$ is indeed an equivalence relation is easy to see.

Isomorphic structures have of course the same cardinality.

*Warning.* A bijective homomorphism need not be an isomorphism!

**Example.** Consider two sets $M$ and $N$ of the same power and a signature $\sigma$ that consists of a unique predicate symbol $R$ only. Turn these sets into $\sigma$-structures $\mathcal{M}$ and $\mathcal{N}$ by setting $R^{\mathcal{M}} = \emptyset$ and $R^{\mathcal{N}} = N$. Any bijection $h : M \to N$ is clearly a homomorphism but not a strong one, hence it cannot be an isomorphism.

**Remark.** In case $\mathbf{R} = \emptyset$, any bijective homomorphism is already an isomorphism, since then every homomorphism is strong.

**Lemma 1.3.1.** *Let $\mathcal{M}$ and $\mathcal{N}$ be $\sigma$-structures and let $h : M \to N$ be a bijection.*
*Then $h : \mathcal{M} \to \mathcal{N}$ and $h^{-1} : \mathcal{N} \to \mathcal{M}$ hold if and only if $h : \mathcal{M} \cong \mathcal{N}$ and $h^{-1} : \mathcal{N} \cong \mathcal{M}$ hold.*

*Proof.* $\Longrightarrow$. The homomorphic condition (iii) for $h^{-1}$ implies that $h$ is an isomorphism.
$\Longleftarrow$. Any isomorphism is a homomorphism.                                              $\square$

**Remark.** Let $\mathcal{M}$ and $\mathcal{N}$ be $\sigma$-structures.
Then $h : \mathcal{M} \to \mathcal{N}$ is an isomorphism if and only if there is $h' : \mathcal{N} \to \mathcal{M}$ such that $hh' = \mathrm{id}_N$ and $h'h = \mathrm{id}_M$.

An **endomorphism** of $\mathcal{M}$ is, by definition, a homomorphism from $\mathcal{M}$ to itself. The endomorphisms of $\mathcal{M}$ form—with respect to composition of maps—a semigroup (see §5.2 below) whose identity element is $\mathrm{id}_M$. An **automorphism** of $\mathcal{M}$ is an isomorphism of $\mathcal{M}$ onto itself. Given $X \subseteq M$, an $X$-isomorphism from $\mathcal{M}$ onto itself is called **automorphism over $X$** or just **$X$-automorphism** of $\mathcal{M}$. The automorphisms of $\mathcal{M}$ form a group with respect to composition of maps, the so-called **automorphism group** of $M$, denoted by $\mathrm{Aut}\,\mathcal{M}$. Given $X \subseteq M$, $\mathrm{Aut}_X \mathcal{M}$ denotes the subgroup formed by the $X$-automorphisms. A monomorphism $h : \mathcal{M} \hookrightarrow \mathcal{N}$ is also called **isomorphic embedding** of $\mathcal{M}$ in $\mathcal{N}$ (cf. the end of §1.4 below). In case $h \upharpoonright X = \mathrm{id}_X$ for some $X \subseteq M$, we speak of **embeddings over** $X$ or $X$-**embeddings**, denoted $h : \mathcal{M} \hookrightarrow_X \mathcal{N}$. $\mathcal{M}$ is said to be (**isomorphically**) **embeddable** (over $X$) in $\mathcal{N}$, if there is an embedding (over $X$) of $\mathcal{M}$ in $\mathcal{N}$. We then write $\mathcal{M} \hookrightarrow \mathcal{N}$ (resp. $\mathcal{M} \hookrightarrow_X \mathcal{N}$).

Thus every automorphism is a surjective and injective endomorphism, and again, the converse need not be true (check!).

As in group theory or other algebraic theories known to the reader, every homomorphic image of a structure $\mathcal{M}$ is (isomorphic to) a factor structure of $\mathcal{M}$ modulo a *congruence relation* on $\mathcal{M}$. In order to have a 1-1 correspondence between the isomorphism types of homomorphic images and factor structures, one needs to restrict oneself to strong homomorphisms (which, of course, is irrelevant if there are no relation symbols around). See §2.4 of Malcev's *Algebraic Systems* for this.

**Exercise 1.3.1.** Given $X \subseteq M$, let $\mathrm{Aut}_{\{X\}}\mathcal{M}$ be the set $\{h \in \mathrm{Aut}\,\mathcal{M} : h[X] = X\}$. Show that $\mathrm{Aut}_X\mathcal{M}$ is a normal subgroup of $\mathrm{Aut}_{\{X\}}\mathcal{M}$. What happens if, instead of $h[X] = X$, we require only $h[X] \subseteq X$?

**Exercise 1.3.2.** Find a structure with a bijective endomorphism that is not an automorphism.

**Exercise 1.3.3.** Find an infinite structure $\mathcal{M}$ with a trivial automorphism group, i. e. $\mathrm{Aut}\,\mathcal{M} = \{\mathrm{id}_M\}$.

## 1.4   Restrictions onto subsets

Often one wants to think of a subset of a given structure as a structure of the same signature, in its own right. This is possible only if the subset meets the following requirement.

Let $\mathcal{M} = (M, \mathbf{C}^{\mathcal{M}}, \mathbf{F}^{\mathcal{M}}, \mathbf{R}^{\mathcal{M}})$ be a $\sigma$-structure and $N$ a subset of $M$. We say $N$ is **closed** in $\mathcal{M}$ **under functions**[1] (from $\sigma$) if $\mathbf{C}^{\mathcal{M}} \subseteq N$ and $\mathbf{F}^{\mathcal{M}}[N] \subseteq N$ (i. e., $c^{\mathcal{M}} \in N$, for all $c \in \mathbf{C}$, and $f^{\mathcal{M}}(\bar{a}) \in N$, for all $f \in \mathbf{F}$ and all $\sigma'(f)$-tuples $\bar{a}$ in $N$). If this is the case, we can turn $N$ into a $\sigma$-structure $\mathcal{N}$ by setting $c^{\mathcal{N}} = c^{\mathcal{M}}$, $f^{\mathcal{N}}(\bar{a}) = f^{\mathcal{M}}(\bar{a})$, and $R^{\mathcal{N}}(\bar{b})$ precisely if $R^{\mathcal{M}}(\bar{b})$, for all $c \in \mathbf{C}$, all $f \in \mathbf{F}$, and all $\sigma'(f)$-tuples $\bar{a}$ from $N$, as well as all $R \in \mathbf{R}$ and all $\sigma'(R)$-tuples $\bar{b}$ from $N$.

**Remark.** If the signature of $\mathcal{M}$ is purely relational, i. e. $\mathbf{C} = \mathbf{F} = \emptyset$, *every* subset $N \subseteq M$ can be made such a uniquely determined structure $\mathcal{N}$. If the signature of $\mathcal{M}$ is without constants, i. e. $\mathbf{C} = \emptyset$, the empty set can be made such a structure (isomorphic to $\emptyset_\sigma$ from §1.2).

---

[1]This terminology is justified by the aforementioned fact that constants can be regarded as nullary functions.

For this relationship between $\mathcal{M}$ and $\mathcal{N}$ we introduce the following terminology.

$\mathcal{N}$ is called a **restriction** (or **relativisation**) of $\mathcal{M}$ onto $N$, denoted $\mathcal{M} \upharpoonright N$. The structure $\mathcal{N}$ is said to be a **substructure** of $\mathcal{M}$, if $N \subseteq M$ and $\mathcal{N}$ is the restriction of $\mathcal{M}$ onto $N$. We write $\mathcal{N} \subseteq \mathcal{M}$ for short.

$\mathcal{M}$ is said to be a **superstructure** or an **extension** of $\mathcal{N}$, if $\mathcal{N}$ is a substructure of $\mathcal{M}$. We write $\mathcal{M} \supseteq \mathcal{N}$ then.

**Remark.** The image $h[M]$ of any homomorphism $h : \mathcal{M} \to \mathcal{N}$ of $\sigma$-structures is closed under functions in $\mathcal{N}$ and thus the universe of a canonical substructure of $\mathcal{N}$. Hence such an image is itself a $\sigma$-structure and as such a homomorphic image of $\mathcal{M}$. This structure on $h[M]$ is denoted by $h(\mathcal{M})$. The homomorphism $h$ is a monomorphism if and only if it is an isomorphism between $\mathcal{M}$ and $h(\mathcal{M})$.

**Exercise 1.4.1.** Describe the difference between substructures of $\mathbb{Z}$ according to whether $\mathbb{Z}$ is considered in the signature $(0; +)$ or in the signature $(0; +, -)$.

## 1.5　Reductions onto subsignatures

We obtain a different kind of canonical structure, if, instead of shrinking the *universe*, we make the *signature* smaller and just 'forget' the interpretation of the symbols left out from the signature. As in the preceding section, we consider also the reverse process—which is no longer canonical—where we have to assign interpretations to symbols added to the signature.

Let $\sigma_0$ and $\sigma_1$ be signatures such that $\sigma_0 \subseteq \sigma_1$ (i. e. with $\mathbf{C}_0 \subseteq \mathbf{C}_1$, $\mathbf{F}_0 \subseteq \mathbf{F}_1$, $\mathbf{R}_0 \subseteq \mathbf{R}_1$, and $\sigma_0' = \sigma_1' \upharpoonright \operatorname{dom} \sigma_0'$). Then every $\sigma_1$-structure $\mathcal{M}$ can be canonically regarded as a $\sigma_0$-structure. More precisely, the **reduct** of $\mathcal{M}$ onto $\sigma_0$ (or the $\sigma_0$-**reduct** of $\mathcal{M}$) is the structure $\mathcal{M} \upharpoonright \sigma_0 =_{\mathrm{def}}$ $(M, \mathbf{C}_0^{\mathcal{M}}, \mathbf{F}_0^{\mathcal{M}}, \mathbf{R}_0^{\mathcal{M}})$ (where $\mathbf{C}_0^{\mathcal{M}} = \{c^{\mathcal{M}} : c \in \mathbf{C}_0\}$ and similarly for $\mathbf{F}_0^{\mathcal{M}}$ and $\mathbf{R}_0^{\mathcal{M}}$). Given a $\sigma_0$-structure $\mathcal{N}$ and a $\sigma_1$-structure $\mathcal{M}$, the structure $\mathcal{M}$ is said to be an **expansion** of the structure $\mathcal{N}$ to $\sigma_1$, if $\mathcal{N}$ is the reduct of $\mathcal{M}$ onto $\sigma_0$.

The relationship between these concepts is illustrated by the following.

restriction $\longleftrightarrow$ extension　　　　—　　　changing universes
reduct $\longleftrightarrow$ expansion　　　　　　—　　　changing signatures

**Exercise 1.5.1.** Given a signature $\sigma$, find a signature $\sigma_1 \supseteq \sigma$ such that all $\sigma$-structures $\mathcal{M}$ and $\mathcal{N}$ with $\mathcal{N} \subseteq \mathcal{M}$ have expansions $\mathcal{M}'$ and $\mathcal{N}'$ to $\sigma_1$ such that $\mathcal{N}' \subseteq \mathcal{M}'$ and $\operatorname{Aut} \mathcal{M}' = \operatorname{Aut}_{\{N\}} \mathcal{M}$ (cf. notation from Exercise 1.3.1).

## 1.6 Products

Here we see how one can, in a canonical way, patch various structures to-gether, provided all of them have the same signature.

Let $I$ be a nonempty set and $(\mathcal{M}_i : i \in I)$ a family of $\sigma$-structures. We define the **direct** (or **cartesian**) **product** $\mathcal{M} = \prod_{i \in I} \mathcal{M}_i$ of this family to be the following $\sigma$-structure $\mathcal{M}$. The universe $M$ of $\mathcal{M}$ is the set of all maps $a : I \to \bigcup_{i \in I} M_i$ that have the property that $a(i) \in M_i$ for all $i \in I$. We often write $a$ as $(a(i) : i \in I)$.

Given $c \in \mathbf{C}$, we let $c^{\mathcal{M}}$ be the element $a \in M$ for which $a(i) = c^{\mathcal{M}_i}$ for all $i \in I$.

Given an $n$-place function symbol $f \in \mathbf{F}$ and a tuple $\bar{a} = (a_0, \dots, a_{n-1})$ from $M$, we let $f^{\mathcal{M}}(\bar{a})$ be the element $b \in M$ such that, for all $i \in I$, we have $b(i) = f^{\mathcal{M}_i}(a_0(i), \dots, a_{n-1}(i))$.

Given an $n$-place relation symbol $R \in \mathbf{R}$ and a tuple $\bar{a} = (a_0, \dots, a_{n-1})$ from $M$, set $R^{\mathcal{M}}(\bar{a})$ in case $R^{\mathcal{M}_i}(a_0(i), \dots, a_{n-1}(i))$ holds for all $i \in I$.

For $\prod_{i < n} \mathcal{M}_i$ we also write $\mathcal{M}_0 \times \dots \times \mathcal{M}_{n-1}$. Given $i \in I$, the structure $\mathcal{M}_i$ is called the $i$th (**direct**) **factor** of the direct product $\mathcal{M}$. If $\mathcal{M}_i = \mathcal{N}$ for all $i \in I$, the product $\prod_{i \in I} \mathcal{M}_i$ is also called $I$th **direct power** of $\mathcal{N}$, in symbols $\mathcal{N}^I$.

**Remark.** The axiom of choice ensures that $\prod_{i \in I} \mathcal{M}_i \neq \emptyset$ if none of the $M_i$ is empty. (In case $I$ is finite, the axiom of choice is not necessary for this.)

In connection with the above direct product we have the following canon-ical homomorphisms.

Given $j \in I$, the map $p_j : \prod_{i \in I} M_i \to M_j$ defined by $p_j(a) = a(j)$ is, by definition, the **projection onto the $j$th factor**.

**Remark.** Any such projection $p_j$ is a homomorphism of $\mathcal{M} = \prod_{i \in I} \mathcal{M}_i$ to $\mathcal{M}_j$, which is surjective if $\mathcal{M} \neq \emptyset$.

**Exercise 1.6.1.** Show that $\mathcal{M} = \prod_{i \in I} \mathcal{M}_i$ is uncountable as soon as no $M_i$ is empty and infinitely many of the $M_i$ have at least two elements.

**Exercise 1.6.2.** Find an embedding $e : \mathcal{M} \to \mathcal{M}^I$ such that $p_i e = \mathrm{id}_M$ for all $i \in I$.

# Chapter 2

# Languages

In the first three sections of this chapter we build a formal language for each signature $\sigma = (\mathbf{C}, \mathbf{F}, \mathbf{R}, \sigma')$. More precisely, we build a *first-order* (also called an *elementary*) language $L = L(\sigma)$, whose building blocks are the symbols from a certain *alphabet*, which depends on the signature $\sigma$, and whose syntactic categories are *terms* and *formulas*. (What we here simply call a **language**, logicians also call an *object language*—as opposed to the *metalanguage*, in which the main text of this book is written and in which we usually argue.)

*Fix an arbitrary signature $\sigma = (\mathbf{C}, \mathbf{F}, \mathbf{R}, \sigma')$.*

## 2.1 Alphabets

The language $L(\sigma)$ we are going to define will consist of certain strings of symbols. The set of these symbols is called the **alphabet** of $L(\sigma)$. It consists of the following.

**Logical symbols**: the **connectives** $\neg$ [read: *not*] for **negation**, $\wedge$ [read: *and*] for **conjunction**, the **existential quantifier** $\exists$ [read: *there exists* or *there is*], and the **equality symbol** $=$ ;

countably many **variables** (see below);

the **non-logical symbols** from the signature $\sigma$ (i. e. the constant, function, and relation symbols from $\sigma$);

the **parentheses** ( and ).

(Thus alphabets can differ only in their non-logical symbols.) Some symbols may seem to be missing in the above, however we will see in due course that this choice suffices.

Although the alphabet contains a fixed (countable) set of variables, there will be no need to know their formal names. We denote them by $x_0, x_1, x_2, \ldots$ or $x, y, z$, or the like. These symbols thus serve as (meta) variables for variables from the alphabet and are formally not part of the alphabet.

## 2.2   Terms

The **terms** of signature $\sigma$ (also called $\sigma$-**terms**) are defined recursively as follows.

(i)    All variables are terms.

(ii)   All constant symbols are terms.

(iii)  If $t_0, \ldots, t_{n-1}$ are terms and $f \in \mathbf{F}$ with $\sigma'(f) = n$, then $f(t_0, \ldots, t_{n-1})$ is a term too.

(iv)   $t$ is a term, if it can be built in finitely many steps using (i)–(iii).

An example of a term in the language with two binary function symbols $f_1$ and $f_2$ is $f_1(f_1(f_2(z, f_2(x, f_2(x, y))), f_2(x, x)), f_2(x, x))$. If $f_1$ is $+$ and $f_2$ is $\cdot$, then this term can be regarded as the polynomial $(z \cdot (x \cdot (x \cdot y)) + (x \cdot x)) + (x \cdot x)$. Be aware that a term is a syntactic object with no meaning attached initially. So we cannot drop the parentheses 'assuming associativity or commutativity of the operations', for terms 'know' nothing about operations, they are just built from function symbols that will later, on the semantic level of Chapter 3, be interpreted as operations in structures. In such a structure any term will then be interpreted as an element of that structure.

Given a set $X$ of variables, the **term algebra** of $L$ over $X$ (or with **basis** $X$, is the $L$-structure $\mathrm{Term}_L(X)$ defined as follows. The domain of $\mathrm{Term}_L(X)$ is the set of all $L$-terms whose variables are in $X$. We interpret the constant and the function symbols of $L$ by themselves, i. e., $c^{\mathrm{Term}_L(X)} = c$ and $f^{\mathrm{Term}_L(X)}(t_1, \ldots, t_n) = f(t_1, \ldots, t_n)$ for each $c \in \mathbf{C}$ and $f \in \mathbf{F}$. The relational symbols of $L$ are interpreted trivially, that is we let them have empty domains, i. e., $R^{\mathrm{Term}_L(X)} = \emptyset$ for all $R \in \mathbf{R}$.

**Exercise 2.2.1.** Show that any map $h_0$ from $X$ to an $L$-structure $\mathcal{M}$ can be uniquely extended to a homomorphism $h$ from $\mathrm{Term}_L(X)$ to $\mathcal{M}$.

**Exercise 2.2.2.** Let $h_0$ and $h$ be as above and let $t(\bar{x})$ be an $L$-term whose variables $\bar{x}$ are in $X$. Prove that $t^{\mathcal{M}}(h_0[\bar{x}]) = h(t(\bar{x}))$.

**Exercise 2.2.3.** (About unique legibility) Prove that no proper initial segment of a term (regarded as a string of symbols of the alphabet) can be a term. Derive that for every term there is a unique way of building it up from its constituents according to the above recursion.

## 2.3 Formulas

Next we build formulas from terms. While terms will later (in Chapter 3) be interpreted as *elements* of structures, formulas will be interpreted as *statements* about these elements. Thus formulas will turn out to be the objects of our languages, whose interpretations have a truth value. But, as for terms, formulas themselves are syntactical objects that 'know' nothing about elements or structures.

We define **formulas** of signature $\sigma$ (or $\sigma$-**formulas**) recursively as follows.

(i)   If $t_1$ and $t_2$ are $\sigma$-terms, then $t_1 = t_2$ is a formula.

(ii)  If $t_0, \dots, t_{n-1}$ are $\sigma$-terms and $R \in \mathbf{R}$ with $\sigma'(R) = n$, then $R(t_0, \dots, t_{n-1})$ is a formula too.

(iii) If $\varphi$ and $\psi$ are formulas and $x$ is a variable, then $\neg\varphi$, $(\varphi \wedge \psi)$, and $\exists x \, \varphi$ are formulas too.

(iv)  $\varphi$ is a formula if it can be obtained from (i)—(iii) in finitely many steps.

The formulas from (i) and (ii) are said to be **atomic**. We denote the class of atomic formulas by **at**. The formulas from (i) are also known as **term equations** and those from (ii) as **relational atomic formulas**. Atomic formulas and their negations are called **literals**. The only proper **subformula** of the formulas $\neg\varphi$ and $\exists x \, \varphi$ is the formula $\varphi$, the only proper **subformulas** of $(\varphi \wedge \psi)$ are its two **conjuncts**, $\varphi$ and $\psi$.

Sometimes we are interested only in atomic formulas in which no variables have been replaced by terms other than variables. These are called **unnested** and defined formally as follows. Variables $x_i$ and constant symbols $c \in \mathbf{C}$ are **unnested terms**, as well as terms of the form $f(x_0, \dots, x_n)$, where the $x_i$ are variables (and $f \in \mathbf{F}$). No other terms are unnested. An atomic formula is said to be **unnested** if it is either an unnested term equation (i. e. an equation of unnested terms) or else a relational formula of the form $R(x_0, \dots, x_{n-1})$, where $R \in \mathbf{R}$ and the $x_i$ are variables.

An example of an atomic formula in $L(+, \cdot)$ is $(z \cdot (x \cdot (x \cdot y))) + y = (z \cdot (x \cdot (x \cdot y)) + (x \cdot x)) + (x \cdot x)$; it is not unnested.

The terms and formulas of the signature $\sigma$ together form the **expressions** of $L(\sigma)$ (or $L(\sigma)$-**expressions**). Formally we define the **language** $L(\sigma)$ to be the set of all $\sigma$-formulas. All concepts from Chapter 1 corresponding to signatures (like **reducts**, signatures **without constants**, **purely relational** signatures etc.) can thus be applied to languages as well.

Note that all the expressions of the language are *finite* strings of symbols, hence one can check in finitely many steps and effectively if a string of symbols is a term or a formula of the given language.

The correspondence between signatures and languages being one-to-one (in fact, uniquely determined by the set of non-logical symbols), we may write $\sigma = \sigma(L)$ instead of $L = L(\sigma)$.

Given a language $L$ without constants, $\emptyset_L$ denotes the empty $L$-structure (cf. §1.2).

**Exercise 2.3.1.** Verify that there are only finitely many unnested atomic sentences in $L$, provided the signature of $L$ is finite.

**Exercise 2.3.2.** (About unique legibility)
Prove that no proper initial segment of a formula (regarded as a string of symbols of the alphabet) can be a formula. Derive that for every formula there is a unique way of building it up from its constituents according to the above recursion.

## 2.4   Abbreviations

We now introduce some common notation as abbreviations, like the **nullary connectives** $\top$ [read: *true* or *verum*] and $\bot$ [read: *false* or *falsum*], the binary connectives $\vee$ [read: *or*] for the **disjunction**, $\rightarrow$ [read: *if, then*] for the **implication** or **subjunction**, and $\leftrightarrow$ [read: *if and only if*] for the **equivalence** or **equijunction**, the many-place connectives $\bigwedge$ and $\bigvee$ for **multiple conjunction** and **disjunction**, and the **universal quantifier** $\forall$ [read: *for all*].

The formal definitions are as follows. Given terms $t_1$ and $t_2$, formulas $\varphi$ and $\psi$, and a natural number $n > 0$ we write
$t_1 \neq t_2$ for $\neg t_1 = t_2$,
$\bot$ for $\exists x\, x \neq x$,
$\top$ for $\neg \bot$,
$\varphi \vee \psi$ for $\neg(\neg\varphi \wedge \neg\psi)$,
$\varphi \rightarrow \psi$ for $\neg\varphi \vee \psi$,
$\varphi \leftrightarrow \psi$ for $(\varphi \rightarrow \psi) \wedge (\psi \rightarrow \varphi)$, $\forall x\, \varphi$ for $\neg\exists x\, \neg\varphi$,
$\exists x_0 \ldots x_{n-1}\, \varphi$ (also $\exists \bar{x}\, \varphi$ in case $\bar{x} = (x_0, \ldots, x_{n-1})$) for $\exists x_0 \ldots \exists x_{n-1}\, \varphi$,
$\bigwedge_{i<n} \varphi_i$ for $(\ldots(\varphi_0 \wedge \varphi_1) \wedge \ldots) \wedge \varphi_{n-1}$ (here $\varphi_0, \ldots, \varphi_{n-1}$ are arbitrary formulas),
$\bigvee_{i<n} \varphi_i$ for $(\ldots(\varphi_0 \vee \varphi_1) \vee \ldots) \vee \varphi_{n-1}$,
$\exists^{\geq n} x\, \varphi$ for $\exists x_0 \ldots x_{n-1}(\bigwedge_{i<j<n} x_i \neq x_j \wedge \bigwedge_{i<n} \varphi(x_i))$,
$\exists^{\leq n-1} x\, \varphi$ for $\neg\exists^{\geq n} x\, \varphi$,
$\exists^{=n} x\, \varphi$ for $\exists^{\geq n} x\, \varphi \wedge \exists^{\leq n} x\, \varphi$.
Sometimes we also write $\exists^{>n}$ for $\exists^{\geq n+1}$ (similarly for $\exists^{<n}$) and $\exists!$ for $\exists^{=1}$.

The **disjuncts** of a disjunction $\varphi \vee \psi$ are the **subformulas** $\varphi$ and $\psi$. The **subformulas** of an implication $\varphi \to \psi$ are the **premise** $\varphi$ and the **conclusion** $\psi$.

Parentheses are used in formulas (and terms) to indicate their syntactic structure. However, in order to avoid too many, we adopt the rule that the unary connective $\neg$ binds more strongly than the binary connectives $\wedge$ and $\vee$, and the latter bind more strongly than $\to$ and $\leftrightarrow$. Further, outer parentheses around formulas may also be omitted (as we have already done in the above abbreviations). On the other hand, parentheses may be added if this serves their readability.

Note that the abbreviations introduced are not formally part of the language. This has the advantage of avoiding many cases when doing proofs by induction on the complexity of a formula, in which case we have to deal only with $\neg$, $\wedge$, and $\exists$. However, we do use the other connectives and quantifier for the following syntactical classification of formulas.

Let $\Sigma$ be a set of formulas. A **boolean combination** of formulas from $\Sigma$ is, by definition, a formula that can be obtained from formulas from $\Sigma$ by using $\vee$, $\wedge$ and $\neg$ only. (Obviously, we may also allow $\to$ and $\leftrightarrow$, and we could do without $\vee$.) A **positive boolean combination** of formulas from $\Sigma$ is a formula that can be obtained from formulas from $\Sigma$ by using only $\wedge$ and $\vee$. The **boolean closure** of $\Sigma$ is the set of all boolean combinations of formulas from $\Sigma$, denoted by $\widetilde{\Delta}$.

A formula is **positive** if it can be obtained from atomic formulas using only $\wedge$, $\vee$, $\exists$, and $\forall$. The class of all positive formulas (of all possible languages) is denoted by $+$.

A **negative** formula is a negated positive formula. The class of all such is denoted by $-$.

A formula is **quantifier-free** if it contains no quantifiers, where, for technical reasons, we assume $\top$ and $\bot$ to be quantifier-free too.[1]

The class of all quantifier-free formulas (of arbitrary signature) is denoted by **qf**.

Thus **qf** is the class of all boolean combinations of atomic formulas. The class of all positive formulas from **qf** is the class of all positive boolean combinations of atomic formulas.

---

[1]This is relevant only in case of quantifier-free *sentences* in languages without constants, cf. Remark (3) in §3.3.

## 2.5   Free and bound variables

A main ingredient of our formal language are the placeholders for elements
of a structure—the variables. (Note that the term *first-order* indicates that
only variables for elements occur, as opposed to *second-order logic*, which
also has variables for sets of elements.) They allow us, as common in mathematics, to formulate in our formal language relations between elements
without naming them concretely. Accordingly, we have to distinguish between two different kinds of occurrences of variables in a formula—an occurrence as such a placeholder and an occurrence as an operator variable for
a quantifier.

More formally, we make the following definition. In the formula $\exists x\,\varphi$,
the subformula $\varphi$ is said to be the **scope** of the quantifier. The occurrence
of $x$ after the quantifier is $x$'s occurrence as **operator   variable**. This
occurrence as well as each occurrence of $x$ in the scope of the quantifier
is a **bound occurrence** of this variable, while any occurrence that is not
bound is said to be a **free occurrence** of this variable. A **free variable**
of a formula is, by definition, a variable that has a free occurrence in this
formula.

**Example.** All occurrences of $x$ in the formula $\forall x(x = y \vee \exists y(x \neq y))$ are
bound, while the first of the occurrences of $y$ is free and the other two are
bound. Hence $y$ is the only free variable of this formula.

A particular role play expressions without free variables. A term $t$ is
said to be **constant** if it contains no variables at all. A formula $\varphi$ is said to
be a **sentence** if it contains no free variables.

**Remark.**
(1)  (i)    Every constant symbol is a constant term.
     (ii)   If $t_0, \ldots, t_{n-1}$ are constant terms and $f \in \mathbf{F}$ with $\sigma'(f) = n$, then
            $f(t_0, \ldots, t_{n-1})$ is also a constant term.
     (iii)  Obviously, a term $t$ is constant iff it can be obtained in finitely
            many steps from (i) and (ii).
(2)  Atomic sentences, i. e. atomic formulas that are sentences, are ei-
     ther equations of constant terms or relational sentences of the form
     $R(t_0, \ldots, t_{n-1})$, where the $t_i$ are constant terms.
(3)  Languages without constants thus have no atomic *sentences*. According
     to our convention about $\top$ and $\bot$ (from §2.4) the only quantifier-free
     sentences in this case are $\top$ and $\bot$ and their boolean combinations.

For technical reasons we introduce the following division among formulas of a given language $L$. Given a tuple $\bar{x}$ of variables, $L_{\bar{x}}$ is to denote the set of $L$-formulas, whose free variable are among the variables from $\bar{x}$. Further, $L_n$ is used to denote the collection of all $L$-formulas that have *precisely* $n$ free variables (no matter which). For $\bigcup_{k<n} L_k$ we also write $L_{\leq n}$.

Thus, $L_0$ is the set of $L$-sentences, and $L = \bigcup_{n \in \mathbb{N}} L_n$. Further, $L_{\bar{x}}$ is the set of all $L$-formulas whose free variables are from $\bar{x}$, while $L_{l(\bar{x})}$ contains only those formulas from $L_{\bar{x}}$ in which *all* entries of $\bar{x}$ occur free (however, $L_{l(\bar{x})}$ also contains all other $L$-formulas with precisely $l(\bar{x})$ free variables). If, for instance, $\bar{x} = (x_0, \ldots, x_7)$, then $x_0 = x_1$ is in $L_{\bar{x}} \subseteq L_{\leq 8}$, but $x_7 = x_{27}$ is not, while for arbitrary variables $x$ and $y$ the formula $x = y$ is in $L_2$, but not in $L_8$. This notational difference is rather technical, for one can always add so-called *dummy variables*, as we will see in §3.2 below.

The **cardinality** or **power** of the language $L$ is, by definition, the cardinal[2] number $|L|$. In §7.6 we will see that $|L| = |L_n| = |L_0| = \max\{\aleph_0, |\sigma|\} = \max\{\aleph_0, |\mathbf{C} \cup \mathbf{F} \cup \mathbf{R}|\}$. Hence a language is countably infinite[3] precisely if the set of symbols from the signature is countable (that is, finite or countably infinite).

**Exercise 2.5.1.** Find a recursive definition of free variable that is according to the syntactical complexity of the formula under consideration.

## 2.6 Substitutions

An important syntactic operation on expressions of the language is that of substitution of variables by terms. For example, in the language $L(+, \cdot)$, substituting $x$ by $z \cdot z$ in the term $x + z$ we obtain the term $(z \cdot z) + z$. Similarly in polynomial equations (which are formulas in that language). Care has to be taken only in the case of bound variables, e. g., substituting $x$ by $z \cdot z$ in the formula $\exists z(x + z = 0)$ would bring the variable $z$ of the term $z \cdot z$ under the scope of $\exists z$. This phenomenon is called **collision of variables**. Renaming is a remedy for this: in the above example replace first $z$ by $y$, say, which yields $\exists y(x + y = 0)$, and only then substitute. The resulting formula would then be $\exists y((z \cdot z) + y = 0)$.

The reader will have no great difficulty to understand how to perform this renaming of bound variables in general (but note, the result is not unique,

---

[2]We introduce cardinals only in Chapter 6. For the time being it suffices to replace every statement about cardinals by a statement about sets having the same power (which is defined, as usual, by the existence of a bijection).

[3]Every language being infinite anyway (due to infinite supply of variables at the least), we may simply say 'countable' here.

for it depends on the choice of new variable used; in a more syntactically aware treatment not necessary here one could make this process determined by choosing always the 'first' variable not yet used). We therefore allow ourselves a certain laxness in what we define next.

Let $x_0, \ldots, x_{n-1}$ be pairwise *distinct* variables and $t_0, \ldots, t_{n-1}$ arbitrary $L$-terms. A **substitution** of $x_i$ by $t_i$ $(i < n)$ in a given $L$-term $t$ consists in replacing in $t$ all occurrences of $x_i$ by $t_i$ (for all $i < n$). The resulting (uniquely determined) term will be denoted by $t_{x_0 \ldots x_{n-1}}(t_0, \ldots, t_{n-1})$. We may simply write $t(t_0, \ldots, t_{n-1})$, but only under the proviso that all variables occurring in $t$ are among $x_0, \ldots, x_{n-1}$. When using this notation we assume that an index $(x_0, \ldots, x_{n-1})$ is tacitly given, and *we adopt the rule that this index is, for a given $t$, always the same*. A **substitution** of $x_i$ by $t_i$ $(i < n)$ in a given $L$-formula $\varphi$ consists in simultaneously replacing in $\varphi$ all *free* occurrences of $x_i$ by $t_i$ $(i < n)$, where, if necessary, one first has to rename variables so that no variable occurring in $t_i$ gets under the scope of a quantifier in $\varphi$. Ignoring the fact that this process of renaming bound variables is not unique, we denote 'the' resulting formula by $\varphi_{x_0 \ldots x_{n-1}}(t_0, \ldots, t_{n-1})$. Similarly to the case of terms, we may simply write $\varphi(t_0, \ldots, t_{n-1})$—provided all free variables of $\varphi$ occur among $x_0, \ldots, x_{n-1}$. As before, an index $\bar{x}$ is assumed to be tacitly given, and *we adopt the rule that this index, for a given $\varphi$, is always the same*.

Let us emphasize that in this notation the $x_i$ are always pairwise distinct. (This applies only to the index $\bar{x}$: in $\varphi_{\bar{x}}(\bar{y})$, for instance, the entries in $\bar{y}$ may all coincide.)

The notation just introduced can also serve to indicate certain variables occurring in a formula. Namely, $\varphi_{\bar{x}}(\bar{x})$ is certainly just $\varphi$ itself, and we may write $\varphi(\bar{x})$ instead of $\varphi_{\bar{x}}(\bar{x})$ precisely if *all* free variables of $\varphi$ occur in $\bar{x}$. Thus the notation $\varphi(\bar{x})$ for $\varphi$ indicates that all free variables of $\varphi$ occur in $\bar{x}$ (however, more variables may occur in $\bar{x}$, be it bound variables of $\varphi$ or those which do not occur in $\varphi$ at all).

Given two terms $t_1$ and $t_2$ (or two formulas $\psi_1$ and $\psi_2$), we can choose a tuple of variables, $\bar{x}$, long enough so that $t_1 = t_1(\bar{x})$ and $t_2 = t_2(\bar{x})$ (or $\psi_1 = \psi_1(\bar{x})$ and $\psi_2 = \psi_2(\bar{x})$, respectively), even if the variables *actually* occurring in $t_1$ and $t_2$ (or $\psi_1$ and $\psi_2$) are quite different. This turns out handy when variables have to be exhibited in general term equations or conjunctions.

Furthermore, given a formula $\varphi$ of the form $\exists x\, \theta$ from $L_n$, we can find an $n$-tuple $\bar{x}$ (of pairwise distinct) variables $x_0, \ldots, x_{n-1}$ (all different from $x$) such that $\varphi = \varphi(\bar{x})$ and the free variables of $\theta$ occur among $x_0, \ldots, x_{n-1}, x$, and hence such that we may write $\varphi = \varphi(\bar{x}) = \exists x\, \theta(\bar{x}, x)$.

As soon as the semantics is understood, syntactic oversubtleties of the sort above will become rather obvious.

**Exercise 2.6.1.** What does the notation $\varphi(y, x)$ mean for a given $\varphi(x, y) \in L_2$?

## 2.7 The language of pure identity

The smallest possible language is that of **pure identity**, $L_=$, in which no constant, function, or relation symbols occur, i. e. $L_= = \mathrm{L}(\sigma_=)$, where $\sigma_=$ denotes the signature with $\mathbf{C} = \mathbf{F} = \mathbf{R} = \emptyset$. (Remember, being a logical symbol, $=$ is always in the alphabet.)

The only terms in this language are the variables. The atomic formulas are the formulas of the form $x = y$, where $x, y$ are arbitrary variables. The language $L_=$ is thus countable. The $L_=$-structures are simply the sets ('pure sets') that have, besides equality and inequality, no other 'structure', i. e. no relations between elements given by non-logical symbols.

# Chapter 3

# Semantics

Now that we have structures and corresponding formal expressions, the crucial step is to establish a connection between the two. More precisely, we want to assign a meaning to each expression in any given structure of corresponding signature. Technically speaking, in every such structure we will *interpret* the expressions of our language in such a way that terms become elements of the structure and sentences of the language turn into statements about these elements so that we can assign a truth value to each sentence (depending only on the structure under consideration). This is straightforward if no quantifiers occur (and will constitute the initial step in our formal definition). When quantifiers do occur, however, we run into problems: we are forced to interpret also formulas containing free variables. Alfred Tarski found the right way to do this, which we will study in the first section.

[Tarski, A. : Der Wahrheitsbegriff in den formalisierten Sprachen, Studia Philosoph. 1 (1935) 261 – 405]

The extra work involving formulas with free variables pays off, as we will see in the second section. Namely, not only are we able to assign a truth value to sentences (formulas *without* free variables) in each structure, but we will see that to any formula *with* free variables, in every structure, we can naturally assign a 'solution set' of elements (or tuples, depending on the number of free variables) of that structure. It is this 'functorial' aspect of formulas that carries model theory beyond mere (logical) axiomatizability questions.

We will interpret the expressions of the language inductively according to their syntactic complexity. The initial step is built into the notion of structure, since a structure comes equipped with the interpretation of the symbols of the signature. This provides a natural interpretation of unnested

terms and unnested atomic formulas in that structure. All this requires a
little care and preparation, which we now begin.

## 3.1  Expansions by constants, truth and satisfaction

As, in general, quantification binds free variables, in the process of building a formula from atomic subformulas the number of free variables often
drops. This is the reason that for most purposes arguments or definitions
by induction cannot be performed within the class of sentences, but lead
automatically to formulas with arbitrarily many free variables. The same
applies to the (technical) concept of truth. A remedy for this is to consider
formulas as *sentences* in a certain extended language.

*For the remainder of this section fix a language $L$ of signature $\sigma = (\mathbf{C}, \mathbf{F}, \mathbf{R}, \sigma')$.*

A **new constant** (**symbol**) for $L$ is any symbol not occurring in the
alphabet of $L$. We use this terminology only in connection with the following
concept. Given a set of new constants, $C$, the **expansion** of $L$ by the new
constants $C$ (or the **expansion** of $L$ **by constants from** $C$, for short),
in symbols $L(C)$, is, by definition, the (uniquely determined) language of
signature $(\mathbf{C} \cup C, \mathbf{F}, \mathbf{R}, \sigma')$.

Most important is the following particular case. Given a set $A$ of elements of an $L$-structure $\mathcal{M}$, we choose a *new* constant for every $a \in A$, which
we denote by $\underline{a}$ (assuming this latter is not part of the alphabet), and set
$\underline{A} = \{\underline{a} : a \in A\}$. Then we define the **expansion** of $L$ **by new constants**
**for** $A$ (or the **expansion** of $L$ **by** $A$ for short) to be the expansion $L(\underline{A})$.

For an arbitrary $L$-structure $\mathcal{M}$ and an arbitrary map $f : A \to M$, we
denote by $(\mathcal{M}, f[A])$ the $L(\underline{A})$-structure in which for all $a \in A$ the constant
$\underline{a}$ is interpreted by $f(a)$. This becomes somewhat simpler when $A \subseteq M$ and
$f = \mathrm{id}_A$: the $L(\underline{A})$-structure thus obtained is denoted by $(\mathcal{M}, A)$. We omit
the underlines if no ambiguity can arise.

Instead of $L(\underline{A})_n$ we write $L_n(\underline{A})$ or even $L_n(A)$, and similarly for $L_{\bar{x}}(A)$.

Generalizing the notation $L(A)$, given any class of formulas $\Delta$ and any
set $A$ of symbols not occurring in $\Delta$, we use the notation $\Delta(A)$ for the class
of all formulas obtained from formulas from $\Delta$ by substitution of variables
by elements from $A$.

Notice, we have $(+ \cap L)(C) = + \cap L(C)$ and $(\mathbf{qf} \cap L)(C) = \mathbf{qf} \cap L(C)$
for the class $+$ of positive formulas and that of quantifier-free formulas, $\mathbf{qf}$
(similarly for any other classes of formulas, like e. g. $\forall$, $\exists$, and $\forall\exists$ introduced

below).

Now we can *interpret* $L(\underline{M})$-terms in $(\mathcal{M}, M)$. Given an $L$-structure $\mathcal{M}$ and a constant term $t$ in $L(\underline{M})$, the **value** (or the **interpretation**) of $t$ in $\mathcal{M}^* =_{\text{def}} (\mathcal{M}, M)$, in symbols $t^{\mathcal{M}^*}$, is defined as follows.

(i)   If $t$ is the constant $c \in \mathbf{C}$, then $t^{\mathcal{M}^*} = c^{\mathcal{M}}$.

(ii)  If $t$ is the constant $\underline{a}$ for some $a \in M$, then $t^{\mathcal{M}^*} = a$.

(iii) If $t$ is the term $f(t_0, \dots, t_{n-1})$, where $n \in \mathbb{N}$, $f \in \mathbf{F}$ with $\sigma'(f) = n$, and $t_0, \dots, t_{n-1}$ are constant $L(\underline{M})$-terms, then $t^{\mathcal{M}^*} = f^{\mathcal{M}}(t_0^{\mathcal{M}^*}, \dots, t_{n-1}^{\mathcal{M}^*})$.

Thus such an evaluation or interpretation of terms is a function from the set of constant $L(\underline{M})$-terms into $M$. The evaluation or interpretation of $L(\underline{M})$-sentences in $\mathcal{M}^*$ will be a function from $L_0(\underline{M})$ to the set {true, false}. Note that due to our working in the expansion $\mathcal{M}^*$ of $\mathcal{M}$ we can, indeed, proceed inductively over *sentences* only.

Given an $L$-structure $\mathcal{M}$, the **truth** of an $L(\underline{M})$-sentence $\varphi$ in $\mathcal{M}^* = (\mathcal{M}, M)$, in symbols $\mathcal{M}^* \models \varphi$, is defined as follows.

Let $R \in \mathbf{R}$ with $\sigma'(R) = n$, let $t_0, \dots, t_{n-1}$ be constant $L(\underline{M})$-terms, let $\psi, \psi_1, \psi_2 \in L_0(\underline{M})$, and let $\theta \in L_{\leq 1}(\underline{M})$. Then we set

(i)   $\mathcal{M}^* \models t_1 = t_2$, if $t_1^{\mathcal{M}^*} = t_2^{\mathcal{M}^*}$,

(ii)  $\mathcal{M}^* \models R(t_0, \dots, t_{n-1})$, if $R^{\mathcal{M}}(t_0^{\mathcal{M}^*}, \dots, t_{n-1}^{\mathcal{M}^*})$,

(iii) $\mathcal{M}^* \models \neg\psi$, if $\mathcal{M}^* \not\models \psi$, i. e. if *not* $\mathcal{M}^* \models \psi$,

(iv)  $\mathcal{M}^* \models \psi_1 \wedge \psi_2$, if $\mathcal{M}^* \models \psi_1$ *and* $\mathcal{M}^* \models \psi_2$,

(v)   $\mathcal{M}^* \models \exists x\, \theta$, if *there is* $a \in M$ such that $\mathcal{M}^* \models \theta_x(\underline{a})$.

If $\mathcal{M}^* \models \varphi$, we say that $\varphi$ is **true** in (or **holds** in or of) $\mathcal{M}^*$, or that $\mathcal{M}^*$ **satisfies** $\varphi$.

Our original concern was the truth of $L$-sentences in $L$-structures. We make a corresponding definition, more generally, for arbitrary $L$-formulas. Let $\mathcal{M}$ be an $L$-structure and $\bar{a} = (a_0, \dots, a_{n-1})$ an $n$-tuple from $M$ (we do not exclude the case $n = 0$, i. e. $\bar{a} = \emptyset$).

Given an $L$-term $t = t(x_0, \dots, x_{n-1})$, the **value** of $t$ at $\bar{a}$ in $\mathcal{M}$, in symbols $t^{\mathcal{M}}(\bar{a})$, is by definition the element $t_{\bar{x}}(\bar{a})^{\mathcal{M}^*}$, where $\mathcal{M}^*$ is as before and $t_{\bar{x}}(\bar{a})$ is the (constant) $L(\underline{M})$-term $t_{x_0 \dots x_{n-1}}(\underline{a}_0, \dots, \underline{a}_{n-1})$.

Given an $L$-formula $\varphi = \varphi(x_0, \dots, x_{n-1})$, the tuple $\bar{a} = (a_0, \dots, a_{n-1})$ is said to **satisfy** $\varphi$ in $\mathcal{M}$, in symbols $\mathcal{M} \models \varphi(a_0, \dots, a_{n-1})$ or $\mathcal{M} \models \varphi(\bar{a})$, if $\mathcal{M}^* \models \varphi_{x_0 \dots x_{n-1}}(\underline{a}_0, \dots, \underline{a}_{n-1})$. We also write $\mathcal{M}^* \models \varphi_{\bar{x}}(\bar{a})$ for the latter.

We extend this notation as follows to arbitrary sets $\Phi = \{\varphi_i(\bar{x}) : i \in I\}$ of $L$-formulas in the same free variables $\bar{x}$. We then write $\Phi = \Phi(\bar{x})$ and set $\mathcal{M} \models \Phi(\bar{a})$ in case $\mathcal{M} \models \varphi_i(\bar{a})$ for all $i \in I$. Correspondingly we say, $\bar{a}$ **satisfies** $\Phi$ in $\mathcal{M}$. In case $\Phi$ is a set of sentences, we say that it is **true** in (or **holds** in or of) $\mathcal{M}^*$, or that $\mathcal{M}^*$ **satisfies** $\Phi$.

An $L$-formula or a set of $L$-formulas is said to be **satisfiable** in $\mathcal{M}$, if some tuple satisfies it in $M$; it is said to be **satisfiable** if it is satisfiable in some *nonempty (!)* $L$-structure. (In §3.3 below we will introduce the concept of *consistency* and show that for sentences this is the same as satisfiability.)

An $L$-formula $\varphi$ is called **valid** or **true** in $\mathcal{M}$, in symbols $\mathcal{M} \models \varphi$, if *every* tuple in $\mathcal{M}$ (of matching length) satisfies $\varphi$ in $\mathcal{M}$. In case $\varphi$ is a sentence, in accordance with the above, we also say that $\varphi$ **holds** in (or of) $\mathcal{M}$. The formula $\varphi$ is called **valid** or **(logically) true**, in symbols $\models \varphi$, if it is valid in *every* nonempty (!) $L$-structure.

In order to check the validity of a formula in a given structure, one has to examine *all* tuples of 'matching length.' We exhibited a certain sloppiness in not specifying *which* matching length. We justify this in the remark below saying that the definition does not depend on the particular length. It turns out that we can take the number of actually occurring free variables for this length. In particular, in case of sentences we may work with length zero, i. e., its satisfiability is witnessed by the empty tuple, whence, in a given structure, there is no difference between the satisfiability and the validity of sentence. Consequently, a sentence is satisfiable in a nonempty structure $\mathcal{M}$ precisely if it is true in $\mathcal{M}$. Hence, a sentence has a model if and only if it is satisfiable.

**Remark.** (Coincidence Lemma for Satisfiability)

It is clear from the definition of substitution (§2.6) that for satisfaction of a formula $\varphi = \varphi(x_0, \ldots, x_{n-1})$ by a tuple $\bar{a} = (a_0, \ldots, a_{n-1})$ only those indices $i$ count for which $x_i$ occurs free in $\varphi$. That is, given tuples $\bar{x} = (x_0, \ldots, x_{n-1})$ and $\bar{y} = (y_0, \ldots, y_{m-1})$ of variables, both containing all the free variables of $\varphi$, and given tuples $\bar{a} = (a_0, \ldots, a_{n-1})$ and $\bar{b} = (b_0, \ldots, b_{m-1})$ from $M$, if $a_i = b_j$ for all those $i < n$ and $j < m$ for which $x_i$ and $y_j$ denote the same variable occurring free in $\varphi$, we have $\mathcal{M} \models \varphi_{\bar{x}}(\bar{a})$ if and only if $\mathcal{M} \models \varphi_{\bar{y}}(\bar{b})$. We leave the proof of this rather obvious fact as an exercise.

Another coincidence of interpretations is expressed in the next lemma. Even though it is rather trivial, we present a proof in order to present, once for all, a complete proof by induction on the complexity of formulas and terms. Later we will omit the rather straightforward arguments and give only hints to the less trivial steps in the induction.

**Lemma 3.1.1.** (Coincidence under Expansion) *Let $L \subseteq L'$ be languages and $\mathcal{N}$ an $L'$-structure.*

*Then for all $L$-formulas $\varphi = \varphi(\bar{x})$ and tuples $\bar{a}$ from $N$ (of matching length) we have*

$\mathcal{N} \models \varphi(\bar{a})$ *if and only if* $\mathcal{N} \restriction L \models \varphi(\bar{a})$.

*Proof.* Let $\sigma(L) = (\mathbf{C}, \mathbf{F}, \mathbf{R}, \sigma')$.

First we prove, for any $L$-term $t = t(\bar{x})$, the equation

(*) $$t^{\mathcal{N}}(\bar{a}) = t^{\mathcal{N} \restriction L}(\bar{a}).$$

(i)   If $t(\bar{x})$ is the variable $x_i$, then $t^{\mathcal{N}}(\bar{a}) = \underline{a}_i^{\mathcal{N}^*} = a_i = \underline{a}_i^{(\mathcal{N} \restriction L)^*} = t^{\mathcal{N} \restriction L}(\bar{a})$.
      (Here $^*$ stands for the corresponding expansion by constants again.)

(ii)  If $t(\bar{x})$ is the constant $c$ (from $L$), then $t^{\mathcal{N}}(\bar{a}) = c^{\mathcal{N}} = t^{\mathcal{N} \restriction L}(\bar{a})$.

(iii) Let $t_0, \dots, t_{n-1}$ be $L$-terms satisfying (*) (induction hypothesis), and
      let $f \in \mathbf{F}$ with $\sigma'(f) = n$. We have to show (*) for $t = f(t_0, \dots, t_{n-1})$.
      By induction hypothesis we have $t_i^{\mathcal{N}}(\bar{a}) = t_i^{\mathcal{N} \restriction L}(\bar{a})$ for $i < n$, and, by
      the definition of reduct, also $f^{\mathcal{N}} = f^{\mathcal{N} \restriction L}$, consequently also $t^{\mathcal{N}}(\bar{a}) = t^{\mathcal{N} \restriction L}(\bar{a})$.

The inductive proof for (*) is thus complete.

Next we prove the assertion of the lemma inductively on the complexity
of the formula $\varphi$.

(i)   If $\varphi$ is a term equation (in $L$), the assertion is immediate from (*).

(ii)  If $\varphi$ is a relational atomic $L$-formula, the assertion follows from (*) and
      the fact that $R^{\mathcal{N}} = R^{\mathcal{N} \restriction L}$ for all $R \in \mathbf{R}$ (by the definition of reduct
      again).

(iii) Suppose the formulas $\psi_0, \psi_1, \psi_2$, and $\theta$ are $L$-formulas satisfying the
      assertion (induction hypothesis). We have to show that any $\varphi$ of the
      form $\neg\psi$, $\psi_1 \wedge \psi_2$, or $\exists x\, \theta$ satisfies the assertion too.
      If $\varphi = \varphi(\bar{x}) = \neg\psi$, then $\psi = \psi(\bar{x})$, hence
      $\mathcal{N} \models \varphi(\bar{a})$ iff $\mathcal{N} \not\models \psi(\bar{a})$ iff $\mathcal{N} \restriction L \not\models \psi(\bar{a})$ iff $\mathcal{N} \restriction L \models \varphi(\bar{a})$.
      If $\varphi = \varphi(\bar{x}) = \psi_1 \wedge \psi_2$, then $\psi_1 = \psi_1(\bar{x})$ and $\psi_2 = \psi_2(\bar{x})$ and hence
      $\mathcal{N} \models \varphi(\bar{a})$ iff $\mathcal{N} \models \psi_1(\bar{a})$ *and* $\mathcal{N} \models \psi_2(\bar{a})$
      iff $\mathcal{N} \restriction L \models \psi_1(\bar{a})$ *and* $\mathcal{N} \restriction L \models \psi_2(\bar{a})$ iff $\mathcal{N} \restriction L \models \varphi(\bar{a})$.
      If, finally, $\varphi = \varphi(\bar{x})$ is of the form $\exists x\, \theta$, then $\theta = \theta(\bar{x}, x)$ and hence
      $\mathcal{N} \models \varphi(\bar{a})$ iff $\mathcal{N} \models \theta(\bar{a}, a)$ for some $a \in N$
      iff $\mathcal{N} \restriction L \models \theta(\bar{a}, a)$ for some $a \in N$ iff $\mathcal{N} \restriction L \models \varphi(\bar{a})$.

This completes the induction. □

**Remark.** (About equality.) The question may arise, why the equality symbol is not considered a relation symbol from $\mathbf{R}$, in which case the definition
of homomorphism and monomorphism would be much more elegant, for its
being a map would be already included in (iii) $a = b \implies h(a) = h(b)$ and
its being injective in the inverse implication. It should be said that there

are logical systems using this option. The drawback (or sometimes wanted effect), however, is that then sets of the same cardinality would not necessarily be isomorphic as $L_=$-structures, since then the isomorphism type of a set would depend on the interpretation of the equality relation $=$. Instead we prefer equality to be 'real' (physical) identity of elements in every structure (not subject to interpretation).

**Exercise 3.1.1.** Prove the coincidence lemma for satisfiability (from the remark before Lemma 3.1.1).

**Exercise 3.1.2.** Prove the following equivalences for all sentences $\varphi$ and $\psi$ and structures $\mathcal{M}$ (of matching signature).
(i)    $\mathcal{M} \models \varphi \lor \psi$ if and only if , $\mathcal{M} \models \varphi$ *or* $\mathcal{M} \models \psi$,
(ii)   $\mathcal{M} \models \varphi \rightarrow \psi$ if and only if, *if* $\mathcal{M} \models \varphi$ *then* $\mathcal{M} \models \psi$,
(iii)  $\mathcal{M} \models \varphi \leftrightarrow \psi$ if and only if , $\mathcal{M} \models \varphi$ *iff* $\mathcal{M} \models \psi$,
(iv)   $\mathcal{M} \models \forall x\, \varphi$ if and only if *for all* $a \in M$, $\mathcal{M} \models \varphi(a)$.

**Exercise 3.1.3.** Show that the sentences $\exists x(x = x)$ and $\forall x(x = x)$ are valid. Which of the two sentences is true in empty structures?

**Exercise 3.1.4.** For any natural number $n$, find an $L_=$-sentence which is true in an $L$-structure precisely if it has cardinality $n$.

**Exercise 3.1.5.** Let $\varphi = \varphi(x)$ be an $L$-formula, and $t = t(\bar{x})$ an $L$-term, $\mathcal{M}$ an $L$-structure, and $\bar{a}$ a tuple from $M$ (of matching length).

Prove the so-called substitution lemma, which says that $\bar{a}$ satisfies the formula $\varphi(t(\bar{x}))$ (in $\mathcal{M}$) if and only if $t^{\mathcal{M}}(\bar{a})$ satisfies $\varphi(x)$. More precisely, letting $\psi$ denote the formula $\varphi_x(t)$ (and hence $\psi = \psi(\bar{x})$), we have $\mathcal{M} \models \varphi(t^{\mathcal{M}}(\bar{a}))$ iff $\mathcal{M} \models \psi(\bar{a})$.

## 3.2   Definable sets and relations

Every $L$-formula with $n$ free variables defines an $n$-place relation in every $L$-structure. Very often in model theory it is this relation that matters rather than the actual formula it comes from. Therefore we care about the syntactic structure of a formula only if it is significant for the relation defined thereby. In particular, two formulas that define the same relation in every structure are model-theoretically the same (the technical term to be defined below is *logically equivalent*).

We fix the following terminology and notation. Let $\mathcal{M}$ be an $L$-structure. Given $\psi \in L_n$ (where $n > 0$), let $\psi(\mathcal{M})$ denote the set **defined** by $\psi$ in $\mathcal{M}$, that is, by definition, the set $\{\bar{a} \in M^n : \mathcal{M} \models \psi(\bar{a})\}$ (which is also known as the **solution set** of $\psi$ in $\mathcal{M}$). A subset $A \subseteq M^n$ is said to be **definable** (in $\mathcal{M}$), if it is defined by some $\psi \in L_n$, i. e. $A = \psi(\mathcal{M})$. An $n$-place relation on $M$ is said to be **definable** (in $\mathcal{M}$), if, as a subset of $M^n$, it is definable in $\mathcal{M}$. The thus defined set of $n$-tuples is called a definable set too. (To distinguish sets definable by 1-place formulas in $\mathcal{M}$ from the latter kind of definable set, we called them definable *sub*set of $\mathcal{M}$, as opposed to sets defined by many-place formulas, which are not subsets of $M$.)

Given $\psi(\bar{x}, \bar{y}) \in L_{n+m}$ (where $\bar{x}$ is an $n$-tuple and $\bar{y}$ an $m$-tuple of variables with no variables in common) and an $m$-tuple $\bar{c}$ from $M$, the formula $\psi(\bar{x}, \bar{c})$ from $L_n(M)$ is called an **instance** of $\psi(\bar{x}, \bar{y})$ (or just a $\psi$-**instance** if the partition of variables is understood). The set defined by it in $\mathcal{M}$, that is the set $\{\bar{a} \in M^n : \mathcal{M} \models \psi(\bar{a}, \bar{c})\}$, is denoted by $\psi(\mathcal{M}, \bar{c})$. The tuple $\bar{c}$ is also known as the **parameter** (**tuple**) of this set. Further, a set $A \subseteq M^n$ is said to be **definable with parameters** or **parametrically definable** (in $\mathcal{M}$), if there is a formula $\psi$ and an $m$-tuple $\bar{c}$ (in $M$) as above such that $\psi(\bar{x}, \bar{c})$ defines $A$ in $\mathcal{M}$.

The easiest examples of definable subsets are the empty set (use the formula $x \neq x$) and the entire structure (use the formula $x = x$). Both of them are definable without parameters. Easy examples of parametrically definable subsets are finite sets and cofinite sets (exercise!). The easiest nontrivial example of set definable by a 2-place formula is the diagonal, defined by the formula $x = y$. Less trivial examples are many curves over the field of real numbers, e. g. the parabola defined by $y = x \cdot x$ (in a signature containing $\cdot$) or, if parameters are allowed, any parabola.

The set $\psi(\mathcal{M})$ defined by a *sentence* $\psi \in L_0(M)$ in $\mathcal{M}$ is, by definition, the entire set $M$ in case $\mathcal{M} \models \psi$, and $\emptyset$ otherwise.

Note that for $n \leq m$ we can embed $L_n$ in $L_m$ by adding what is known as **redundant** or **dummy** variables. This means that we associate with any formula $\varphi = \varphi(x_0, \ldots, x_{n-1}) \in L_n$ the formula $\varphi' \in L_m$ defined as $\varphi \wedge (x_n = x_n) \wedge \ldots \wedge (x_{m-1} = x_{m-1})$. For the corresponding definable relations we then have $\varphi'(\mathcal{M}) = \varphi(\mathcal{M}) \times M^{m-n}$.

Sometimes (when we are not so much interested in the arity) we identify $\varphi$ and $\varphi'$ as above, assuming that redundant variables have been added. This usually does not lead to problems, as both formulas are logically equivalent in the sense of the next section. So, in a way, we always may assume that arities match. Just as well, all substitutions are always assumed to be matching. *In general, in all notation, lengths of tuples are assumed to be the correct ones (if necessary, after tacitly adding dummy variables).*

**Exercise 3.2.1.** Show that every finite subset of a structure $\mathcal{M}$ is parametrically definable in $\mathcal{M}$. Derive the same for every **cofinite** subset of $\mathcal{M}$ (i. e. a subset $X \subseteq M$ such that $M \setminus X$ is finite).

**Exercise 3.2.2.** Consider a given set $M$ as an $L_=$-structure $\mathcal{M}$. Describe all sets defined (with parameters) by atomic $L_=$-formulas in $\mathcal{M}$.

The next exercise shows that for many purposes it suffices to consider purely relational languages.

**Exercise 3.2.3.** Let $L$ be a language of signature $\sigma = (\mathbf{C}, \mathbf{F}, \mathbf{R}, \sigma')$. For all $c \in \mathbf{C}$, choose a new unary relation symbol (a *new predicate*) $P_c$, and for all $f \in \mathbf{F}$ with $\sigma'(f) = n$ a new $n + 1$-place relation symbol $R_f$.

Set $\mathbf{R}^* = \mathbf{R} \cup \{P_c : c \in \mathbf{C}\} \cup \{R_f : f \in \mathbf{F}\}$ and let $L^*$ be the language with non-logical symbols $\mathbf{R}^*$. Given an $L$-structure $\mathcal{M}$, let $\mathcal{M}^*$ be the $L^*$-structure with the same underlying set $M$ such that

$R^{\mathcal{M}} = R^{\mathcal{M}^*}$, for all $R \in \mathbf{R}$,

$\mathcal{M}^* \models P_c(d)$ iff $c^{\mathcal{M}} = d$, for all $c \in \mathbf{C}$,

$\mathcal{M}^* \models R_f(\bar{a}, b)$ iff $f^{\mathcal{M}}(\bar{a}) = b$, for all $f \in \mathbf{F}$.

Prove that $\mathcal{M}$ and $\mathcal{M}^*$ have the same definable sets.

## 3.3   Models and entailment

We now turn to the notion which gave the field its name.

Let $\Sigma$ be a set of $L$-sentences. We defined in §3.1 what it meant for $\Sigma$ to be true (or to hold) in an $L$-structure $\mathcal{M}$, or, equivalently, for $\mathcal{M}$ to satisfy $\Sigma$. We introduced, remember, the notation $\mathcal{M} \models \Sigma$ for this. If, in addition, $\mathcal{M}$ is not empty, we say $\mathcal{M}$ is a **model** of $\Sigma$. We use $\mathrm{Mod}_L\Sigma$, or, if the context $L$ is understood, simply $\mathrm{Mod}\,\Sigma$, to denote the class of $L$-structures that are models of $\Sigma$ (i. e. the class of all nonempty $L$-structures satisfying $\Sigma$)—the **model class** of $\Sigma$.

Clearly, $\Sigma$ can be regarded as a set of sentences in any language $L'$ extending $L$, and so $\mathrm{Mod}_{L'}\Sigma$ is defined. But note, this class is essentially different from $\mathrm{Mod}_L\Sigma$ if $L' \neq L$. For example, $\mathrm{Mod}_L\emptyset$ is the class of all nonempty $L$-structures, while $\mathrm{Mod}_{L'}\emptyset$ is the class of all nonempty $L'$-structures.

Turning to the central concept of logic, we say that the $L$-sentence $\varphi$ is a **consequence** of $\Sigma$, or that $\Sigma$ **entails** $\varphi$, in symbols[1] $\Sigma \models_L \varphi$ or just $\Sigma \models \varphi$, if every model of $\Sigma$ is also a model of $\varphi$. For $\{\psi\} \models \varphi$ we simply write $\psi \models \varphi$. We sometimes use the term **logical consequence** to emphasize that we are talking about consequences in the above technical sense—as opposed to consequences in the every day sense, that is, in the metalanguage.

---

[1] The symbol $\models$ thus stands for two different relations, that of being a model and that of entailment. Which of the two is under consideration is evident from the context: the former is a relation between structures and (sets of) sentences, while the latter is a relation between sets of sentences and sentences.

Note that in contrast to the relation of being a model, entailment does not depend on the context $L$ (this justifying the omission of the subscript); more precisely, $\Sigma \models_L \varphi$ holds if and only if $\Sigma \models_{L'} \varphi$ holds for any (or equivalently the smallest) language $L'$ containing both $\Sigma$ and $\varphi$: the direction from right to left follows immediately from the coincidence lemma (3.1.1), for the converse one needs in addition that if $L' \subset L$, every $L'$-structure can be trivially expanded to an $L$-structure (the new symbols can, in fact, be interpreted in an arbitrary way, as is easily seen). We can thus say that $\Sigma \models \varphi$ if and only if every model of $\Sigma$ is also a model of $\varphi$.

$\Sigma_L^\models$ is to denote the set of all consequences of $\Sigma$ in $L$, i. e. $\Sigma_L^\models = \{\varphi \in L_0 : \Sigma \models \varphi\}$. This set is called the **deductive closure** of $\Sigma$ in $L$ (usually we omit the subscript $L$ when it is understood.) A set of $L$-sentences $\Sigma$ is said to be **deductively closed** if $\Sigma^\models = \Sigma$. Two sets of $L$-sentences $\Sigma_0$ and $\Sigma_1$ are **equivalent** modulo a set of $L$-sentences $\Sigma$ (or just $\Sigma$**-equivalent**), in symbols $\Sigma_0 \sim_\Sigma \Sigma_1$, if $\Sigma \cup \Sigma_0$ and $\Sigma \cup \Sigma_1$ have the same deductive closure in $L$. We say (**logically equivalent**) **equivalent** instead of $\emptyset$-equivalent.

[Tarski, A. : Über einige fundamentale Begriffe der Mathematik, Comptes Rendus Séances Soc. Sci. Lettres Varsovie Cl. III **23** (1930) 22 – 29]

Clearly, $\Sigma_0$ and $\Sigma_1$ are $\Sigma$-equivalent if and only if $\Sigma \cup \Sigma_0$ and $\Sigma \cup \Sigma_1$ are logically equivalent. Further, it is not hard to see that $(\Sigma^\models)^\models = \Sigma^\models$, whence the deductive closure of $\Sigma$ is deductively closed.

Next we verify that we may write $\models \varphi$ instead of $\emptyset \models \varphi$ without causing any confusion.

**Lemma 3.3.1.** *An $L$-sentence $\varphi$ is a consequence of the empty set if and only if $\varphi$ is valid (i. e., the deductive closure $\emptyset^\models$ of $\emptyset \subseteq L_0$ is precisely the set of valid $L$-sentences).*

*Proof.* We have $\emptyset \models \varphi$ if and only if $\mathcal{M} \models \emptyset$ implies $\mathcal{M} \models \varphi$ for all nonempty $L$-structures $\mathcal{M}$. This is the case if and only if $\mathcal{M} \models \varphi$ for *all* nonempty $L$-structures $\mathcal{M}$ (as every $L$-structure is a model of the empty set of sentences), hence if and only if $\models \varphi$. $\qquad\qquad\square$

Suppose $\Sigma \subseteq L_0$, $\bar{x}$ is an $n$-tuple of variables ($n > 0$), $\Phi(\bar{x}) \subseteq L_{\bar{x}}$ and $\Psi(\bar{x}) \subseteq L_{\bar{x}}$. The sets $\Phi(\bar{x})$ and $\Psi(\bar{x})$ are called **equivalent** in the $L$-structure $\mathcal{M}$, or just $\mathcal{M}$**-equivalent**, in symbols $\Phi \sim_\mathcal{M} \Psi$, if for all $\bar{a} \in M^n$ we have $\mathcal{M} \models \Phi(\bar{a})$ if and only if $\mathcal{M} \models \Psi(\bar{a})$. They are called **equivalent** modulo $\Sigma$, or just $\Sigma$**-equivalent**, in symbols $\Phi \sim_\Sigma \Psi$, if they are $\mathcal{M}$-equivalent for all models (i. e. nonempty $L$-structures) $\mathcal{M} \models \Sigma$.

We extend these definitions to $L$-formulas $\varphi(\bar{x})$ and $\psi(\bar{x})$ in the obvious way by applying it to the sets $\{\varphi\}$ and $\{\psi\}$, so that the notations $\varphi \sim_{\mathcal{M}} \psi$ and $\varphi \sim_{\Sigma} \psi$ make sense.

The $L$-terms $t(\bar{x})$ and $s(\bar{x})$ are said to be $\mathcal{M}$- or $\Sigma$-**equivalent** if the formulas $y = t(\bar{x})$ and $y = s(\bar{x})$ are.

Instead of $\emptyset$-equivalent we say (**logically**) **equivalent** and we write $\sim$ instead of $\sim_{\emptyset}$.

As in the coincidence lemma, in this definition only those entries of $\bar{a}$ are relevant which are substituted for a free variable in $\Phi$ or $\Psi$. Thus this definition does not depend on $n$.

Clearly, $\varphi(\bar{x})$ and $\psi(\bar{x})$ are $\Sigma$-equivalent if and only if $\Sigma \cup \{\varphi\}$ and $\Sigma \cup \{\psi\}$ are logically equivalent (if and only if $\Sigma \models \forall \bar{x}(\varphi \leftrightarrow \psi)$). If these formulas have the same number of free variables, then they are $\mathcal{M}$-equivalent if and only if they define the same sets in $\mathcal{M}$. In general, they are $\mathcal{M}$-equivalent if and only if the latter is true after adding dummy variables appropriately.

It is now easy to write and verify the well-known logical laws as equivalences, e. g. $\neg(\varphi \wedge \psi) \sim \neg\varphi \vee \neg\psi$ or $(\exists x\, \varphi \wedge \exists x\, \psi) \sim \exists xy(\varphi \wedge \psi_x(y))$, in case $y$ occurs free in neither $\varphi$ nor $\psi$.

We leave it to the reader to find and prove the equivalences that make the next remarks evident.

**Remarks.**

(1)   (Disjunctive Normal Forms) Let $\Sigma$ be a set of formulas.
      Every boolean combination $\varphi$ of formulas from $\Sigma$ is logically equivalent to a formula (in the same free variables) of the form $\bigvee_{i<n} \bigwedge_{j<m} \varphi_{ij}$, a so-called **disjunctive normal form** of $\varphi$, where each $\varphi_{ij}$ is from $\Sigma$ or a negation of a formula from $\Sigma$.

(2)   Every quantifier-free formula is logically equivalent to $\top$, $\bot$, or to a formula of the form $\bigvee_{i<n} \bigwedge_{j<m} \varphi_{ij}$, where the $\varphi_{ij}$ are literals (i. e. atomic or negations of atomic formulas). If we want the same free variables to occur, then, in the case of quantifier-free *sentences* in languages without constants, we *have* to allow $\top$ and $\bot$, see next remark.

(3)   The formulas $x = x$ and $x \neq x$ are equivalent to $\top$ and $\bot$, respectively (note, the latter uses that models are never empty). If there is $c \in \mathbf{C}$ then $\top$ and $\bot$ are logically equivalent to $c = c$ and $c \neq c$, respectively. Therefore we do not need $\top$ and $\bot$ in this case (cf. previous remark).

(4)   (Prenex Normal Forms) Every formula $\psi$ is logically equivalent to a formula (with the same free variables) of the form $Q_0 x_0 \ldots Q_{n-1} x_{n-1} \varphi$, the so-called **prenex normal form** of $\psi$, where the $Q_i$ stand for quantifiers (that is $\exists$ or $\forall$) and $\varphi$ is a quantifier-free formula as in the second

remark above. (By the first half of the previous remark, we can avoid $\top$ and $\bot$ in the formula $\varphi$ by introducing an extra quantifier.) These normal forms are of course far from being unique.

Given a formula $\psi$ in prenex normal form $Q_0 x_0 \ldots Q_{n-1} x_{n-1} \varphi$, the formula $\varphi$ is also known as a (**quantifier-free**) **matrix** of $\varphi$, while the (**quantifier**) **prefix** of $\psi$ is $Q_0 x_0 \ldots Q_{n-1} x_{n-1}$. In case the quantifier $\forall$ (resp. the quantifier $\exists$) does not occur in $\psi$, this formula is called **existential** or an $\exists$-**formula** [read: e-formula] (resp. **universal** or $\forall$-**formula** [read: a-formula]). The corresponding classes of formulas are denoted by $\exists$ and $\forall$. Correspondingly, $\forall\exists$ denotes the class of all formulas of the form $\forall \bar{x} \, \varphi$, where $\varphi \in \exists$, while $\exists\forall$ denotes that of formulas $\exists \bar{x} \, \varphi$, where $\varphi \in \forall$, and so on.

**Remark.**
(5) Every positive formula is logically equivalent to a formula in prenex normal form whose matrix is a positive boolean combination of atomic formulas.

A **contradiction** in $L$ is an $L$-sentence of the form $\varphi \wedge \neg\varphi$. A set of $L$-sentences or a single such sentence is said to be **consistent** if no consequence of it (in $L$) is a contradiction, i. e., if its deductive closure contains no contradictions. Otherwise the set is said to be **inconsistent**.

Since no nonempty structure can satisfy a contradiction, the deductive closure of a contradiction is the set of *all* sentences (of the language under consideration). Conversely, if the deductive closure of a set of sentences is the set of *all* sentences (of the language under consideration), it certainly contains (many) contradictions. Consequently, a set of sentences (or a single sentence) is consistent if and only if its deductive closure is *not* the set of *all* sentences (of the language under consideration).

Similarly, no contradiction has a model, and, conversely, a set of sentences without a model obviously entails *every* sentence (of the language under consideration).

The next remarks are now immediate.

**Remarks.**
(6) Every contradiction is inconsistent and logically equivalent to $\bot$. Every valid sentence is consistent and logically equivalent to $\top$.
(7) A set of sentences is consistent if and only if it has a model.

Next we derive an important semantic property of new constants.

**Lemma 3.3.2.** (On New Constants) *Let* $\Sigma \subseteq L_0$, $\varphi \in L_n$, *and let* $\bar{c}$ *be an n-tuple of constants not occurring in L.*

*Then* $\Sigma \models_{L(\bar{c})} \varphi(\bar{c})$ *implies* $\Sigma \models_L \forall \bar{x}\, \varphi(\bar{x})$.

*Proof.* For the nontrivial direction, let the $L$-structure $\mathcal{M}$ be a model of $\Sigma$. We have to show that $\mathcal{M} \models \varphi(\bar{a})$ for all $n$-tuples $\bar{a}$ from $M$. Now, the $L$-structure $\mathcal{M}$ becomes an $L(\bar{c})$-structure $\mathcal{M}^*$ when setting $\bar{c}^{\mathcal{M}^*} = \bar{a}$, i. e., when $\mathcal{M}^* = (\mathcal{M}, \bar{a})$. By the coincidence lemma (3.1.1), $\mathcal{M} \models \Sigma$ implies $\mathcal{M}^* \models \Sigma$. Then the hypothesis yields $\mathcal{M}^* \models \varphi(\bar{c})$, which is the same as $\mathcal{M} \models \varphi(\bar{a})$.                                                                                                □

**Remark.**

(8)  Consider the recurrent special case, where $\varphi(\bar{x})$ is of the form $\psi(\bar{x}) \to \theta$ with $\theta \in L_0$. Then $\Sigma \models_{L(\bar{c})} \varphi(\bar{c})$ implies $\Sigma \models_L \exists \bar{x}\, \psi(\bar{x}) \to \theta$, since the sentences $\forall \bar{x}(\psi(\bar{x}) \to \theta)$ and $\exists \bar{x}\, \psi(\bar{x}) \to \theta$ are logically equivalent.

**Exercise 3.3.1.** (The Deduction Theorem) Given $L$-sentences $\varphi$ and $\psi$ and a set of $L$-sentences $\Sigma$, show that $\Sigma \models \varphi \to \psi$ if and only if $\Sigma \cup \{\varphi\} \models \psi$.

**Exercise 3.3.2.** Prove that a finite conjunction of existential (resp. universal) formulas is logically equivalent to an existential (resp. universal) formula. Prove the same for finite disjunctions. In which of the four statements to prove you cannot in general simply exchange the quantifiers with the logical connectives?

**Exercise 3.3.3.** Verify the assertions made in all the above remarks, in particular, prove the theorems on normal forms.

**Exercise 3.3.4.** Let $\Sigma$ be a set of $L$-sentences, and $\Phi(\bar{x})$ and $\Psi(\bar{x})$ sets of $L$-formulas in the free variables $\bar{x} = (x_0, \ldots, x_{n-1})$. Then $\Phi$ and $\Psi$ are $\Sigma$-equivalent iff for some (every) expansion $L(\bar{c})$ by an $n$-tuple of new constants $\bar{c}$, the sets of $L(\bar{c})$-sentences $\Phi(\bar{c})$ and $\Psi(\bar{c})$ are $\Sigma$-equivalent.

**Exercise 3.3.5.** Show that adding dummy variables preserves logical equivalence.

**Exercise 3.3.6.** Prove by induction on the complexity of formulas that every formula is equivalent to a formula obtained from *unnested* atomic formulas using $\neg$, $\wedge$, and $\exists$.

## 3.4  Theories and axiomatizable classes

An *L*-**theory** (or a **theory** in *L*) is a consistent and deductively closed set of *L*-sentences. The **cardinality** or **power** of an *L*-theory $T$, in symbols $|T|$, is by definition the cardinality of *L*.

Since an $L$-theory $T$ is deductively closed, it contains all valid $L$-sentences, in particular the sentence $\exists x(x = x)$. Hence, no structure satisfying a theory is empty. In other words, every such structure is a model, i. e., given a *theory* $T$, the class $\mathrm{Mod}_L T$ is exactly the class of $L$-structures satisfying $T$ (no empty structures have to be excluded).

The next remarks are immediate from what was said about consistency in the preceding section.

**Remarks.**
(1) $L_0$ is the only deductively closed and inconsistent set of $L$-sentences.
(2) Let $T$ and $T'$ be $L$-theories.
   $T' \subseteq T$ if and only if for all $L$-structures $\mathcal{M}$, if $\mathcal{M} \models T$ then $\mathcal{M} \models T'$.
   (This follows directly from the definition of $\models$ and from the fact that $\varphi \in T$ iff $T \models \varphi$.)

Given a class $\mathbf{K}$ of $L$-structures, the $L$-**theory** of $\mathbf{K}$, or simply the **theory** of $\mathbf{K}$, is, by definition, the set $\mathrm{Th}\,\mathbf{K}$ (or $\mathrm{Th}_L\mathbf{K}$) of those $L$-sentences that are true in all *nonempty* structures from $\mathbf{K}$, i. e.

$$\mathrm{Th}\,\mathbf{K} = \{\varphi \in L_0 : \mathcal{M} \models \varphi, \text{ for all nonempty } \mathcal{M} \in \mathbf{K}\}.$$

We write $\mathrm{Th}\,\mathcal{M}$ instead of $\mathrm{Th}\,\{\mathcal{M}\}$ and call this set the $L$-**theory** of $\mathcal{M}$. Note that $\mathrm{Th}\,\mathbf{K} = \mathrm{Th}_L(\mathbf{K} \smallsetminus \{\emptyset_L\})$, in particular, $\mathrm{Th}\,\emptyset_L = \mathrm{Th}_L\{\emptyset_L\} = \mathrm{Th}_L\emptyset = L_0$.

Note that in the pathological cases when $\mathbf{K} = \emptyset$ or $\mathbf{K} = \{\emptyset_L\}$, that is when $\mathrm{Th}\,\mathbf{K} = L_0$, the set $\mathrm{Th}\,\mathbf{K}$ is not a *theory* in the sense of our definition, for then it is inconsistent. However, it is easy to see that otherwise it indeed is a theory (in any case, $\mathrm{Th}\,\mathbf{K}$ contains all valid $L$-sentences). In particular, if $T$ is a theory, so is $\mathrm{Th}\,\mathrm{Mod}\,T$. Next we convince ourselves that, conversely, every theory is of this form.

**Lemma 3.4.1.** *Let $T$ be a set of sentences.*
(1) $T \subseteq \mathrm{Th}\,\mathrm{Mod}\,T$.
(2) $T^\models = \mathrm{Th}\,\mathrm{Mod}\,T$.
(3) $T$ *is deductively closed iff* $T = \mathrm{Th}\,\mathrm{Mod}\,T$.
(4) $T$ *is a theory if and only if* $T$ *is consistent and* $T = \mathrm{Th}\,\mathrm{Mod}\,T$.

*Proof.* The first assertion is immediate from the definitions. The second is immediate from the definition of consequence (and the fact that $\mathrm{Mod}\,T$ contains only models). The remaining two assertions now follow easily. $\square$

Given a theory $T$, denote by $T^\infty$ the theory of the class of all infinite models of $T$.

The simplest example of a theory is the $L_=$-theory of all sets (regarded as $L_=$-structures), which we call the **theory of pure identity** and denote by $T_=$. Clearly, $T_=$ is the set of all valid $L_=$-sentences and $T_=^\infty$ is the theory of all infinite sets.

Given an $L$-theory $T$ and an arbitrary class of formulas $\Delta$, we call $T_\Delta =_{\mathrm{def}} (T \cap \Delta)^\models$ the $\Delta$-**part** of $T$. We write $\mathrm{Th}_\Delta \mathbf{K}$ for $(\mathrm{Th}\,\mathbf{K})_\Delta$ and call this the $\Delta$-**theory** of $\mathbf{K}$. By this, notation like $T_{\mathsf{qf}}$, $T_+$, $T_-$, $T_\forall$, $T_\exists$, $T_{\forall\exists}$, etc. should be clear. The first five of these are also called **quantifier-free**, **positive**, **negative**, **universal**, and **existential part** of $T$.

We conclude this section with another basic notion. A class $\mathbf{K}$ of $L$-structures is said to be **axiomatizable** or **elementary** (in $L$) if there is a set $\Sigma \subseteq L_0$ such that $\mathbf{K}$ is the class of $L$-structures satisfying $\Sigma$. We then say that $\Sigma$ **axiomatizes** $\mathbf{K}$. In case $\Sigma$ consists of a single sentence $\varphi$, i. e. $\Sigma = \{\varphi\}$, we omit the braces and say, $\varphi$ **axiomatizes** $\mathbf{K}$. A class is said to be **finitely axiomatizable** (in $L$) if it can be axiomatized by a single $L$-sentence (or, equivalently, by a finite set of those). A theory $T$ is **axiomatized** by $\Sigma$ if $\Sigma \subseteq T \subseteq \Sigma^\models$ or, equivalently, the model classes of $T$ and $\Sigma$ are the same.

It is easily seen that the empty set axiomatizes (in $L$) the class of all $L$-structures. Further, a class $\mathbf{K}$ is axiomatized by a set $\Sigma$ if and only if the class $\mathbf{K} \cup \{\emptyset_L\}$ is axiomatized by $\{\sigma \vee \neg \exists x\, (x = x) \; : \; \sigma \in \Sigma\}$, and also if and only if $\mathbf{K} \setminus \{\emptyset_L\}$ is axiomatized by $\Sigma \cup \{\exists x\, (x = x)\}$. This allows us to exclude the empty structure easily from discussions related to axiomatizability. It also shows that a class is axiomatizable if and only if it is of the form $\mathbf{K} = \mathrm{Mod}_L \Sigma$ or $\mathbf{K} = \mathrm{Mod}_L \Sigma \cup \{\emptyset_L\}$.

Given an arbitrary class of formulas, $\Delta$, we call an $L$-theory $\Delta$-**axiomatizable** (or simply a $\Delta$-**theory**) if it can be axiomatized by a subset of $\Delta \cap L_0$.

$\forall$-theories [read: a-theories] are also called **universal** or **universally axiomatizable**. Similarly, $\exists$-theories [read: e-theories] are called **existential** or **existentially axiomatizable**. +-theories are also called **positive** or **positively axiomatizable**.

**Remarks.**

(3)  Obviously, $\Sigma$ axiomatizes $T$ if and only if $\Sigma$ and $T$ are (logically) equivalent, i. e., if and only if $\Sigma^\models = T$.

(4)  The theory $T$ is a $\Delta$-theory if and only if $T \subseteq T_\Delta$ (if and only if $T = T_\Delta$).

**Example.**  $T^\infty$ is axiomatized by $T \cup \{\exists^{\geq n} x\, (x = x) \; : \; n \in \mathbb{N}\}$.

**Lemma 3.4.2.** *A class* **K** *of L-structures is axiomatizable if and only if* **K** $\smallsetminus \{\emptyset_L\} = \text{Mod Th}\,\mathbf{K}$.

*Proof.* We saw above that **K** is axiomatizable if and only if **K** $\smallsetminus \{\emptyset_L\}$ is. So we may restrict to the case where **K** does not contain the empty structure.

Then, if **K** has the form Mod Th **K**, it clearly is axiomatized by Th **K**. Let, conversely, **K** be axiomatized by $\Sigma \subseteq L_0$. The inclusion from left to right holds for any class **K** not containing the empty structure. Since **K** = Mod$\Sigma$, clearly $\Sigma$ is contained in Th **K** and so Mod Th **K** $\subseteq$ Mod$\Sigma$ = **K**, whence the innverse inclusion also holds. $\qquad \square$

In Chapter 5 we will see a number of concrete axiomatizations (of algebraic classes).

It is advisable to understand well the dualism between axiomatizable classes (as described above) and deductively closed sets of sentences (as described in Lemma 3.4.1(3)). This is treated in a more systematic way in the last two exercises (and only for the case without empty structures).

**Exercise 3.4.1.** Prove that isomorphic $L$-structures have the same $L$-theory and hence, if $\mathcal{M}$ is isomorphic to a structure from an axiomatizable class **K**, then $\mathcal{M}$ itself is in **K**.

**Exercise 3.4.2.** Axiomatize the class of all vector spaces over a field $K$ (in the language of Exercise 1.2.1).

**Exercise 3.4.3.** (In the notation of Exercise 3.2.3.)
Let $\Sigma = \{\exists^{=1} x\, P_c(x) : c \in \mathbf{C}\} \cup \{\forall \bar{x}\, \exists^{=1} y\, R_f(\bar{x}, y) : f \in \mathbf{F}\} \subseteq L_0^*$. Prove that the class of structures satisfying $\Sigma$ is $\{\mathcal{M}^* : \mathcal{M}$ is an $L$-structure$\}$, hence Mod $\Sigma = \{\mathcal{M}^* : \mathcal{M}$ is a nonempty $L$-structure$\}$.

Let $X$ and $Y$ be any two classes. A **Galois correspondence** between $X$ and $Y$ is a pair of maps $\alpha : \mathfrak{P}(X) \to \mathfrak{P}(Y)$ and $\beta : \mathfrak{P}(Y) \to \mathfrak{P}(X)$ such that for all $X_0, X_1 \subseteq X$ and $Y_0, Y_1 \subseteq Y$, if $X_0 \subseteq X_1$ then $\alpha(X_0) \supseteq \alpha(X_1)$, if $Y_0 \subseteq Y_1$ then $\beta(Y_0) \supseteq \beta(Y_1)$, and $X_0 \subseteq \beta\alpha(X_0)$ and $Y_0 \subseteq \alpha\beta(Y_0)$. A subset $X_0 \subseteq X$ (respectively, $Y_0 \subseteq Y$) is said to be **closed** with respect to the above Galois correspondence, if $X_0 = \beta\alpha(X_0)$ (respectively, $Y_0 = \alpha\beta(Y_0)$).

**Exercise 3.4.4.** Prove that $\Sigma \mapsto \text{Mod}_L\Sigma$ and $\mathbf{K} \mapsto \text{Th}_L\mathbf{K}$ defines a Galois correspondence between $L_0$, the set of all $L$-sentences, and $\text{Mod}_L\emptyset$, the class of all nonempty $L$-structures. Derive that, given theories $S$ and $T$ (in the same language), we have $S = T$ if and only if Mod $S$ = Mod $T$.

**Exercise 3.4.5.** Show that, under the Galois correspondence of the preceding exercise, the deductively closed sets of $L$-sentences (including the inconsistent one) are exactly the closed subsets of $L_0$, while the axiomatizable classes of nonempty $L$-structures (including the empty class) are exactly the closed subsets of $\text{Mod}_L\emptyset$.

## 3.5   Complete theories

Let us consider maximal consistent sets of sentences. These clearly are
deductively closed and thus theories. In fact, it is easy to see that these are
maximal theories, that is, if $L$ is the language we are working in, theories
that are not properly contained in another $L$-theory. This leads to the
following definition.

An $L$-theory $T$ is said to be **complete** if for every $L$-sentence $\varphi$, either
$\varphi \in T$ or $\neg\varphi \in T$.

**Lemma 3.5.1.** *The following are equivalent for every L-theory T.*
(i)   *$T$ is complete.*
(ii)  *$T$ is a maximal L-theory.*
(iii) *$T$ is a maximal consistent set of L-sentences.*
(iv)  *$T = \operatorname{Th}\mathcal{M}$ for all $\mathcal{M} \models T$.*
(v)   *$T = \operatorname{Th}\mathcal{M}$ for some $\mathcal{M}$.*

*Proof.* If $T$ is complete, every $L$-sentence $\varphi$ that is consistent with $T$ must
be contained in $T$, hence $T$ is a maximal theory. Similarly, a maximal theory
is a maximal consistent set.

If $\mathcal{M}$ is a model of $T$, then $T \subseteq \operatorname{Th}\mathcal{M}$. If $T$ is in addition maximal, we
obtain equality here.

We are left with the implication from the last assertion to the first. But
this follows easily from the fact that, for every $\varphi \in L_0$, either $\varphi$ or $\neg\varphi$ is
true in $\mathcal{M}$.                                                            □

As every consistent set of sentences has a model, we readily derive

**Corollary 3.5.2.** (Lindenbaum's Theorem)
*Every consistent set of L-sentences is contained in a complete L-theory, a
so-called* **completion**.                                                     □

This assertion is also known as the Lindenbaum-Tarski theorem. Lindenbaum
was a colleague of Tarski, who was killed by the Nazis and never published this
result, cf.
[Tarski, A. : Über einige fundamentale Begriffe der Mathematik, Comptes Rendus
Séances Soc. Sci. Lettres Varsovie Cl. III **23** (1930) 22 – 29].

Not every theory is complete, e. g. the set of all valid $L$ sentences is a
theory by 3.3.1. But this is far from being complete, as we are going to
see next. Note first that there are $L$-structures of every finite cardinality,
for on *every* nonempty set one can define an $L$-structure, for instance by

giving all terms the same value and interpreting the rest of the language arbitrarily. Further, let $\varphi_n$ denote the sentence $\exists^{=n}x\,(x\,=\,x)$, which is contained in every language and says that there are precisely $n$ elements. Then, as mentioned, $\emptyset^\models \cup \{\varphi_n\}$ is consistent for all $n > 0$, while $\varphi_n \wedge \varphi_m$ is inconsistent for $n \neq m$. Hence none of the sentences $\varphi_n$ is contained in $\emptyset^\models$, whence the latter is an incomplete theory. Moreover, by Lindenbaum's theorem, it has infinitely many completions.

Later we will see less trivial examples, like the theory of all groups.

A very nontrivial example of an incomplete theory is the following theory, the so-called **Peano arithmetic**, denoted by PA.

**Example.** Consider the language $L$ of signature $(0, 1; +, \cdot)$. The Axioms of Peano arithmetic are the following $L$-sentences.
(P1)   $\forall x(x + 1 \neq 0)$,
(P2)   $\forall x(x \neq 0 \rightarrow \exists y(x = y + 1))$,
(P3)   $\forall xy(x + 1 = y + 1 \rightarrow x = y)$,
(P4)   $\forall x(x + 0 = x)$,
(P5)   $\forall xy(x + (y + 1) = (x + y) + 1)$,
(P6)   $\forall x(x \cdot 0 = 0)$,
(P7)   $\forall xy(x \cdot (y + 1) = (x \cdot y) + x)$,
and for all $\varphi \in L_1$ the axiom of induction
(P$\varphi$)   $(\varphi(0) \wedge \forall x(\varphi(x) \rightarrow \varphi(x + 1))) \rightarrow \forall x\,\varphi(x)$.

Denote by PA the deductive closure of $\{(\text{P1}), \dots, (\text{P7})\} \cup \{(\text{P}\varphi) : \varphi \in L_1\}$. This is consistent and thus a theory, for $(\text{N}; 0, 1; +, \cdot)$ is a model. This model is called the **standard model** of PA.

Gödel's first incompleteness theorem says that this theory is incomplete. More precisely, it says that there are true number-theoretic statements $\psi$ in $L$ (which means nothing more than true in $(\text{N}; 0, 1; +, \cdot)$) that are not provable in PA, i. e., PA $\not\models \psi$. Since $(\text{N}; 0, 1; +, \cdot)$ itself is a model of PA, we have on the other hand PA $\not\models \neg\psi$. Hence PA is incomplete and thus a proper subtheory of the $L$-theory $\text{Th}\,(\text{N}; 0, 1; +, \cdot)$. The latter is also called **true arithmetic** (or **complete first-order number theory**). Gödel's theorem is a deep result in mathematical logic, for which we refer the reader to the standard literature.

**Exercise 3.5.1.** An $L$-theory $T$ is complete iff $\varphi \vee \psi \in T$ implies $\varphi \in T$ or $\psi \in T$.

**Exercise 3.5.2.** Consider a language $L$ without relations. Prove that there is, up to isomorphism, precisely one $L$-structure of power 1. Does this remain true if $L$ has also relation symbols? E. g., how many $L$-structures of power 1 are there if $L$ has precisely two relation symbols both of which are unary (i. e. 1-place)?

**Exercise 3.5.3.** Find a formula defining the set of prime numbers in the standard model of Peano arithmetic.

## 3.6  Empty structures in languages without constants

As in the preceding chapter, we close with a side remark about a 'trivial' topic. We mentioned in §1.2 that there is an empty $\sigma$-structure $\emptyset_\sigma$ (or an empty $L$-structure $\emptyset_L$) if and only if $\sigma$ (or $L = L(\sigma)$) has no constants. Fix such a language $L$. Although also in $\emptyset_L$ an $L$-*sentence* is satisfiable if and only if it is true in $\emptyset_L$ (since 'there is' an empty tuple, i. e. a 0-tuple in $\emptyset$), due to the lack of tuples of positive length in $\emptyset$ *no* formula with free variables is satisfiable in $\emptyset$, while *all* such formulas are true in $\emptyset_L$. For the same reason $\forall x\, \varphi$ is true in $\emptyset_L$ for *all* $\varphi \in L$, while $\exists x\, \varphi$ is not. Further, *all* $L$-formulas are $\emptyset_L$-equivalent even though some are true in $\emptyset_L$ and some are not (note that also in $\emptyset_L$, the sentence $\top$ is true, while $\bot$ is not).

These anomalies are likely to be the reason that in most logic texts empty structures are not even considered and that $\exists x(x = x)$ became a logical axiom. However, this exclusion leads to other anomalies, namely in connection with the notion of *generated substructure*, cf. §6.3, and sometimes even to incorrect statements of certain theorems, see the example after Theorem 9.2.2. We do allow empty structures, but for traditional reasons and in order to avoid the aforementioned semantic singularities, we exclude empty *models* (cf. §3.3) so that all models satisfy $\exists x(x = x)$. In the definition of truth (§3.1), however, we did not exclude empty structures so that the statement $\emptyset_L \models \Sigma$ makes sense. (Note that this statement is false if $\Sigma$ is a theory, see introduction to §3.4.) In any case, all this is no real problem, and Hodges therefore says, *in practice these points never matter* (see p.42 of his *Model Theory*).

**Exercise 3.6.1.** Prove that the set of $L$-sentences true in $\emptyset_L$ is neither consistent nor deductively closed.

**Exercise 3.6.2.** Determine the set of $L$-sentences true in $\emptyset_L$ precisely.

**Exercise 3.6.3.** Find an algorithm to produce normal forms that respects not only logical equivalence but also equivalence modulo the empty structure.

## Part II
# Beginnings of model theory

The main objective of model theory is to describe the model classes of theories, more precisely, the relationship between the models of a theory on the one hand, and that between the definable sets of a given model on the other. We start, in Chapter 4, with the most fundamental statement about the existence of models, the finiteness (or compactness) theorem. We deal with immediate consequences in Chapter 5, where we also introduce our main algebraic examples. In Chapter 6 we present Malcev's local theorems of group theory of the early 40s, apparently the first non-trivial application of model theory altogether. In Chapter 7 we conclude this part with some theory of orderings, ordinals, and cardinals.

*Throughout we work in a language $L = L(\sigma)$, unless it is fixed otherwise.*

# Chapter 4

# The finiteness theorem

In this chapter we prove the fundamental result saying that a set of sentences has a model whenever every finite subset has a model.

## 4.1   Filters and reduced products

Let us start with an example. Consider a family $\mathcal{G}_i$ ($i \in \mathbb{N}$) of abelian groups in the signature $(0; +, -)$ (cf. §5.2). Let $\mathcal{G}$ be the abelian group $\prod_{i \in \mathbb{N}} \mathcal{G}_i$. The **support** supp $a$ of an element $a$ of $\mathcal{G}$ is, by definition, the set $\{i \in \mathbb{N} : a(i) \neq 0\}$. It is easy to see that the set $\{a \in G : \text{supp}\, a$ is finite$\}$ is closed under addition and inverses and that it contains the neutral element $0^{\mathcal{G}}$. Hence this set is a substructure, that is, a subgroup of $\mathcal{G}$, which is called the **direct sum** of the $\mathcal{G}_i$, in symbols $\bigoplus_{i \in \mathbb{N}} \mathcal{G}_i$. In many cases, the factor group $\prod_{i \in \mathbb{N}} \mathcal{G}_i / \bigoplus_{i \in \mathbb{N}} \mathcal{G}_i$ is algebraically an interesting object. We are interested in it, because it is a prototype for the general construction to be presented here. Let's have a closer look at it.

Two elements $a = (a_i : i \in \mathbb{N}) + \bigoplus_{i \in \mathbb{N}} \mathcal{G}_i$ and $b = (b_i : i \in \mathbb{N}) + \bigoplus_{i \in \mathbb{N}} \mathcal{G}_i$ of this factor group are equal if and only if the set supp $(a_i - b_i : i \in \mathbb{N}) = \{i \in \mathbb{N} : a_i \neq b_i\}$ is finite, or, in other words, if $a_i = b_i$ for all but finitely many $i \in \mathbb{N}$. A subset of $\mathbb{N}$ is said to be **cofinite** if its complement in $\mathbb{N}$ is finite. The set of all cofinite subsets of $\mathbb{N}$ is called the **Fréchet filter** (on $\mathbb{N}$). In this terminology, the above elements $a$ and $b$ are equal if and only if the set $\{i \in \mathbb{N} : a_i = b_i\}$ is contained in the Fréchet filter. This leads to the following generalization.

Given a direct product $\mathcal{M} = \prod_{i \in I} \mathcal{M}_i$ of arbitrary nonempty $L$-structures indexed by a nonempty set $I$, we are going to define certain factor structures of $\mathcal{M}$ by way of identification of some of its elements. The measure of identification will be given by certain subsets of the power set of $I$, the so-called filters, which we define next.

Let $I$ be a nonempty set. A nonempty subset $F$ of its power set $\mathfrak{P}(I)$ is said to be a **filter** (**on** $I$) if the following conditions hold true.[1]

(i)   $\emptyset \notin F$
(ii)  If $A, B \in F$, then $A \cap B \in F$.
(iii) If $A \in F$ and $A \subseteq B \subseteq I$, then $B \in F$.

Since $F \neq \emptyset$, condition (iii) implies $I \in F$.

If $I$ is an infinite set, the **cofinite** subsets of $I$ (i. e. the sets $A \subseteq I$ whose complement $I \smallsetminus A$ is finite) constitute a filter on $I$, the so-called **Fréchet filter** on $I$. For every $B \subseteq I$, the set $\mathrm{F}(B) =_{\mathrm{def}} \{A \subseteq I : B \subseteq A\}$ is a filter on $I$, which is known as the filter **generated** by $B$. Filters generated by a single element are said to be **principal**. For principal filters we simplify the above notation by omitting braces: given $b \in I$, we write $\mathrm{F}(b)$ instead of $\mathrm{F}(\{b\})$. Note that filters containing a singleton are automatically generated by that singleton.

We now turn to the aforementioned factoring. For this, let $I$, $\mathcal{M}_i$ ($i \in I$), and $\mathcal{M} = \prod_{i \in I} \mathcal{M}_i$ as before, in particular, $\mathcal{M}$ is not empty.

Given $\varphi \in L_n$ and an $n$-tuple $\bar{a}$ in $M$, the set $\|\varphi(\bar{a})\| =_{\mathrm{def}} \{i \in I : \mathcal{M}_i \models \varphi(\bar{a}(i))\}$ is called the **boolean extension** of $\varphi(\bar{a})$; here $\bar{a}(i)$ is a shorthand for $(a_0(i), \ldots, a_{n-1}(i))$, where $\bar{a} = (a_0, \ldots, a_{n-1})$ and $a_j = (a_j(i) : i \in I)$ with $a_j(i) \in M_i$.

**Remark.** Suppose $\varphi, \psi \in L_n$ and $\bar{a}$ is an $n$-tuple in $M$.
(1)  $\|\varphi(\bar{a}) \wedge \psi(\bar{a})\| = \|\varphi(\bar{a})\| \cap \|\psi(\bar{a})\|$.
(2)  $\|\varphi(\bar{a}) \vee \psi(\bar{a})\| = \|\varphi(\bar{a})\| \cup \|\psi(\bar{a})\|$.
(3)  $\|\neg\varphi(\bar{a})\| = I \smallsetminus \|\varphi(\bar{a})\|$.
(4)  For all $(n-1)$-tuples $\bar{a}$ in $M$ and all $b \in M$, we have $\|\varphi(b, \bar{a})\| \subseteq \|\exists x\, \varphi(x, \bar{a})\|$, and there is $b \in M$ such that $\|\varphi(b, \bar{a})\| = \|\exists x\, \varphi(x, \bar{a})\|$.

*Proof.* (1), (2), (3), and the first half of (4) are immediate from the definition. For the other half of (4), choose $b$ in $M$ as follows.

For every $i \in \|\exists x\, \varphi(x, \bar{a})\|$ pick $c_i$ in $M_i$ such that $\mathcal{M}_i \models \varphi(c_i, \bar{a}(i))$, and set $b(i) = c_i$. For every $j \in I \smallsetminus \|\exists x\, \varphi(x, \bar{a})\|$ let $b(j)$ be arbitrary in $M_j$.

By choice of $b$, we have $\|\exists x\, \varphi(x, \bar{a})\| \subseteq \|\varphi(b, \bar{a})\|$.              $\square$

To every filter $F$ on $I$ we can associate a relation $\sim_F$ on $M$ by setting $a \sim_F b$ if $\|a = b\| \in F$. It is easily seen that this is an equivalence relation on

---

[1]When being in a filter is thought of as being big, then these conditions become very natural.

$M$. Given $a \in M$, let $a/F$ denote its equivalence class modulo $\sim_F$. Given $\bar{a} = (a_0, \ldots, a_{n-1})$, we write $\bar{a}/F$ for $(a_0/F, \ldots, a_{n-1}/F)$.

The **reduced product** $\mathcal{M}/F$ (or $\prod_{i \in I} \mathcal{M}_i /F$) of the (nonempty) $\mathcal{M}_i$ with respect to (or modulo) the filter $F$ is defined to be the following $L$-structure.

The universe of $\mathcal{M}/F$ is the set $M/F$ of all equivalence classes $a/F$, where $a \in M$.

Given a constant symbol $c \in \mathbf{C}$, set $c^{\mathcal{M}/F} = (c^{\mathcal{M}_i} : i \in I)/F$ (i. e., $c^{\mathcal{M}/F} = c^{\mathcal{M}}/F$).

Given an $n$-place function symbol $f \in \mathbf{F}$ and a tuple $\bar{a} = (a_0, \ldots, a_{n-1})$ in $M$, set $f^{\mathcal{M}/F}(\bar{a}/F) = (f^{\mathcal{M}_i}(a_0(i), \ldots, a_{n-1}(i)) : i \in I)/F$ (that is, $f^{\mathcal{M}/F}(\bar{a}/F) = f^{\mathcal{M}}(\bar{a})/F$).

Given an $n$-place relation symbol $R \in \mathbf{R}$ and a tuple $\bar{a}$ in $M$, set $R^{\mathcal{M}/F}(\bar{a}/F)$ just in case $\|R(\bar{a})\| \in F$.

It is easily checked that these definitions do not depend on the choice of representatives (exercise!). Thus the reduced product $\prod_{i \in I} \mathcal{M}_i /F$ is well defined by this.

If all the $\mathcal{M}_i$ are equal to a single structure $\mathcal{N}$, the reduced product is called **reduced power** of $\mathcal{N}$ with respect to $F$, in symbols $\mathcal{N}^I/F$.

**Example.** The factor group $\prod_{i \in \mathbb{N}} \mathcal{G}_i / \bigoplus_{i \in \mathbb{N}} \mathcal{G}_i$ considered above is the reduced product of the $\mathcal{G}_i$ modulo the Fréchet filter on $\mathbb{N}$.

**Exercise 4.1.1.** Verify that the definition of reduced product does not depend on the representatives chosen.

**Exercise 4.1.2.** Show that every structure is embeddable in each of its reduced powers.

**Exercise 4.1.3.** Let $F$ be the Fréchet filter on $\mathbb{N}$ and $\mathcal{N}$ the standard model of Peano arithmetic. Prove that $\mathcal{N}^{\mathbb{N}}/F$ contains no prime divisors,[2] and hence the embedding from the previous exercise does not preserve prime divisors.

**Exercise 4.1.4.** Prove that every reduced power of a structure $\mathcal{N}$ modulo a principal filter is isomorphic to a (direct) power of $\mathcal{N}$.

## 4.2 Ultrafilters and ultraproducts

Reduced products modulo so-called ultrafilters behave very nicely with respect to the satisfiability of formulas, as we will see now.

---

[2] An element $a$ different from 0 and 1 is a **prime divisor** if it divides one of the factors of every product it divides, cf. Exercise 3.5.3

Let $I$ be a nonempty set. An **ultrafilter** on $I$ is a filter $F$ on $I$ such that for all $A \subseteq I$, either $A \in F$ or $I \smallsetminus A \in F$. Reduced products (resp. powers) modulo ultrafilters are called **ultraproducts** (resp. **ultrapowers**).

**Example.** If $I$ has at least two elements ($a$ and $b$, say), the singleton $\{I\}$ is a filter that is not an ultrafilter. The filter generated by $\{a, b\}$ is not an ultrafilter either. Another example of a filter that is not an ultrafilter is the Fréchet filter on an infinite set. The filters generated by a singleton (or just containing a singleton) are precisely the principal ultrafilters (exercise!). Nonprincipal ultrafilters are harder to come by, for the ultrafilters are precisely the maximal filters (exercise!). In general their existence rests on Zorn's lemma (i. e. on the axiom of choice), see Lemma 4.3.1 below, whence there is no constructive way of describing them (for more on this and the connection with the compactness theorem see §6.2 of Hodges' *Model Theory*). Note that every (ultra)filter on a set $I$ is principal if and only if $I$ is finite (exercise!).

The next result is often called the fundamental theorem about ultraproducts.

**Theorem 4.2.1.** (Łoś [pronounce roughly like 'wash'])
   *Suppose $I$ is a nonempty set, $\mathcal{M}_i \, (i \in I)$ are nonempty L-structures, $\mathcal{M} = \prod_{i \in I} \mathcal{M}_i$ is their direct product, and $U$ is an ultrafilter on $I$.*
   *Then, for every formula $\varphi \in L_n$ and every n-tuple $\bar{a}$ from $M$,*
$\mathcal{M}/U \models \varphi(\bar{a}/U)$ *iff* $\|\varphi(\bar{a})\| \in U$.

[Łoś, J. : Quelques remarques, théorèmes et problèmes sur les classes définissables d'algèbres, in **Mathematical interpretations of formal systems** (L. e. J. Brouwer et al., ed.), North-Holland, Amsterdam, 1955, pp. 98 – 113]

*Proof.* by induction on the complexity of $\varphi$.
   First of all we have to show that $t^{\mathcal{M}/U}(\bar{a}/U) = (t^{\mathcal{M}_i}(\bar{a}(i)) : i \in I)/U$ (hence $t^{\mathcal{M}/U}(\bar{a}/U) = t^{\mathcal{M}}(\bar{a})/U$) for every term $t(\bar{x})$ with $n$ free variables and every n-tuple $\bar{a}$ from $M$. The proof of this by induction too, this time on the complexity of the term $t$. We leave it to the reader.
   From this one easily derives that a term equation holds in $\mathcal{M}/U$ if and only if its boolean extension lies in $U$; similarly for relational atomic formulas. Thus the assertion of the theorem is true for all atomic formulas $\varphi$.
   Let now $\psi$, $\psi_1$, $\psi_2$, and $\theta$ be formulas satisfying the assertion of the theorem (induction hypothesis). We are to show that also $\neg\psi$, $\psi_1 \wedge \psi_2$, and $\exists x \, \theta(x)$ satisfy this assertion.

(i)   $\mathcal{M}/U \models \neg\psi(\bar{a}/U)$ iff $\mathcal{M}/U \not\models \psi(\bar{a}/U)$ iff $\|\psi(\bar{a})\| \notin U$ iff $I \smallsetminus \|\psi(\bar{a})\| \in U$
      (by the ultrafilter axiom). However, $I \smallsetminus \|\psi(\bar{a})\| = \|\neg\psi(\bar{a})\|$, cf. Remark
      (3) of §4.1.

(ii)  $\mathcal{M}/U \models \psi_1(\bar{a}/U) \wedge \psi_2(\bar{a}/U)$ iff $\mathcal{M}/U \models \psi_1(\bar{a}/U)$ and $\mathcal{M}/U \models \psi_2(\bar{a}/U)$.
      Hence one direction of the assertion follows from the finite intersection
      property (taking into account Remark (1) from §4.1), while the other
      follows from the extension axiom (iii) in the definition of filter.

(iii) $\mathcal{M}/U \models \exists x\, \theta(x, \bar{a}/U)$ iff $\mathcal{M}/U \models \theta(b, \bar{a}/U)$ for some $b \in M/U$ iff
      $\mathcal{M}/U \models \theta(b/U, \bar{a}/U)$ for some $b \in M$ iff $\|\theta(b, \bar{a})\| \in U$ for some $b \in M$.
      By Remark (4) of §4.1, the latter is equivalent to $\|\exists x\, \theta(x, \bar{a})\| \in U$.

$\square$

**Exercise 4.2.1.** Prove the assertions about filters made above.

**Exercise 4.2.2.** Prove the statement left to the reader in the above proof.

**Exercise 4.2.3.** Let $U$ be an ultrafilter containing the Fréchet filter on $\mathbb{N}$, and let
$\mathcal{N}$ be the standard model of Peano arithmetic. Prove that $\mathcal{N}^{\mathbb{N}}/U$ contains elements
with infinitely many prime divisors (cf. Exercise 4.1.3).

A **positive primitive** formula is a formula of the form $\exists \bar{x}\, \varphi$, where $\varphi$ is a finite
conjunction of atomic formulas.

**Exercise 4.2.4.** Prove that Łoś' theorem restricted to positive primitive formulas
$\varphi$ holds in arbitrary reduced products.

**Exercise 4.2.5.** Show that the ultrapowers of a structure $\mathcal{N}$ modulo a principal
ultrafilter are isomorphic to $\mathcal{N}$. [cf. 4.1.4 and 4.2.1.]

## 4.3   The finiteness theorem

We already mentioned that not every filter is an ultrafilter and that, however,
every filter can be extended to an ultrafilter. We need the latter property
for our proof of the finiteness theorem. We prove a slight generalization of
it next, for which we make the following definition.

Let $I$ be a nonempty set. A subset $F$ of $\mathfrak{P}(I)$ is said to have the **finite intersection property** if $F \neq \emptyset$ and no intersection of finitely many
members of $F$ is empty.

**Remark.** $F$ has the finite intersection if and only if the set $F'$ of all finite
intersections of sets from $F$ satisfies the filter axioms (i) and (ii).

**Lemma 4.3.1.** Let $I$ be a nonempty set and $F$ a subset of $\mathfrak{P}(I)$ with the
finite intersection property.
      Then there is an ultrafilter $U$ on $I$ containing $F$.

[Tarski, A. : Une contribution à la théorie de la mesure, Fundamenta Math. **15** (1930) 42 – 50]

*Proof.* Suppose $F'$ is as in the preceding remark and $U \subseteq \mathfrak{P}(I)$ is a maximal extension of $F'$ with the finite intersection property. The existence of such a set $U$ is guaranteed by Zorn's lemma. Let us show that $U$ is an ultrafilter on $I$.

The finite intersection property implies (i) ($\emptyset \notin U$). To derive (ii), the intersection axiom, just note that if $A, B \in U$, then also $U \cup \{A \cap B\}$ has the finite intersection property and is thus equal to $U$ by maximality. Considering $U \cup \{B\}$ for any $A \in U$ and $A \subseteq B$, one obtains also (iii), the extension axiom. Thus $U$ is a filter.

In order to derive the ultrafilter axiom, let $A \subseteq I$ and consider the set $U \cup \{A\}$. If $C \cap A = \emptyset$ for some $C \in U$, then $C \subseteq I \smallsetminus A$ and hence $I \smallsetminus A \in U$ by (ii). Otherwise $U \cup \{A\}$ has the finite intersection property and hence $A$ must lie in $U$ by maximality. Consequently, $U$ is an ultrafilter on $I$.  $\square$

We now turn to the fundamental theorem of model theory. It is often called compactness theorem, and we will see in the next chapter why.

**Theorem 4.3.2.** (The Finiteness Theorem) *A set of L-sentences, $\Sigma$, has a model if and only if every finite subset of $\Sigma$ has a model.*

[Gödel, K. : Die Vollständigkeit der Axiome des Akiome des Funktionenkalküls, Monatshefte f. Math. u. Phys. **37** (1930) 349 – 360][3]

[Malcev, A. I. : Untersuchungen aus dem Gebiet der mathematischen Logik, Rec. Math. N.S. (Mat. Sbornik) **1** (1936) 323 – 336, (In German, English translation in Malcev (1971).)][4]

*Proof.* One direction is trivial, for if $\Sigma$ has a model, so does every subset.

Suppose now that every finite subset of $\Sigma$ has a model, and let $I$ be the set of all nonempty finite subsets of $\Sigma$. By hypothesis, there is a nonempty $L$-structure $\mathcal{M}_i$ for every $i \in I$ such that $\mathcal{M}_i \models i$. Let $\mathcal{M}$ be their direct product and set $i^* = \{j \in I : i \subseteq j\}$. Clearly, $i_0^* \cap i_1^* = (i_0 \cup i_1)^*$. Hence the set $\{i^* : i \in I\}$ has the finite intersection property, whence Lemma 4.3.1 yields an ultrafilter $U$ on $I$ containing it.

We show that $\mathcal{M}/U \models \Sigma$. So let $\varphi \in \Sigma$. Then $\{\varphi\} \in I$ and $\mathcal{M}_i \models \varphi$ for all $i \in I$ with $\varphi \in i$, hence $\{\varphi\}^* = \{i \in I : \varphi \in i\} \subseteq \{i \in I : \mathcal{M}_i \models \varphi\}$. By

---

[3]For countable theories.

[4]In general.

the choice of $U$, we have $\{\varphi\}^* \in U$. Then filter axiom (iii) yields $\{i \in I : \mathcal{M}_i \models \varphi\} = \|\varphi\| \in U$. Finally, Łoś' theorem 4.2.1 implies $\mathcal{M}/U \models \varphi$. Since $\varphi$ was arbitrary in $\Sigma$, we see that $\mathcal{M}/U \models \Sigma$, as desired. $\square$

In courses of first-order logic one usually derives the finiteness theorem from Gödel's completeness theorem for first-order logic, which says that every logical consequence can be derived in a certain effective formal proof system. The finiteness theorem then follows immediately from the finite character of that proof system ('every formal proof is finite'!). Since it is model-theoretically irrelevant, we avoid this proof theory and refer the reader to any of the standard logic texts for this. We can derive the finite character of logical entailment now from the finiteness theorem.

**Corollary 4.3.3.** *The notion of logical consequence is finitary, that is, every consequence of a set of sentences, $\Sigma$, is a consequence of some finite subset of $\Sigma$. Consequently, a set of sentences is consistent if and only if every finite subset is.*

*Proof.* Note that $\Sigma \models \varphi$ if and only if $\Sigma \cup \{\neg\varphi\}$ has no model. $\square$

This result is often identified with the finiteness theorem.

**Exercise 4.3.1.** Show that a nonempty set of subsets of a nonempty set $I$ is contained in a filter on $I$ if and only if it has the finite intersection property.

**Exercise 4.3.2.** Prove that if a set of sentences, $\Sigma$, axiomatizes a finitely axiomatizable class of structures, then this class is axiomatized already by a finite subset of $\Sigma$.

**Exercise 4.3.3.** Show that a class **K** of $L$-structures is finitely axiomatizable if and only if **K**, as well as its complement (within the class of all $L$-structures), is axiomatizable.

**Exercise 4.3.4.** Verify that the class of all infinite sets (as $L_=$-structures) is axiomatizable, but not finitely axiomatizable.

# Chapter 5

# First consequences of the finiteness theorem

As a first application of the finiteness theorem we derive a fundamental result about arbitrarily large models. Then we give some simple applications to classical structures. In order to apply the finiteness theorem to these, we have to first axiomatize them in an appropriate (first-order) language. Finally, we explain why the finiteness theorem is also called compactness theorem. This leads to certain topological spaces, the so-called Stone spaces, which we will return to later in Part IV.[1]

## 5.1 The Löwenheim-Skolem Theorem Upward

**Theorem 5.1.1.** (Löwenheim-Skolem Upward) *Let* $\Sigma \subseteq L_0$.

*If* $\Sigma$ *has arbitrarily large finite models or an infinite model, then* $\Sigma$ *has models of arbitrarily large power.*

[Löwenheim, L. : Über Möglichkeiten im Relativkalkül, Mathematische Annalen **76** (1915) 447 – 470]

[Skolem, Th. : Logisch-kombinatorische Untersuchungen über die Erfüllbarkeit oder Beweisbarkeit mathematischer Sätze nebst einem Theorem über dichte Mengen, Skrifter, Videnskabsakademie i Kristiania I. Mat.-Nat. Kl. no. 4 (1920) 1 – 36]

---

[1]The use of topology in this text is only sporadic (for the most part in §§5.6–7 and §11.3) and could usually be avoided (see e. g. Exercise 9.1.1). Consult any topology text for the rudiments of the theory, e. g., Ch. 1 of Jänich's *Topology*.

The theorem is contained in these two papers in one form or another. However, the full strength for arbitrary languages was formulated (but never published) by A. Tarski, whence the alternative name Löwenheim-Skolem-Tarski theorem. See also

[Skolem, Th. : **Selected works in logic**, p. 366, Universitetsforlaget, Oslo, 1970, Bem. 3].

*Proof.* For any given set $C$ of new constants, we are going to find a model of $\Sigma$ of power at least the power of $C$.

For this, consider the following set of $L(C)$-sentences,

$$\Sigma_C = \Sigma \cup \{c \neq c' \ : \ c, c' \in C \text{ and } c \neq c'\}.$$

Since $\Sigma$ has arbitrarily large finite models or an infinite model, every finite subset $\Sigma' \Subset \Sigma_C$ has a model, for in any sufficiently large structure we can find pairwise distinct interpretations for the finitely many $c$ occurring in $\Sigma'$. The finiteness theorem yields a model of $\Sigma_C$. Its $L$-reduct is then a model of $\Sigma$ of power $\geq |C|$.    □

This was a typical, albeit easy application of the finiteness theorem. As a consequence we see that finiteness is not axiomatizable. In particular, the class of finite sets is not axiomatizable (compare this with Exercise 4.3.4, showing that the class of infinite sets, though axiomatizable, is not *finitely* axiomatizable). Similarly, the theorem above immediately yields—simply for cardinality reasons—results like the following, due to Thoralf Skolem, who derived it directly by some precursor of the ultraproduct construction. (It is ironic that these results are due to Skolem, who had the most serious methodological doubts about the usefulness of the uncountable—a historical phenomenon that is, though still existing, quite obsolete.)

**Corollary 5.1.2.** *The complete first-order number theory (or true arithmetic, cf. §3.5) has* nonstandard *models, i. e. models not isomorphic to that of the natural numbers.*    □

[Skolem, Th. : Über die Nichtcharakterisierbarkeit der Zahlenreihe mittels endlich oder abzählbar unendlich vieler Aussagen mit ausschließlich Zahlenvariablen, Fundamenta Math. **23** (1934) 150 – 161]

**Exercise 5.1.1.** Using the finiteness theorem, prove that Peano arithmetic (cf. §3.5) has models with nonzero elements having infinitely many prime divisors.

## 5.2 Semigroups, monoids, and groups

In the next few sections we find axiomatizations of some well-known classes of algebraic structures and see what immediate consequences of the finiteness theorem we can derive about them.

We choose appropriate languages, first for semigroups, monoids, and groups, in which we are able to axiomatize these classes.

Consider the signature $(\cdot)$, where $\cdot$ is a binary function symbol (whose intended interpretation is the multiplication). In the corresponding language we formulate the sentence

(1) $\forall xyz\,((x \cdot y) \cdot z = x \cdot (y \cdot z))$,

the axiom of associativity of the operation $\cdot$. Its models are precisely what is known to be **semigroups**. We denote by SG the $L(\cdot)$-theory of all semigroups, i. e. the deductive closure of axiom (1).

We simplify the expressions in this language by introducing the following abbreviations.

Given a nonzero $n \in \mathbb{N}$, the symbol $x^n$ stands for the term $(\ldots (x \cdot x) \cdot \ldots) \cdot x$, where $x$ occurs $n$ times. We also write $xy$ instead of $x \cdot y$.

Let 1 be a constant symbol. If we add to (1) the $(1; \cdot)$-sentence

(2) $\forall x\,(x1 = x \wedge 1x = x)$,

saying that the interpretation of 1 is the neutral element of (the interpretation of) $\cdot$, we obtain the system of axioms $\{(1), (2)\}$ in the language $L(1; \cdot)$, whose models are called **monoids**. We denote by TM the $L(1; \cdot)$-theory of all **monoids**, i. e. the deductive closure of $\{(1), (2)\}$.

In order to describe inverses we pass to a richer signature $(1; \cdot,^{-1})$, which has in addition a unary function symbol $^{-1}$ (with the obvious intended interpretation). Again we introduce some handy (and common) abbreviations.

We write $x^{-n}$ for the $L(1; \cdot,^{-1})$-term $(x^n)^{-1}$ ($n$ as before), and $xy^{-1}$ or $x/y$ for the term $x \cdot (y^{-1})$.

If we extend the previous axioms by the $L(1; \cdot,^{-1})$-sentence

(3) $\forall x\,(xx^{-1} = 1 \wedge x^{-1}x = 1)$,

saying that (the interpretation of) $^{-1}$ is the operation of taking inverses with respect to (the interpretations of) $\cdot$ and 1, we get $\{(1), (2), (3)\}$, which axiomatizes the **groups**. That is, a group is an $L(1; \cdot,^{-1})$-structure satisfying this set of axioms. We denote by TG the $L(1; \cdot,^{-1})$-theory of all **groups**, i. e. the deductive closure of $\{(1), (2), (3)\}$.

If we add also the commutativity axiom

(4) $\forall xy\,(xy = yx)$

for the operation $\cdot$, we obtain the axioms of **abelian** groups. Usually one writes abelian groups additively, that is, as a structure for the signature $(0; +, -)$, where $0$ is a constant symbol, $+$ is a binary function symbol, and $-$ is, as $^{-1}$, a *unary* function symbol.

Let $L_{\mathbb{Z}}$ denote the language of this signature $(0; +, -)$ and AG the $L_{\mathbb{Z}}$-theory of all **abelian** groups, i. e. the deductive closure of the following collection of axioms.

$(1)^+$  $\forall xyz \, ((x + y) + z) = x + (y + z))$
$(2)^+$  $\forall x \, (x + 0 = x \wedge 0 + x = x)$
$(3)^+$  $\forall x \, (x + (-x) = 0 \wedge (-x) + x = 0)$
$(4)^+$  $\forall xy \, (x + y = y + x)$

The theory of **torsionfree** abelian groups, i. e., the $L_{\mathbb{Z}}$-theory with the axioms AG $\cup \{\forall x \, (nx = 0 \longrightarrow x = 0) : 1 < n \in \mathbb{N}\}$, is denoted by AG$_{\mathrm{tf}}$.

Similarly as in the case of multiplication, we write $nx$ for the term $(\ldots (x + x) + \ldots) + x$ ($n$ times), where $n \in \mathbb{N}$ is not zero. Further, we use $(-n)x$ to denote $-(nx)$, and $x - y$ to denote $x + (-y)$.

Now that we have these axiomatizations, we can apply the finiteness theorem to derive interesting properties of axiomatizable classes of semigroups, monoids, or groups. As an example we have the following result, and the reader is asked to state and prove similar ones (see also the exercises).

**Proposition 5.2.1.** *Let $\Sigma$ be a set of sentences containing the above sentences (1) and (2).*

*Suppose $\Sigma$ has, for arbitrarily large $n \in \mathbb{N}$, a model containing an element of **order** at least $n$ (i. e., an element $a$ such that $a^k \neq 1$ for all $k < n$). Then $\Sigma$ has a model containing an element of **infinite order** (i. e., an element $a$ such that $a^k \neq 1$ for all $k \in \mathbb{N}$).*

*Proof.* Let $L$ be a language containing the language of signature $(1; \cdot)$. Set $L' = L(c)$ and $\Sigma' = \Sigma \cup \{c^n \neq 1 : n \in \mathbb{N}\}$, where $c$ is a new constant. Any model of $\Sigma'$ is a model of $\Sigma$, and the interpretation of $c$ therein is the desired element. Therefore it suffices to find a model of $\Sigma'$. By the finiteness theorem this is equivalent to finding a model for every finite subset of $\Sigma'$. Now, every such finite subset is contained in a set of the form $\Sigma \cup \{c^k \neq 1 : k < n\}$ for some $n \in \mathbb{N}$, hence it suffices to find a model for the latter. This is possible by hypothesis, for given $\mathcal{M}'_n = (\mathcal{M}_n, a_n)$, where $\mathcal{M}_n \models \Sigma$ and $a_n \in M_n$ has order $\geq n$, setting $c^{\mathcal{M}'_n} = a_n$ shows that $\mathcal{M}'_n$ a model of $\Sigma'$. $\qquad\square$

**Corollary 5.2.2.** *The class of **periodic** groups, that is groups all of whose elements have finite order, is not axiomatizable.* $\qquad\square$

The congruence classes of integers modulo a natural number $n$ form an abelian group of cardinality (more group-theoretically, of **order**) $n$, which we denote by $\mathbb{Z}_n$.

$\mathbb{Z}_n$ is thus the cyclic group $(\mathbb{Z}/n\mathbb{Z}; 0; +, -)$, hence homomorphic image of $(\mathbb{Z}; 0; +, -)$ under the group homomorphism $k \mapsto k + n\mathbb{Z}$.

**Exercise 5.2.1.** Prove that the class of finite cyclic groups is not axiomatizable.

**Exercise 5.2.2.** Prove that the class of cyclic groups is not axiomatizable.

**Exercise 5.2.3.** Show that the class of groups can also be axiomatized in the language of signature $(1; \cdot)$. What are then the substructures of a group? What about the same question for the signature $(1; \cdot, :)$, where $:$ is a binary function symbol (which is to be interpreted so that $x : y = x \cdot y^{-1}$)?

**Exercise 5.2.4.** Prove that every reduced product of groups is a factor group of the corresponding direct product.

**Exercise 5.2.5.** Find a direct product of infinitely many groups and a factor group thereof which is not a reduced product of these groups.

**Exercise 5.2.6.** Prove that every factor group of a product of simple groups is a reduced product of these simple groups.

## 5.3 Rings and fields

We consider rings in the signature of additive groups extended by that of monoids, that is in the signature $(0, 1; +, -, \cdot)$, where $0$ and $1$ are constant symbols, $-$ is a unary function symbol, and $+$ and $\cdot$ are binary function symbols. Then

$$\mathrm{AG} \cup \mathrm{TM} \cup \{\forall xyz \, (x \cdot (y+z) = (x \cdot y) + (x \cdot z) \wedge (y+z) \cdot x = (y \cdot x) + (z \cdot x))\}$$
$$\cup \{0 \neq 1\}$$

axiomatizes the class of **rings**.[2] We denote the theory of all rings by TR.

A ring is said to be **commutative** if its multiplication is commutative, i. e., if it satisfies axiom (4). The theory of all commutative rings is denoted by CR. Note that CR is thus the deductive closure of $\mathrm{TR} \cup \{(4)\}$.

The set of $\mathrm{L}(0, 1; +, -, \cdot)$-sentences

$$\mathrm{TR} \cup \{\forall x \exists y \, (x \neq 0 \longrightarrow x \cdot y = 1 \wedge y \cdot x = 1)\}$$

---

[2] Note that it is built into this definition of ring that it be associative with a 1.

axiomatizes the **division rings**. We denote the theory of all division rings by DR. Commutative division rings, that is division rings that are, as rings, also commutative, are called **fields**. The theory of all fields is denoted by TF. Thus TF is the deductive closure of DR $\cup$ {(4)}, which is the same as the deductive closure of DR $\cup$ CR.

Division rings are also known as **skew fields**, since they are like fields, except they need not satisfy the commutativity axiom. We prefer the positive term *division ring*, because *skew* could also suggest that the ring be *non*commutative, which is not intended.

For the above axiomatization we chose not to include the operation of taking multiplicative inverses into the signature. The reason is that it would not be total, as $0^{-1}$ is not defined, while, in our setting, all function symbols have to be interpreted by total functions. One remedy would be to assign to $0^{-1}$ an arbitrary, but fixed (and irrelevant) value, like e. g. 0. We chose not to do so (which seems to be closer to the usual mathematical practice) and to use $x^{-1}$ as an abbreviation instead. More precisely, given a division ring $\mathcal{K}$, as an L(0, 1; +, −, ·)-structure, and an element $0 \neq a \in K$, we use $a^{-1}$ to denote the uniquely determined $b$ satisfying $a \cdot b = 1 \wedge b \cdot a = 1$.

Note that due to this choice of signature every substructure of a division ring is a subring (that need not be a division ring).[3] The same applies to fields, of course, which means that a substructure of a field need not be a field. We therefore introduce the term **subfield** to mean a subring which is a field in its own right. (Note that the inverses of nonzero elements are the same, whether taken in a subfield or the initial field.)

An important invariant of a division ring is its characteristic, as defined next. The **characteristic** of a division ring $\mathcal{K}$ is the smallest natural number $p$ such that $\mathcal{K} \models p1 = 0$, if such a number exists, and 0 otherwise. Hence $\mathcal{K}$ has characteristic 0 if and only if $\mathcal{K} \models \{p1 \neq 0 : p > 0\}$.

Thus the characteristic of $\mathcal{K} \models$ DR is simply the order (in the sense of Proposition 5.2.1) of the element 1 in the *additive group* of $\mathcal{K}$, i. e., in the (0; +, −)-reduct of $\mathcal{K}$. Recall that this number, if not 0, is always a prime (see Exercise 5.3.3 below).

The class of division rings of fixed characteristic is also axiomatizable. It is obvious that in the case of positive characteristic we need to add to DR only finitely many—in fact just one—extra axiom for this purpose. We will see shortly, as an easy application of the finiteness theorem, that in the characteristic zero case this is impossible.

---

[3] The abstinence we impose on the signature will pay off later in the quantifier elimination for algebraically closed fields, cf. §9.4

Let $q$ be a prime number or 0. Then $DR_q$ denotes the theory of all division rings of characteristic $q$; similarly for $TF_q$.

Recall from basic algebra that every field of characteristic 0 contains a subfield isomorphic to $\mathbb{Q}$, the field of rational numbers.[4] Similarly, every field of characteristic $p$, with $p$ a prime number, contains a subfield isomorphic to $\mathbb{F}_p$, where $\mathbb{F}_p$ denotes the field with universe $\mathbb{Z}_p$, that is the $(0, 1; +, -, \cdot)$-expansion of $\mathbb{Z}_p$ by multiplication modulo $p$ (more precisely, the expansion obtained by interpreting $\cdot$ as multiplication modulo $p$). These subfields are known as the **prime fields**. To summarize, the prime field of a field $\mathcal{F}$ is isomorphic to $\mathbb{Q}$ or $\mathbb{Z}_p$, depending on whether $\mathcal{F}$ has characteristic 0 or $p\ (> 0)$.

Note that $\mathbb{Z}_p$ is the $(0; +, -)$-reduct, in more common terms, the additive group of $\mathbb{F}_p$.

In analogy to the proposition about orders in (expansions of) groups in the previous section, the finiteness theorem yields the following about characteristics of (expansions of) division rings. By the **characteristic** of an expansion of a division ring we here simply mean the characteristic of the underlying division ring.

**Proposition 5.3.1.** *Suppose $\Sigma$ is a set of sentences in a language $L$ whose signature contains $(0, 1; +, -, \cdot)$ and $DR \subseteq \Sigma^\vDash$.*

*If $\Sigma$ has models of arbitrarily high characteristic, then $\Sigma$ has a model of characteristic 0.*

[Robinson, A. : **On the metamathematics of algebra**, Studies in Logic and the Foundations of Mathematics, North-Holland, Amsterdam, 1951]

*Proof.* We have to show that $\Sigma' = \Sigma \cup \{p1 \neq 0 : p > 0\}$ is consistent, for which, by the finiteness theorem, it suffices to show that every finite subset is consistent. Every such subset, $\Sigma'$, is contained in some set of the form $\Sigma \cup \{n1 \neq 0 : n < p\}$, where $p$ is a prime. But the latter set has, by hypothesis, a model (namely one of characteristic $\geq p$).  $\square$

**Remark.** Note that the set $\Sigma$ may contain sentences of the possibly larger language $L$. However, important only is that it has, for every natural number $n$, a model whose $(0, 1; +, -, \cdot)$-reduct is a division ring of characteristic $\geq n$.

**Corollary 5.3.2.** *Let $L$ be as before and $\varphi$ an $L$-sentence.*

*If $DR_0 \vDash_L \varphi$, then there is a prime number $p_\varphi$ such that $DR_p \vDash_L \varphi$ for all primes $p > p_\varphi$.*

---

[4]We identified the field of rational numbers with its universe, $\mathbb{Q}$, here. This laxness is very common in mathematics when the signature is understood.

*Proof.* Otherwise $\Sigma = \mathrm{DR} \cup \{\neg\varphi\}$ would have models of arbitrarily high characteristic, hence, by Proposition 5.3.1, also one of characteristic 0, contradicting the hypothesis.                                                                    □

**Remarks.** (1)  Corollary 5.3.2 means: if an $(0, 1; +, -, \cdot)$-sentence $\varphi$ holds in every division ring of characteristic 0, then there is a prime number $p_\varphi$ such that $\varphi$ holds in all division rings of characteristic $\geq p_\varphi$. In particular, $\mathrm{DR}_0$ is not finitely axiomatizable.

(2)  A similar proof yields analogous results for any other axiomatizable class of division rings of characteristic 0, which thus are not finitely axiomatizable. In particular, $\mathrm{TF}_0$, the theory of all fields of characteristic 0, is not finitely axiomatizable.

*Warning.* One cannot replace $\varphi$ by an infinite set of sentences: consider the counterexample $\{p1 \neq 0 : p > 0\}$.

We conclude this section with some observations about terms and formulas of signature $(0, 1; +, -, \cdot)$. A priori, terms can be quite complicated, e. g. $x(y + zx) + x(y + zx) + myy(z - y)x$, where $m \in \mathbb{Z}$ is already used to abbreviate an $m$-fold sum. However, modulo CR, the theory of commutative rings, such a term is equivalent to a much simpler expression, namely to a sum of monomials, where in addition occurrences of the same variable can be joined and, using the abbreviations of §5.2, written as powers. What we obtain are polynomials with coefficients from $\mathbb{Z}$. The above term, for instance, is CR-equivalent to $2xy + 2x^2z + mxy^2z - mxy^3$, which is a polynomial in $\mathbb{Z}[x, y, z]$. Clearly, we can consider this term also as a polynomial in $x$ with coefficients from $\mathbb{Z}[y, z]$. Since every term equation is, via subtraction, CR-equivalent to a term equation of the form $t = 0$, we may summarize this discussion as follows.

**Lemma 5.3.3.** *Let $L$ be the language of signature $(0, 1; +, -, \cdot)$.*

(1)  *Every atomic $L$-formula in the variables $\bar{x}$ is CR-equivalent to a polynomial equation $t = 0$, where $t \in \mathbb{Z}[\bar{x}]$. If $\bar{x}$ is the union of the tuples $\bar{y}$ and $\bar{z}$, then the term $t$ may be regarded as a polynomial in $\bar{z}$ with coefficients from $\mathbb{Z}[\bar{y}]$, i. e., $t \in (\mathbb{Z}[\bar{y}])[\bar{z}]$.*

(2)  *Let $\mathcal{A}$ be a commutative ring.*
     *Every atomic $L(\mathcal{A})$-formula in the free variables $\bar{x}$ is $\mathcal{A}$-equivalent to a polynomial equation $t(\bar{x}) = 0$, where $t \in \mathcal{A}[\bar{x}]$.*                    □

**Remark.** Atomic $L$-*sentences* are, by (1), CR-equivalent to (trivial) polynomial equations of the form $k = 0$, where $k$ is an integer. Similarly, atomic $L(\mathcal{A})$-*sentences* are, by (2), CR-equivalent to (trivial) polynomial equations of the form $k = 0$, where $k \in \mathcal{A}$.

**Exercise 5.3.1.** Justify the above Remark (2) (after the corollary).

A **zero-divisor** of a ring is a nonzero element whose product with a certain nonzero element is 0.

**Exercise 5.3.2.** Find a finite system of axioms for the class of rings without zero-divisors. Do the same for the class of commutative rings without zero-divisors (such rings are known as **integral domains**).

**Exercise 5.3.3.** Define the characteristic of an arbitrary ring (as before) and show that, for rings without zero-divisors, this number is 0 or a prime.

An element $r \neq 0$ from a ring $\mathcal{R}$ is called **nilpotent**, if there exists an $n \in \mathbb{N}$ such that $r^n = 0$.

**Exercise 5.3.4.** Show that the class of all rings without nilpotent elements is finitely axiomatizable.

**Exercise 5.3.5.** Show that the reduced product of a given family of rings is a factor ring of the direct product of this family.

**Exercise 5.3.6.** Prove the converse of the preceding statement for division rings (in contrast to groups, cf. Exercise 5.2.5, or arbitrary rings (check!)).

**Exercise 5.3.7.** Show that the natural ordering $\leq$ on $\mathbb{R}$ is definable in the language of fields.

# 5.4 Vector spaces

Vector spaces over a *fixed* division ring $\mathcal{K}$ are usually regarded as structures of the signature $(0; +, -) \cup \{f_k : k \in K\}$, where the $f_k$ are unary function symbols (for the multiplication by the scalars $k \in \mathcal{K}$). We denote the corresponding language by $L_{\mathcal{K}}$.

Consider the following sets of $L_{\mathcal{K}}$-sentences.

(1) $\{\forall x \, (f_0(x) = 0 \land f_1(x) = x)\}$
(2) $\{\forall x y \, (f_k(x + y) = f_k(x) + f_k(y)) : k \in K\}$
(3) $\{\forall x \, (f_{k+k'}(x) = f_k(x) + f_{k'}(x)) : k, k' \in K\}$
(4) $\{\forall x \, (f_k(f_{k'}(x)) = f_{k \cdot k'}(x)) : k, k' \in K\}$

Together with AG these axiomatize the class of **left vector spaces over** $\mathcal{K}$ (or **left $\mathcal{K}$-vector spaces**). We denote the $L_{\mathcal{K}}$-theory of all such spaces by $T_{\mathcal{K}}$. Thus $T_{\mathcal{K}}$ is the deductive closure of AG and the above four sets.

Recall that every vector space has a uniquely determined dimension and that two vector spaces over the same division ring are isomorphic if and only if they have the same dimension (if the reader knows this only for vector spaces over fields, he or she may want to go over the proofs again and check that they never use the commutativity of the field).

**Exercise 5.4.1.** How would a similar $L_K$-axiomatization of right $K$-vector spaces have to look?

We always consider vector spaces in the above languages, which depend on the given division ring. One can axiomatize *all* left vector spaces in a single language by adding the scalars to the structure. This can be established in a so-called **two-sorted** setting, where one has two sorts of variables, one for the scalars and one for the vectors, and where a structure is regarded as a pair $(K, V)$, where $V$ is a vector space and $K$ is the division ring of scalars. There is also a standard way of coding this kind of two-sorted structure in a **one-sorted** structure (which is our setting): one simply introduces a predicate (i. e., a unary relation symbol) for the scalars, $K(x)$, and one for the vectors, $V(x)$, (or one uses $\neg K(x)$ instead), and adds the axiom that every element is either a vector or a scalar.

**Exercise 5.4.2.** Find the corresponding signatures and axiomatizations (if the two-sorted setting sounds too unfamiliar, do this only for the one-sorted 'coding').

One of the reasons for not using this language in this book lies in the phenomenon to be described in the next two exercises.

**Exercise 5.4.3.** What are the substructures of such a two-sorted vector space (be it in the literally two-sorted or the one-sortedly coded setting)?

**Exercise 5.4.4.** Given any infinite division ring $K$, can you axiomatize the class of all left vector spaces over $K$ in the two-sorted (or 'coded') setting?

**Exercise 5.4.5.** Axiomatize the class of all left vector spaces over a finite field $K$ in the two-sorted (or 'coded') setting.

Note that we would gain nothing by allowing finite division rings in the preceding, for, by a classical theorem of Wedderburn, finite division rings are fields.

## 5.5   Orderings and ordered structures

All we need in the signature for orderings is a binary relation symbol, $<$. We denote the corresponding language by $L_<$.

Consider the following $L_<$-sentences and their names (leading to corresponding names of orderings).

| | | |
|---|---|---|
| (1) | $\forall x \, \neg x < x$ | **irreflexivity** (or **strictness**) |
| (2) | $\forall xyz \, (x < y \wedge y < z \longrightarrow x < z)$ | **transitivity** |
| (3) | $\forall xy \, (x < y \vee x = y \vee y < x)$ | **linearity** |
| (4) | $\forall xy \, (x < y \longrightarrow \exists z \, (x < z \wedge z < y))$ | **density** |
| (5) | $\exists xy \, x < y$ | **nontriviality** |

$\{(1),(2)\}$ axiomatizes the class of **partial orderings**, and $\{(1),(2),(3)\}$ axiomatizes that of **linear orderings** or **chains**, while $\{(1),(2),(3),(4),(5)\}$ axiomatizes the class of (nontrivial) **dense linear orderings**. A **chain** in a partial ordering $(X,<)$ is, by definition, a subordering of $X$ which is a chain in its own right (i. e. which is linearly ordered w.r.t. $<$).

Note that (1) and (2) imply

(6)  $\forall xy\,(x < y \longrightarrow \neg y < x))$ **antisymmetry**

**Remark.** Because of the irreflexivity, (1), homomorphisms of linear orderings are automatically embeddings.

Consider the following endpoint axioms, whose names are to suggest the stated existence or lack of left or right endpoints.

$(--)$  $\forall x\exists yz\,(y < x \wedge x < z)$
$(+-)$  $\exists x\forall y\,(x < y \vee x = y) \wedge \forall x\exists y\, x < y$
$(-+)$  $\forall x\exists y\, y < x \wedge \exists x\forall y\,(y < x \vee x = y)$
$(++)$  $\exists x\forall y\,(x < y \vee x = y) \wedge \exists x\forall y\,(y < x \vee x = y)$

We use $T_<$ to denote the $L_<$-theory of all partial orderings, LO to denote that of all linear orderings, and DLO to denote that of all (nontrivial) dense linear orderings. Thus these theories are the deductive closures of the corresponding sets of axioms used to define the respective classes of orderings above. We denote the deductive closure of DLO $\cup \{(-+)\}$ by DLO$_{-+}$, and call this the theory of dense linear orderings **with a right, but no left endpoint**; similarly for the other three endpoint axioms.

We simplify notation by introducing some common abbreviations: $x > y \Longleftrightarrow_{\text{def}} y < x$ and $x \leq y \Longleftrightarrow_{\text{def}} x < y \vee x = y$ and also $x \not< y \Longleftrightarrow_{\text{def}} \neg x < y$.

We will consider orderings in more detail in Chapter 7.

Often one can order structures in such a way that the operations of the signature are in some sense compatible with the order. A few cases are collected next.

A **partially ordered group** is a $(1;\cdot;<)$-structure satisfying

$$\forall xyuv\,(x < y \longrightarrow uxv < uyv)$$

whose $(1;\cdot)$-reduct is a group and whose $(<)$-reduct is a partial ordering. This partial ordering is also called a **partial ordering** of the underlying group. If this ordering is linear, we speak of a **linear ordering** and correspondingly of a **linearly ordered group**.

A **partially ordered ring** is a $(0,1;+,-,\cdot;<)$-structure satisfying

$$\forall xyuv\,(x < y \wedge u > 0 \wedge v > 0 \longrightarrow uxv < uyv)$$

whose $(0,1;+,-,\cdot)$-reduct is a ring and whose $(0;+,-;<)$-reduct is a partially ordered (abelian) group.

A **partially ordered division ring** is a partially ordered ring whose $(0,1;+,-,\cdot)$-reduct is a division ring. A **partially ordered field** is a commutative partially ordered division ring, etc. etc.

**Exercise 5.5.1.** Show that every dense linear ordering is infinite. What would happen if the nontriviality axiom, (5), were dropped?

**Exercise 5.5.2.** Prove that every linearly ordered group is torsionfree. Derive that every linearly ordered division ring has characteristic 0.

**Exercise 5.5.3.** Prove that homomorphic images of linear orderings need not be partial orderings (in our sense). Show that, nevertheless, every homomorphic image of a given linear ordering $X$ which is itself a partial ordering is isomorphic to $X$.

## 5.6 Boolean algebras

We consider boolean algebras in the signature $(0,1;+,\cdot,^{-})$, where 0 and 1 are constant symbols (for the smallest and greatest elements), $+$ and $\cdot$ are binary function symbols (for supremum and infimum), and $^{-}$ is a unary function symbol (for the complement). In the corresponding language, **boolean algebras** are axiomatized by the following sentences.

(1) $\forall xyz\, (x + (y + z) = (x + y) + z \wedge x \cdot (y \cdot z) = (x \cdot y) \cdot z)$ **associativity**

(2) $\forall xy\, (x + y = y + x \wedge x \cdot y = y \cdot x)$ **commutativity**

(3) $\forall x\, (x + x = x \wedge x \cdot x = x)$ **idempotence**

(4) $\forall xyz\, (x + (y \cdot z) = (x + y) \cdot (x + z) \wedge x \cdot (y + z) = (x \cdot y) + (x \cdot z))$ **distributivity**

(5) $\forall xy\, (x + (x \cdot y) = x \wedge x \cdot (x + y) = x)$ **absorption**

(6) $\forall xy\, (\overline{x + y} = \overline{x} \cdot \overline{y} \wedge \overline{x \cdot y} = \overline{x} + \overline{y})$ **De Morgan's laws**

(7) $\forall x\, (x + 0 = x \wedge x \cdot 0 = 0 \wedge x + 1 = 1 \wedge x \cdot 1 = x \wedge 0 \neq 1 \wedge x + \overline{x} = 1$
$\wedge x \cdot \overline{x} = 0 \wedge \overline{\overline{x}} = x)$ **laws about 0, 1, and** $^{-}$

Denote the $(0,1;+,\cdot,^{-})$-theory of all boolean algebras by BA. Thus BA is the deductive closure of $\{(1),(2),(3),(4),(5),(6),(7)\}$.

Every boolean algebra can be partially ordered by stipulating that $x \leq y$ if and only if $x + y = y$, and $x < y$ if and only $x \leq y$ and $x \neq y$. Obviously, in this ordering 1 is the greatest element and 0 is the smallest element. Further, it is not hard to see that, with respect to this partial ordering, $x + y$ is the smallest upper bound (the *supremum*) while $x \cdot y$ is the greatest lower bound (the *infimum*) of the set $\{x, y\}$ for any $x, y$.

**Example.** Let $X$ be a nonempty set and $S$ a subset of the power set of $X$ that contains $\emptyset$ and $X$ and is closed under taking (finite) unions and intersections as well as complements. Then $(S; \emptyset, X; \cup, \cap, X\setminus)$ is called a **boolean algebra of sets** or just a **set algebra**. It is easy to see that this is indeed a boolean algebra. In such an algebra the ordering defined above turns out to be set inclusion. It is an important result of the theory of boolean algebras that, conversely, every boolean algebra is isomorphic to ('can be represented as') a set algebra. More precisely, one can embed every boolean algebra $\mathcal{B}$ in the boolean algebra of *all* subsets of a certain set. (Hence, when looking for a boolean algebra with certain properties, one may as well confine oneself to set algebras.) We will state this famous result without proof (as Theorem 5.6.1). However, the right choice of this set, which is the so-called Stone space of $\mathcal{B}$, is half way down the road to a complete proof. In order to define this properly we have to extend the concept of filter and ultrafilter (from 4.1 and 4.2) to arbitrary boolean algebras, which we do next.

Let $\mathcal{B} \models \mathrm{BA}$. A nonempty subset $F \subseteq B$ is called a **filter** of the boolean algebra $\mathcal{B}$ if the following conditions are satisfied for all $a, b \in B$.
(i)  $0 \notin F$
(ii)  If $a, b \in F$, then $a \cdot b \in F$.
(iii)  If $a \in F$ and $a \leq b$, then $b \in F$.

Note that all we did in this generalization is replace (the algebra of all subsets of) the set $I$ and the partial order of inclusion by an arbitrary boolean algebra $\mathcal{B}$ and its canonical partial ordering, respectively. In the same way we can generalize **principal filter, generation, ultrafilter**, and **principal ultrafilter**. E. g., given an element $b \neq 0$ in $\mathcal{B}$, the principal filter generated by $b$ is the set $F(b) = \{a \in B : b \leq a\}$; and an ultrafilter of $\mathcal{B}$ is a filter, $F$, such that $a \in F$ or $\bar{a} \in F$ for every $a \in B$. It is common to use $\mathrm{S}(\mathcal{B})$ to denote the set of all ultrafilters of $\mathcal{B}$, where 'S' stands for (M. H.) Stone.

**Remarks.** Let $\mathcal{B} \models \mathrm{BA}$.
(1)  $1 \in F$ for every filter $F$ of $\mathcal{B}$.
(2)  Given $F \subseteq B$ satisfying filter axiom (iii), $F$ satisfies axiom (i) if and only if $F \neq B$.
(3)  The sets $\langle a \rangle =_{\mathrm{def}} \{x \in \mathrm{S}(\mathcal{B}) : a \in x\}$, where $a$ runs through $B$, form a basis of a topology on $\mathrm{S}(\mathcal{B})$, the so-called **Stone topology**. The corresponding space is known as the **Stone space** of the boolean algebra $\mathcal{B}$ and denoted by $\mathrm{S}(\mathcal{B})$ again.

(4) As in the previous chapter, Zorn's lemma can be used to show that every filter of a boolean algebra can be extended to an ultrafilter. In general some form of the axiom of choice is necessary for this. If the boolean algebra happens to be well-ordered, however, in particular, if it is countable, there is a way around this (exercise!).

A subset of a topological space is said to be **clopen** if it is at the same time *clo*sed and *open*. A **Stone space** is a nonempty topological space which has a basis of clopen sets and which is **compact** (in the sense that every covering by open subsets contains a finite subcovering) and **hausdorff** (in the sense that any two distinct points can be separated by disjoint open subsets).

**Theorem 5.6.1.** (The Stone Representation Theorem)
(1) *If $\mathcal{B} \models$ BA then* S($\mathcal{B}$) *is a Stone space, the so-called* **Stone space** *of the boolean algebra $\mathcal{B}$.*
(2) *If $S$ is a Stone space, then the clopen subsets of $S$ form a boolean set algebra* B($S$).
(3) *Every boolean algebra $\mathcal{B}$ is isomorphic to the boolean algebra* B(S($\mathcal{B}$)) *via the map $a \mapsto \langle a \rangle$. Hence $\mathcal{B}$ is isomorphic to a subalgebra of the boolean algebra of all subsets of* S($\mathcal{B}$).
(4) *Every Stone space $S$ is homeomorphic to the Stone space* S(B($S$)) *via the map $x \mapsto \{a \in $ B($S$) $: x \in a\}$.* □

[Stone, M. H. : The representation theorem for boolean algebra, Transactions of the American Mathematical Society **40** (1936) 37 – 111]

We leave the proof as an exercise (see also the next section or some of the algebraic literature listed in the bibliography).

An **atom** in a boolean algebra is an element $x \neq 0$ such that there's no element between 0 and $x$, i. e., for all $y$ with $0 \leq y \leq x$ we have either $y = 0$ or $y = x$. A boolean algebra is called **atomic** if for every $x \neq 0$ there is an atom $y$ with $y \leq x$. A boolean algebra is said to be **atomless** if it contains no atoms.

**Remark.** There are boolean algebras that are neither atomic nor atomless (exercise!).

We obtain axiomatizations for the class of atomic boolean algebras by adding the axiom

$$\forall x \, (x \neq 0 \longrightarrow \exists y \, (y \leq x \wedge y \neq 0 \wedge \forall z \, (z < y \longrightarrow z = 0))),$$

while adding the axiom

$$\forall y \, (y \neq 0 \longrightarrow \exists z \, (0 < z \wedge z < y))$$

yields an axiomatization of the class of atomless boolean algebras.

Important (and maybe historically the first) examples of boolean algebras are the so-called Lindenbaum-Tarski algebras, which we conclude with. For this, first partition the set $L_0$ of all $L$-sentences into equivalence classes with respect to logical equivalence $\sim$, cf. §3.3. Let $\varphi/\sim$ denote the equivalence class of $\varphi \in L_0$ and $L_0/\sim$ the set of all such equivalence classes. On this latter set one can canonically define the operations $\wedge$, $\vee$, and $\neg$ (by passing to representatives; exercise!).

Then $\mathcal{B}_L = (L_0/\sim; \bot/\sim, \top/\sim; \vee, \wedge, \neg)$ forms a boolean algebra, the **Lindenbaum-Tarski algebra** of the language $L$.

One can extend this as follows to $T$-equivalence modulo an arbitrary $L$-theory $T$ as well as to arbitrary formulas.

Suppose $T$ is an $L$-theory (or even just a subset of $L_0$) and $\bar{x}$ is an $n$-tuple of variables.

Given $\varphi, \psi \in L_{\bar{x}}$, we write $\varphi \leq_T \psi$ if $T \models \forall \bar{x} \, (\varphi \longrightarrow \psi)$, and we write $\varphi \sim_T \psi$ if $\varphi \leq_T \psi$ and $\psi \leq_T \varphi$. Then $(L_{\bar{x}}/\sim_T; \bot/\sim_T, \top/\sim_T; \vee, \wedge, \neg)$ forms a boolean algebra whose isomorphism type depends on $n$ and $T$ only, whence the notation $\mathcal{B}_n(T)$ for it. Its canonical ordering, again denoted by $\leq_T$, satisfies $\varphi/\sim_T \leq_T \psi/\sim_T$ iff $T \models \forall \bar{x} \, (\varphi \longrightarrow \psi)$. The algebra $\mathcal{B}_n(T)$ is also known as the $n$th **Lindenbaum-Tarski algebra** of $T$.

The 0th Lindenbaum-Tarski algebra $\mathcal{B}_0(\emptyset^\models)$ is thus just $\mathcal{B}_L$. The $L$-theories are (up to logical equivalence) the filters of this algebra, while its ultrafilters are the complete $L$-theories (exercise!).

Note that, by Exercise 3.3.5, the class $\varphi/\sim_T$ contains every formula from $L_{\bar{x}}$ that differs from $\varphi$ by redundant variables.

**Exercise 5.6.1.** Prove—without using the axiom of choice (or any equivalent, like Zorn's lemma)—that in a well-ordered boolean algebra, every filter can be extended to an ultrafilter. (Note that this applies to any countable boolean algebra, in particular to the Lindenbaum-Tarski algebras of countable languages (or theories).)

**Exercise 5.6.2.** Prove Theorem 5.6.1.

**Exercise 5.6.3.** Find a boolean algebra which is neither atomic nor atomless.

**Exercise 5.6.4.** Show that the principal ultrafilters of a boolean algebra are precisely the (ultra)filters generated by atoms.

**Exercise 5.6.5.** Verify the statements made in the text that $\mathcal{B}_L$ and $\mathcal{B}_n(T)$ are well defined.

**Exercise 5.6.6.** Prove the aforementioned fact that the $L$-theories are (up to logical equivalence) the filters of the algebra $\mathcal{B}_L$, while its ultrafilters are the complete $L$-theories (so the Stone space of $\mathcal{B}_L$ is $S_L$, as introduced in the beginning of the next section).

## 5.7  Some topology (or why the finiteness theorem is also called compactness theorem)

This section serves among other things as preparation for §9.1 (in particular Lemma 9.1.1) and §11.3. Another objective is to show that the finiteness theorem is equivalent to the compactness theorem saying that the set of all complete $L$-theories, equipped with a certain topology, is compact (in the sense of the previous section). The proof of the direction that finiteness follows from compactness is left as an exercise. However, we are going to set the stage for this. Note that we may not use the finiteness theorem then!

Consider the set of all complete $L$-theories, and denote it by $S_L$. Given $\varphi \in L_0$, set $\langle \varphi \rangle =_{\text{def}} \{T \in S_L \,:\, \varphi \in T\}$. As $\varphi \wedge \psi$ is in $T$ if and only if $\varphi$ and $\psi$ are in $T$, we see that $\langle \varphi \rangle \cap \langle \psi \rangle = \langle \varphi \wedge \psi \rangle$. Hence $\{\langle \varphi \rangle \,:\, \varphi \in L_0\}$ constitutes a basis for a topology on $S_L$. Denote the resulting topological space again by $S_L$. This reminds us of Remark (3) before Theorem 5.6.1 and might deceive us into inferring that this is the Stone space of the Lindenbaum-Tarski algebra $\mathcal{B}_L$. As a matter of fact, it is (Exercise 5.6.6)! What's wrong then? The point is that this rests on the finiteness theorem, which we must not use, at least not in the proof that compactness implies finiteness. We could use the Stone space argument, of course, in the proof of the other direction (that finiteness implies compactness), which would then be a trivial consequence of the Stone representation theorem. However, we want to avoid this latter theorem and give a straightforward proof instead, which is interesting in its own right, see Theorem 5.7.1 below. Finally, this is a good exercise in understanding Stone spaces.

So let's get back to the topological space $S_L$ as defined above. Its basic open sets $\langle \varphi \rangle$ are also closed, for $S_L \smallsetminus \langle \varphi \rangle = \langle \neg \varphi \rangle$. Hence $S_L$ has a basis of clopen sets. It is also hausdorff, since for all $T, T' \in S_L$ with $T \neq T'$ there is $\varphi \in L_0$ such that $T \in \langle \varphi \rangle$ and $T' \in \langle \neg \varphi \rangle$. (That $S_L$ is a Stone space will follow—without the use of Stone's theorem—once we have compactness).

The open sets of $S_L$ are the sets of the form $\bigcup_{\varphi \in \Sigma} \langle \varphi \rangle$, where $\Sigma \subseteq L_0$. Hence the closed sets of $S_L$ are the sets of the form $\bigcap_{\varphi \in \Sigma} \langle \varphi \rangle = \{T \in S_L \,:\, \Sigma \subseteq T\}$, where $\Sigma \subseteq L_0$.

It is not hard to verify (exercise!) that $\bigcap_{\varphi \in \Sigma} \langle \varphi \rangle \mapsto \Sigma^{\vDash}$ defines a bijection between the nonempty closed sets in $S_L$, on the one hand, and the $L$-theories on the other. Therefore the space $S_L$ reflects not only the complete, but *all* $L$-theories.

**Theorem 5.7.1.** (The Compactness Theorem) *The space $S_L$ as defined above is compact, that is every open covering of $S_L$ contains a finite sub-covering.*

*Proof.* Suppose $S_L = \bigcup_{i \in I} U_i$ is a covering such that $U_i = \bigcup_{\varphi \in \Sigma_i} \langle \varphi \rangle$ for some $\Sigma_i \subseteq L_0$ (i. e., every $U_i$ is open). Set $\Sigma = \bigcup_{i \in I} \Sigma_i$. We will find $\varphi_0, \ldots, \varphi_{n-1} \in \Sigma$ such that $S_L = \bigcup_{i<n} \langle \varphi_i \rangle$.

We have

$$\emptyset = S_L \smallsetminus \bigcup_{i \in I} U_i = S_L \smallsetminus \bigcup_{\varphi \in \Sigma} \langle \varphi \rangle = \bigcap_{\varphi \in \Sigma} S_L \smallsetminus \langle \varphi \rangle = \bigcap_{\varphi \in \Sigma} \langle \neg \varphi \rangle.$$

This means that the set of sentences $\{\neg \varphi : \varphi \in \Sigma\}$ has no completion, hence no model either. The finiteness theorem then yields finitely many $\varphi_0, \ldots, \varphi_{n-1} \in \Sigma$ such that $\{\neg \varphi_0, \ldots, \neg \varphi_{n-1}\}$ has no model, hence no completion either. This in turn means that $\bigcap_{i<n} \langle \neg \varphi_i \rangle = \emptyset$. Thus

$$S_L = S_L \smallsetminus \bigcap_{i<n} \langle \neg \varphi_i \rangle = \bigcup_{i<n} S_L \smallsetminus \langle \neg \varphi_i \rangle = \bigcup_{i<n} \langle \varphi_i \rangle,$$

and consequently there are $j_i \in I$ $(i < n)$ such that $S_L = \bigcup_{i<n} U_{j_i}$. □

**Corollary 5.7.2.** *The clopen sets of $S_L$ are precisely the sets of the form $\langle \varphi \rangle$, where $\varphi \in L_0$.*

*Proof.* $\Longleftarrow$ is clear.
$\Longrightarrow$. Obviously, the empty set $\emptyset = \langle \varphi \wedge \neg \varphi \rangle$ is of this form. Let now $\emptyset \neq U = \bigcup_{\varphi \in \Sigma} \langle \varphi \rangle$ be closed for any $\Sigma \subseteq L_0$. Being a closed subset of a compact space, $U$ is compact. Hence $U = \bigcup_{i<n} \langle \varphi_i \rangle = \langle \bigvee_{i<n} \varphi_i \rangle$ for some $\varphi_i$ $(i < n)$. Thus $U$ is of the desired form. □

**Example.** Consider the theory of pure identity, $T_=$. Let $T_=^n$ be the $L_=$-theory containing the sentence $\exists^{=n} x \, (x = x)$. There is only one such theory, as there is only one model satisfying this sentence. The theory $T_=^n$ is therefore complete. The theory $T_=^\infty$ of the infinite models of $T_=$ (cf. p. 33) has in every infinite cardinality up to isomorphism exactly one model. That it is in fact complete will then follow from Theorem 8.5.1 below. Then $S_{L_=} = \{T_=^n : n \in \mathbb{N}\} \cup \{T_=^\infty\}$, where the $T_=^n$ are finitely axiomatizable and isolated in $S_{L_=}$.

**Exercise 5.7.1.** Verify the statement made about the bijection given before the compactness theorem.

**Exercise 5.7.2.** Derive the finiteness theorem from the compactness theorem.

**Exercise 5.7.3.** Show that a complete $L$-theory is finitely axiomatizable if and only if it is isolated as a point in $S_L$.

**Exercise 5.7.4.** Derive from the previous exercise that $T_\cong^\infty$ is not finitely axiomatizable (a result we already know from Exercise 4.3.4).

# Chapter 6

# Malcev's applications to group theory

In this chapter we present a method for the transfer of certain properties of finitely generated subgroups of a given group to the group itself developed by Anatolij Ivanovich Malcev [read: 'Maltsev'] in his classical paper

[Мальцев, А. И. (Malcev, A. I.): Об одном общем методе получения локальных теорем теории групп, Uchenye Zapiski Ivanov. Ped. Inst. 1, 1 (1941) 3 – 9]

whose English translation appeared as

[A. I. Malcev : A general method for obtaining local theorems in group theory]

only in the 1971 volume of his collected papers (pp. 15 – 21) cited in Appendix H.

We start with an account of the method of diagrams, introduced independently by Malcev and Abraham Robinson, and proceed with the method of interpretation, which seems to have been some sort of folklore as early as in the thirties.

## 6.1 Diagrams

Recall that if a structure $\mathcal{M}$ is a substructure of a structure $\mathcal{N}$, then all symbols of the signature are interpreted in $\mathcal{M}$ in the same way as they are (on the subset $M$) in $\mathcal{N}$; in other words, $\mathcal{M} \models \varphi(\bar{a})$ if and only if $\mathcal{N} \models \varphi(\bar{a})$ for every unnested atomic formula $\varphi$ and every tuple $\bar{a}$ from $M$. In the first lemma we extend this to arbitrary quantifier-free formulas $\varphi$.

**Lemma 6.1.1.** *Let $\mathcal{M}$ and $\mathcal{N}$ be L-structures.*

$\mathcal{M} \subseteq \mathcal{N}$ (*i. e.,* $\mathcal{M}$ *is a substructure of* $\mathcal{N}$) *if and only if* $M \subseteq N$, $M$ *is closed in* $\mathcal{N}$ *under functions from the signature (cf. §1.4), and*
(*)    $\mathcal{M} \models \varphi(\bar{a})$ *iff* $\mathcal{N} \models \varphi(\bar{a})$
*for all quantifier-free L-formulas* $\varphi$ *and matching tuples* $\bar{a}$ *from* $M$.

*Proof.* The direction from right to left is trivial (it suffices to consider unnested atomic formulas for that).

For the converse, we need only show that condition (*) for unnested formulas—as given by the definition of substructure—implies the full condition (*). The first step for this is to prove (*) for *all* atomic formulas, which can be easily established by an induction on the complexity of terms occurring in the atomic formulas. The second step, an induction on the complexity of the quantifier-free formulas, is even easier, since the condition (*) easily 'extends' to conjunctions, disjunctions and negations. The details are left as an exercise.    □

Next we introduce some useful technical notation. Suppose $\mathcal{M}$ and $\mathcal{N}$ are $L$-structures and $\Delta$ is an arbitrary class of formulas (not necessarily from $L$).

We write $\mathcal{M} \Rightarrow_\Delta \mathcal{N}$ or $\mathcal{N} \Leftarrow_\Delta \mathcal{M}$ if $\mathcal{M} \models \varphi$ implies $\mathcal{N} \models \varphi$, for all $\varphi \in \Delta \cap L_0$ (or, equivalently, $\mathrm{Th}_\Delta \mathcal{M} \subseteq \mathrm{Th}_\Delta \mathcal{N}$). The notation $\mathcal{M} \equiv_\Delta \mathcal{N}$ is used to denote that both, $\mathcal{M} \Rightarrow_\Delta \mathcal{N}$ and $\mathcal{N} \Rightarrow_\Delta \mathcal{M}$ hold. We omit the subscript $\Delta$ in case $\Delta \supseteq L_0$.

The notation $f : \mathcal{M} \xrightarrow{\Delta} \mathcal{N}$ means, by definition, that $f : M \to N$ and that $\mathcal{M} \models \varphi(\bar{a})$ implies $\mathcal{N} \models \varphi(f[\bar{a}])$, for all $\varphi \in \Delta \cap L$ (hence also for $L$-formulas $\varphi$ containing free variables) and all matching tuples $\bar{a}$ from $M$. For reasons of tradition we write $f : \mathcal{M} \xrightarrow{\equiv} \mathcal{N}$ instead of $f : \mathcal{M} \xrightarrow{L} \mathcal{N}$. If $\Delta = \{\varphi\}$, we omit the braces.

For technical reasons we have built some redundancy into these definitions: in $\mathcal{M} \Rightarrow_\Delta \mathcal{N}$ (or $f : \mathcal{M} \xrightarrow{\Delta} \mathcal{N}$) only those formulas in $\Delta$ count that are also in $L$. For instance, if $\Delta$ contains all of $L$, then $f : \mathcal{M} \xrightarrow{\Delta} \mathcal{N}$ says nothing more than just $f : \mathcal{M} \xrightarrow{\equiv} \mathcal{N}$.

As an illustration, we state a few facts whose proof we leave as an exercise.

**Remark.**
(1)    $f : \mathcal{M} \xrightarrow{\Delta} \mathcal{N}$ iff $(\mathcal{M}, M) \Rightarrow_{\Delta(M)} (N, f[M])$.

(2)    Let $\Delta \subseteq L_0$. Then $f : \mathcal{M} \xrightarrow{\Delta} \mathcal{N}$ if and only if $\mathcal{M} \Rightarrow_\Delta \mathcal{N}$ and $f : M \to N$.

(3)    $f : \mathcal{M} \to \mathcal{N}$ if and only if $f : \mathcal{M} \xrightarrow{\text{at}} \mathcal{N}$.

(4) Suppose $\Delta \subseteq L$ contains **at**, the set of all atomic formulas, as well as all negations of unnested relational atomic formulas, i. e., all formulas $\neg R(\bar{x})$ where $R \in \mathbf{R}$.

Then $f : \mathcal{M} \overset{\Delta}{\to} \mathcal{N}$ implies that $f$ is a strong homomorphism (while the converse is not true).

(5) $f : M \to N$ is injective if and only if $f : \mathcal{M} \overset{\Delta}{\to} \mathcal{N}$ for the set $\Delta = \{x \neq y\}$.

(6) If $\Delta \subseteq L_0$ contains, along with every sentence, also its negation, then $\mathcal{M} \Rrightarrow_\Delta \mathcal{N}$ implies $\mathcal{M} \equiv_\Delta \mathcal{N}$.

(7) If $\Delta \subseteq L$ contains, along with every formula, also its negation, then $f : \mathcal{M} \overset{\Delta}{\to} \mathcal{N}$ implies that for all $\varphi \in \Delta$ and tuples $\bar{a}$ from $M$ we have $\mathcal{M} \models \varphi(\bar{a})$ iff $\mathcal{N} \models \varphi(f[\bar{a}])$.

Remark (2) shows that, in case $\Delta$ consists of sentences only, the map $f$ is redundant in the notation $f : \mathcal{M} \overset{\Delta}{\to} \mathcal{N}$.

If $x \neq y \in \Delta$, Remark (5) allows us to write $f : \mathcal{M} \overset{\Delta}{\hookrightarrow} \mathcal{N}$ instead of $f : \mathcal{M} \overset{\Delta}{\to} \mathcal{N}$.

The relation $\equiv_\Delta$, is an equivalence relation. In the particular case where $\Delta = L_0$ this relation, that is $\equiv$, is known as **elementary equivalence**. It will be investigated in more detail in §8.1. Similarly, the maps $f : \mathcal{M} \overset{\equiv}{\hookrightarrow} \mathcal{N}$ are known as **elementary maps** and will be investigated in more detail in §8.2.

Let $\mathcal{M}$ be an $L$-structure. The **diagram** of $\mathcal{M}$, in symbols $\mathrm{D}(\mathcal{M})$, is, by definition, the set of all atomic and negated atomic $L(M)$-sentences that are true in $\mathcal{M}$; i. e.

$$\mathrm{D}(\mathcal{M}) = \{\varphi(\bar{a}) : \mathcal{M} \models \varphi(\bar{a}), \varphi \in L \text{ is atomic}, \bar{a} \text{ in } M\} \cup$$
$$\{\neg\varphi(\bar{a}) : \mathcal{M} \models \neg\varphi(\bar{a}), \varphi \in L \text{ is atomic}, \bar{a} \text{ in } M\}.$$

The diagram of $\mathcal{M}$ is thus the set of all $L(M)$-literals that are sentences and true in $\mathcal{M}$. The diagram of an empty structure is empty, as empty structures exist only in languages without constant symbols, which, by Remark (3) in §2.5 have no atomic sentences.

**Example.** The diagram of a group is TG-equivalent to the Cayley diagram (i. e. its multiplication table) joined by all negations of term equations that are *not* true in the group (see also the remark at the end of §6.4*).

Lemma 6.1.1 can now be restated like this: $\mathcal{M} \subseteq \mathcal{N}$ if and only if $M \subseteq N$, $M$ is closed under functions, and $(\mathcal{N}, M) \models \mathrm{D}(\mathcal{M})$. The latter is equivalent to $\mathrm{id}_M : \mathcal{M} \overset{\mathbf{qf}}{\to} \mathcal{N}$ (where **qf** denotes the class of all quantifier-free formulas, as introduced in §2.4). We have, more generally,

**Lemma 6.1.2.** (The Diagram Lemma)
*Suppose $\mathcal{M}$ and $\mathcal{N}$ are L-structures and $f : M \to N$.*
(1) $f : \mathcal{M} \hookrightarrow \mathcal{N}$ (*i. e., f is a monomorphism from $\mathcal{M}$ to $\mathcal{N}$*) *if and only if*
   $f : \mathcal{M} \overset{\mathbf{qf}}{\to} \mathcal{N}$ *if and only if* $(\mathcal{N}, f[M]) \models D(\mathcal{M})$.
   *In particular, $f : \mathcal{M} \overset{\equiv}{\hookrightarrow} \mathcal{N}$ implies $f : \mathcal{M} \hookrightarrow \mathcal{N}$.*
(2) $\mathcal{M} \hookrightarrow \mathcal{N}$ *if and only if $\mathcal{N}$ has an L(M)-expansion which is a model of* $D(\mathcal{M})$.

*Proof.* Ad (1). Let $\Delta$ be the set of all literals. As in the preceding lemma, we have $f : \mathcal{M} \hookrightarrow \mathcal{N}$ iff $f : \mathcal{M} \overset{\mathbf{qf}}{\to} \mathcal{N}$ iff $f : \mathcal{M} \overset{\Delta}{\to} \mathcal{N}$. But the latter is obviously equivalent to $(\mathcal{N}, f[M]) \models D(\mathcal{M})$.

(2) follows from (1), for if $\mathcal{N}'$ is an $L(M)$-expansion of $\mathcal{N}$ satisfying $D(\mathcal{M})$, then, on setting $f(a) = \underline{a}^{\mathcal{N}'}$ for all $a \in M$, we obtain $f : \mathcal{M} \hookrightarrow \mathcal{N}$. □

This implies the important feature of isomorphic structures that—also with respect to their theory—they are indistinguishable.

**Proposition 6.1.3.** *If $f : \mathcal{M} \cong \mathcal{N}$, then $f : \mathcal{M} \overset{\equiv}{\hookrightarrow} \mathcal{N}$, hence also $\mathcal{M} \equiv \mathcal{N}$ (i. e. $\mathrm{Th}\,\mathcal{M} = \mathrm{Th}\,\mathcal{N}$).*

*Proof.* We already know from Lemma 6.1.1 that $f : \mathcal{M} \overset{\mathbf{qf}}{\to} \mathcal{N}$, i. e. that
(*)   $\mathcal{M} \models \varphi(\bar{a})$ iff $\mathcal{N} \models \varphi(f[\bar{a}])$
holds for all $\varphi \in \mathbf{qf}$ and matching tuples $\bar{a}$ from $M$.

By induction on the complexity of $\varphi$ we are going to verify (*) for all $\varphi \in L$. The inductive steps for conjunction and negation are trivial. So we are left with the existential quantifier step.

Let $\varphi$ be of the form $\exists x\, \psi(x, \bar{a})$, where $\psi \in L_{n+1}$ and $\bar{a} \in M^n$. Suppose that for all $b \in M$ we have $\mathcal{M} \models \psi(b, \bar{a})$ iff $\mathcal{N} \models \psi(f(b), f[\bar{a}])$. We have to show that then $\mathcal{M} \models \exists x\, \psi(x, \bar{a})$ iff $\mathcal{N} \models \exists x\, \psi(x, f[\bar{a}])$.

Note that $\mathcal{M} \models \exists x\, \psi(x, \bar{a})$ if and only if $\mathcal{M} \models \psi(b, \bar{a})$ for some $b \in M$. By the above supposition the latter is, in turn, equivalent to the existence of some $b \in M$ such that $\mathcal{N} \models \psi(f(b), f[\bar{a}])$. Since $f$ is surjective, this is equivalent to the existence of $c \in N$ such that $\mathcal{N} \models \psi(c, f[\bar{a}])$, hence to $\mathcal{N} \models \exists x\, \psi(x, f[\bar{a}])$, as desired.

This concludes the proof of $f : \mathcal{M} \overset{\equiv}{\hookrightarrow} \mathcal{N}$, and $\mathcal{M} \equiv \mathcal{N}$ follows as a special case. □

**Remark.** (8)  This can be used to find, for any structure $\mathcal{M}$, a disjoint structure $\mathcal{N} \models \mathrm{Th}\,\mathcal{M}$: consider a bijection of $f$ of $M$ onto an arbitrary disjoint set $N$ and define thereon a structure $\mathcal{N}$ according to the preimages. Then $f : \mathcal{M} \cong \mathcal{N}$ and hence, by the preceding result, $\mathcal{N} \models \mathrm{Th}\,\mathcal{M}$.

(9)  Another important consequence is that sets definable without parameters are invariant under automorphisms, i. e., given an $L$-structure $\mathcal{M}$ and an automorphism $f \in \mathrm{Aut}\,\mathcal{M}$, we have $f[\psi(\mathcal{M})] = \psi(\mathcal{M})$ for all $\psi \in L$.

**Exercise 6.1.1.** Write out a detailed proof of Lemma 6.1.1.

**Exercise 6.1.2.** Verify Remarks (1) through (9).

**Exercise 6.1.3.** Prove $\mathrm{Th}_{\mathbf{qf}}(\mathcal{M}, M) \subseteq \mathrm{D}(\mathcal{M})^{\models}$ (see notation in §3.4).

**Exercise 6.1.4.** Show that if $\mathcal{M} \subseteq \mathcal{N}$, then $\mathrm{Th}_{\mathbf{qf}}(\mathcal{M}, M) = \mathrm{Th}_{\mathbf{qf}}(\mathcal{N}, M)$, hence also $\mathrm{Th}_{\mathbf{qf}}(\mathcal{N}, M) \subseteq \mathrm{D}(\mathcal{M})^{\models}$.

**Exercise 6.1.5.** Verify that the diagram of a structure is logically equivalent to the subset of all its *unnested* atomic formulas and all its negations of *unnested* atomic formulas.

**Exercise 6.1.6.** Show that the diagram of a finite structure in a finite signature is finitely axiomatizable.

**Exercise 6.1.7.** Write down the diagram of a field with three elements (up to TF-equivalence).

**Exercise 6.1.8.** Generalize Remark (9) appropriately to parametrically defined sets.

## 6.2  Simple preservation theorems

We start with a simple consequence of the diagram lemma 6.1.2.

**Lemma 6.2.1.** *Let $f : \mathcal{M} \hookrightarrow \mathcal{N}$.*
*Then $\mathcal{M} \Lleftarrow_{\forall} \mathcal{N}$ and even $(\mathcal{M}, M) \Lleftarrow_{\forall} (\mathcal{N}, f[M])$.*

*Proof.* Let $\psi(\bar{x})$ be an arbitrary $\forall$-formula, i. e., $\psi$ has the form $\forall \bar{y}\, \varphi(\bar{x}, \bar{y})$, where $\varphi \in \mathbf{qf}$. By Lemma 6.1.2(1) (and Remark (7) of the preceding section) we have $\mathcal{M} \models \varphi(\bar{a}, \bar{b})$ iff $\mathcal{N} \models \varphi(f[\bar{a}], f[\bar{b}])$ for all tuples $\bar{a}$ and $\bar{b}$ from $M$. Hence $\mathcal{N} \models \forall \bar{y}\, \varphi(f[\bar{a}], \bar{y})$ implies $\mathcal{M} \models \forall \bar{y}\, \varphi(\bar{a}, \bar{y})$, i. e., $\mathcal{N} \models \psi(f[\bar{a}])$ implies $\mathcal{M} \models \psi(\bar{a})$. $\qquad\square$

Let $T$ be an $L$-theory. Mod $T$ (or just $T$) is said to be **closed** under (or **preserved** in) (nonempty) **substructures** if $\mathcal{N} \models T$ implies $\mathcal{M} \models T$ for all nonempty $L$-structures $\mathcal{M}$ and $\mathcal{N}$ with $\mathcal{M} \subseteq \mathcal{N}$. For 'closed under substructures' one often says **hereditary**.

**Remark.** If $T$ is preserved in substructures, then it is also preserved in embeddings (downward), i. e., if $\mathcal{M} \hookrightarrow \mathcal{N} \models T$ and $M \neq \emptyset$ then $\mathcal{M} \models T$ (for then there is $\mathcal{N}' \subseteq \mathcal{N}$ isomorphic to $\mathcal{M}$; and the former yields $\mathcal{N}' \models T$, which, together with the latter, implies $\mathcal{M} \models T$ by Proposition 6.1.3).

Lemma 6.2.1 implies that $\forall$-theories are preserved in substructures. We are going to prove the converse, for which we need the next lemma. Before that consider the following application of the previous lemma.

**Example.** One cannot universally axiomatize the theory of groups in the signature $(1; \cdot)$, since here, 'substructure' means only 'submonoid'.

**Lemma 6.2.2.** *The following conditions are equivalent for any $L$-theories $T$ and $T'$.*
(i)   *Every model of $T'$ can be embedded in a model of $T$.*
(ii)  $T_\forall \subseteq T'$.

*Proof.* (i)$\Longrightarrow$(ii) is immediate from the preceding lemma (and Remark (2) of §3.4).

(ii)$\Longrightarrow$(i). Let $T_\forall \subseteq T'$ and $\mathcal{M} \models T'$. We have to find some $\mathcal{N} \models T$ with $\mathcal{M} \hookrightarrow \mathcal{N}$. By Lemma 6.1.2(2) this is equivalent to the consistency of $T \cup D(\mathcal{M})$.

Assume $T \cup D(\mathcal{M})$ is inconsistent. Then there is a finite conjunction, $\varphi(\bar{a})$, of formulas from $D(\mathcal{M})$ such that $T \cup \{\varphi(\bar{a})\}$ is inconsistent, hence $T \models \neg\varphi(\bar{a})$. Since the constants $\bar{a}$ do not occur in $T$, the lemma on new constants (3.3.2) then implies $T \models \forall \bar{x} \neg\varphi(\bar{x})$, hence $\forall \bar{x} \neg\varphi(\bar{x}) \in T$. But $\forall \bar{x} \neg\varphi(\bar{x}) \in \forall$ (as $\varphi \in \mathbf{qf}$), whence $\forall \bar{x} \neg\varphi(\bar{x}) \in T_\forall$. Then $\mathcal{M} \models T'$ and $T_\forall \subseteq T'$ imply $\mathcal{M} \models \forall \bar{x} \neg\varphi(\bar{x})$, which contradicts the fact that $\varphi(\bar{a})$ holds in $\mathcal{M}$.                                                    □

**Remark.** The case $T' = T_\forall$ shows that

$$\operatorname{Mod} T_\forall = \{\mathcal{M} : \mathcal{M} \hookrightarrow \mathcal{N} \models T\},$$

i. e., the models of the $\forall$-part, $T_\forall$, of a theory $T$ are (up to isomorphism) *precisely* the substructures of models of $T$. (In Lemma 8.3.1 we will see that we may omit the clause 'up to isomorphism'.) From this one can derive that

$$T_\forall = \operatorname{Th} \{\mathcal{M} : \mathcal{M} \hookrightarrow \mathcal{N} \models T\},$$

see Exercise 6.2.1.

**Proposition 6.2.3.** *The following are equivalent for any L-theory T and any set of formulas,* $\Phi(\bar{x}) \subseteq L_n$ *(cf. notation in §3.3).*

(i)   $\Phi(\bar{x})$ *is T-equivalent to a set of* $\forall$*-formulas from L in the same free variables.*

(ii)  *For all models* $\mathcal{M}$ *and* $\mathcal{N}$ *of T and all* $\bar{a} \in M^n$, *if* $\mathcal{M} \subseteq \mathcal{N}$ *and* $\mathcal{N} \models \Phi(\bar{a})$, *then* $\mathcal{M} \models \Phi(\bar{a})$. *In other words, for all models* $\mathcal{M}$ *and* $\mathcal{N}$ *of T, if* $\mathcal{M} \subseteq \mathcal{N}$, *then* $(\mathcal{N}, M) \Rrightarrow_{\Phi(M)} (\mathcal{M}, M)$.

[Łoś, J. : On extending of models I, Fundamenta Math. **42** (1955) 38 – 54]

[Tarski, A. : Contributions to the theory of models I, II, Koninkl. Ned. Akad. Wetensch. Proc. Ser. A **57** (1954) 572 – 588]

*Proof.* In view of Exercise 3.3.4 it suffices to show that (ii) holds if and only if $\Phi(\bar{c})$ is T-equivalent to a set of $\forall$-sentences in $L(\bar{c})$ (where $\bar{c}$ are new constants). If $\mathcal{M}^*$ and $\mathcal{N}^*$ are $L(\bar{c})$-structures with $\mathcal{M}^* \subseteq \mathcal{N}^*$, then, for their L-reducts $\mathcal{M}$ and $\mathcal{N}$, we have $\mathcal{M} \subseteq \mathcal{N}$, and there is $\bar{a} \in M^n$ such that $\mathcal{M}^* = (\mathcal{M}, \bar{a})$ and $\mathcal{N}^* = (\mathcal{N}, \bar{a})$. Conversely, if $\bar{a} \in M^n$ and $\mathcal{M} \subseteq \mathcal{N}$, then $(\mathcal{M}, \bar{a}) \subseteq (\mathcal{N}, \bar{a})$ as $L(\bar{c})$-structures. Hence (ii) is equivalent to the fact that the set of $L(\bar{c})$-sentences, $\Phi(\bar{c})$, is preserved in substructures whose L-reducts are themselves models of T.

If now $\Phi(\bar{c})$ is T-equivalent to a set of $\forall$-sentences, then (ii) follows from Lemma 6.2.1 for $L(\bar{c})$.

For the converse, let $\Phi(\bar{c})$ be preserved in substructures whose L-reducts are themselves models of T. Let $\Psi(\bar{c})$ be the $\forall$-part of $(T \cup \Phi(\bar{c}))^{\models}$ (in $L(\bar{c})$). By the above remark, the models of $\Psi(\bar{c})$ are precisely the $L(\bar{c})$-substructures of models of $T \cup \Phi(\bar{c})$. Hence, under the above preservation hypothesis, the models of $T \cup \Psi(\bar{c})$ are also models of $T \cup \Phi(\bar{c})$. Consequently, $T \cup \Phi(\bar{c})$ and $T \cup \Psi(\bar{c})$ are equivalent, i. e., $\Phi(\bar{c})$ and $\Psi(\bar{c})$ are T-equivalent. $\square$

**Corollary 6.2.4.** *Let T be an L-theory.*

*An L-formula* $\varphi(\bar{x})$ *is T-equivalent to an* $\forall$*-formula if and only if* $\mathcal{N} \models \varphi(\bar{a})$ *implies* $\mathcal{M} \models \varphi(\bar{a})$, *for all* $\mathcal{M} \models T$ *and* $\mathcal{N} \models T$ *with* $\mathcal{M} \subseteq \mathcal{N}$ *and all* $\bar{a}$ *in* $\mathcal{M}$.

*Proof.* The second clause is, by the the preceding proposition, equivalent to the above (i) for $\Phi = \{\varphi\}$. Invoking the finiteness theorem this shows that $\varphi$ is T-equivalent to (the conjunction of) *finitely* many $\forall$-formulas. But a conjunction of $\forall$-formulas is equivalent to an $\forall$-formula. So (i) says, in this case, that $\varphi$ is T-equivalent to a single $\forall$-formula, as desired. $\square$

**Corollary 6.2.5.** (The Łoś-Tarski Preservation Theorem) *A theory is preserved in substructures if and only if it is an ∀-theory.*

*Proof.* Apply Proposition 6.2.3, where $\Phi$ is the theory under consideration and $T = \emptyset^{\models}$ (and $n = 0$). $\qquad\square$

Let us turn to the inverse situation of passing from substructures to bigger structures. Suppose $T$ is an $L$-theory. $\mathrm{Mod}\,T$ (or just $T$) is said to be **closed** under (or **preserved** in) **extensions** if $\mathcal{M} \models T$ implies $\mathcal{N} \models T$, for all nonempty $L$-structures $\mathcal{M}$ and $\mathcal{N}$ with $\mathcal{M} \subseteq \mathcal{N}$.

**Remark.** As in the previous case, this preservation also holds true for general embeddings: if $T$ is preserved in extensions, then it is also preserved in embeddings (upward), i. e., if $\mathcal{M} \hookrightarrow \mathcal{N}$ and $M \models T$ then $\mathcal{N} \models T$ (by the same argument as in the corresponding remark after Lemma 6.2.1).

The contrapositive of Lemma 6.2.1 shows that if $f : \mathcal{M} \hookrightarrow \mathcal{N}$ then also $f : \mathcal{M} \xrightarrow{\exists} \mathcal{N}$, for the $\exists$-formulas are precisely (equivalent to) the negations of the $\forall$-formulas. In particular, $\mathcal{M} \hookrightarrow \mathcal{N}$ implies $\mathcal{M} \Rightarrow_\exists \mathcal{N}$, hence $\exists$-sentences are preserved under taking extensions. Again there is a corresponding preservation theorem, i. e., the converse is true too, as we will show in the remainder of this section.

If we restrict ourselves to single sentences (or finitely axiomatizable theories), this is almost immediate. First of all, it is easy to see that Corollary 6.2.4 implies the following dual (exercise!).

**Corollary 6.2.6.** *Let $T$ be an $L$-theory.*

*An $L$-formula $\varphi(\bar{x})$ is $T$-equivalent to an $\exists$-formula if and only if, for all $\mathcal{M} \models T$ and $\mathcal{N} \models T$ with $\mathcal{M} \subseteq \mathcal{N}$, we have $\mathcal{M} \xrightarrow{\varphi} \mathcal{N}$ (i. e. $\mathcal{M} \models \varphi(\bar{a})$ implies $\mathcal{N} \models \varphi(\bar{a})$, for all $\bar{a}$ from $\mathcal{M}$).* $\qquad\square$

As a special case, we have

**Corollary 6.2.7.** *A sentence $\varphi$ (or a finitely axiomatizable theory $T$) is preserved in extensions if and only if $\varphi$ is equivalent to an $\exists$-statement (respectively, $T$ is an $\exists$-theory).* $\qquad\square$

The reason that the direction $\Longrightarrow$ cannot be derived so easily for arbitrary (non-finitely axiomatizable) theories $T$ is that in general $\mathcal{M} \not\models T$ is not the same as $\mathcal{M} \models \{\neg\varphi : \varphi \in T\}$ (in other words, it is impossible in general to take the negation of an infinite set of sentences). Nevertheless, the theorem can be proved, albeit by a more complicated argument.

**Lemma 6.2.8.** *Let $\Delta$ be a set of L-sentences that is closed under $\bigvee$ (i. e., if $\varphi_0, \dots, \varphi_{n-1} \in \Delta$, then also $\bigvee_{i<n} \varphi \in \Delta$).*
*An L-theory $T$ is a $\Delta$-theory (i. e. $T \subseteq T_\Delta$) if and only if, for all $\mathcal{M} \models T$ and all $\mathcal{N} \models \mathrm{Th}_\Delta \mathcal{M}$, we have $\mathcal{N} \models T$.*
*Moreover, one can restrict oneself to disjoint $\mathcal{M}$ and $\mathcal{N}$ in this.*

**Remark.** The assertion states that $T$ is a $\Delta$-theory if and only if $T \subseteq T'_\Delta$ for all completions $T'$ of $T$. Thus the lemma holds for complete theories per definitionem.

*Proof.* The direction $\Longrightarrow$ is trivial. For the nontrivial direction, let $\mathcal{N} \models T_\Delta$. We have to verify $\mathcal{N} \models T$ (cf. Remark (2) in §3.4). By hypothesis, for this we only have to find $\mathcal{M} \models T$ such that $\mathcal{N} \models \mathrm{Th}_\Delta \mathcal{M}$.

Set $\neg\Delta = \{\neg\delta : \delta \in \Delta\}$. If $T \cup \mathrm{Th}_{\neg\Delta} \mathcal{N}$ is consistent, such an $\mathcal{M}$ exists. Therefore we are left with showing the consistency of $T \cup \mathrm{Th}_{\neg\Delta}\mathcal{N}$.

So let $\delta_i \in \Delta$ $(i < n)$ and $\mathcal{N} \models \bigwedge_{i<n} \neg\delta_i$. Denote the sentence $\bigvee_{i<n} \delta_i$ by $\varphi$. Then $\mathcal{N} \models \neg\varphi$ and, by hypothesis, $\varphi$ is in $\Delta$. If $T \cup \{\neg\delta_i : i < n\}$ were inconsistent, we would have $T \models \varphi$, hence $\varphi \in T_\Delta$ and thus also $\mathcal{N} \models \varphi$, a contradiction.

Consequently there is a model $\mathcal{M} \models T \cup \mathrm{Th}_{\neg\Delta}\mathcal{N}$, which, by Remark (8) at the end of §6.1, may be chosen disjoint from $N$. □

Noting that this lemma applies to the case $\Delta = \exists$, we are now able to give a proof of the preservation theorem for extensions.

**Theorem 6.2.9.** (Łoś' Preservation Theorem) *A theory is preserved in extensions if and only if it is an $\exists$-theory.*

*Proof.* We proved already that $\exists$-theories are preserved in extensions. For the converse, suppose the theory $T$ is preserved in extensions. Let us first prove the following assertion.
(*) $\mathrm{Th}\,\mathcal{N} \cup \mathrm{D}(\mathcal{M})$ is consistent for every structure $\mathcal{M}$ and all models $\mathcal{N} \models \mathrm{Th}_\exists \mathcal{M}$.
To prove (*), let $\varphi(\bar{a})$ be a finite conjunction of sentences from $\mathrm{D}(\mathcal{M})$. Then $\varphi$ is quantifier-free , $\bar{a}$ is from $M$, and $\mathcal{M} \models \varphi(\bar{a})$. Clearly, $\mathcal{M} \models \exists\bar{x}\,\varphi(\bar{x})$ and hence $\exists\bar{x}\,\varphi(\bar{x}) \in \mathrm{Th}_\exists\mathcal{M}$. Since $\mathcal{N} \models \mathrm{Th}_\exists\mathcal{M}$, we also have $\mathcal{N} \models \exists\bar{x}\,\varphi(\bar{x})$. Hence there is a tuple $\bar{b}$ in $N$ such that $\mathcal{N} \models \varphi(\bar{b})$. Then $(\mathcal{N}, \bar{b}) \models \mathrm{Th}\,\mathcal{N} \cup \{\varphi(\bar{a})\}$, and (*) follows on applying the finiteness theorem.

In order to prove that $T$ is an $\exists$-theory, by the preceding lemma, we need only show that $\mathcal{N} \models T$, whenever $\mathcal{M} \models T$ and $\mathcal{N} \models \mathrm{Th}_\exists\mathcal{M}$. Suppose $\mathcal{M}$ and $\mathcal{N}$ are of that kind. Condition (*) yields a model of $\mathrm{Th}\,\mathcal{N} \cup \mathrm{D}(\mathcal{M})$, for whose $L$-reduct, $\mathcal{N}'$ say, we have $\mathcal{M} \hookrightarrow \mathcal{N}'$ and $\mathcal{N} \equiv \mathcal{N}'$. The former implies $\mathcal{N}' \models T$ by preservation, hence the latter implies $\mathcal{N} \models T$.   $\square$

**Exercise 6.2.1.** Prove $T_\forall = \mathrm{Th}\{\mathcal{M} : \mathcal{M} \hookrightarrow \mathcal{N} \models T\}$.

**Exercise 6.2.2.** Derive Corollary 6.2.5 directly from (the remark after) Lemma 6.2.2.

**Exercise 6.2.3.** Prove Corollary 6.2.6.

**Exercise 6.2.4.** Formulate and prove the dual of Proposition 6.2.3 for $\exists$-formulas (that is, the generalization of Corollary 6.2.6 to arbitrary sets of formulas $\Phi(\bar{x})$).

In the next exercises we deal with the dual of the remark after Lemma 6.2.2, i. e. with the models of $T_\exists$.

**Exercise 6.2.5.** Given theories $S$ and $T$, show that $S_\forall \cup T$ is consistent if and only if there are $\mathcal{M} \models T$ and $\mathcal{N} \models S$ such that $\mathcal{M} \hookrightarrow \mathcal{N}$ (if and only if there is a sentence in $S_\forall$ whose negation is in $T_\exists$).

**Exercise 6.2.6.** Prove the following two equations.
(1)   $\mathrm{Mod}\, T_\exists = \{\mathcal{N} : \text{there are } \mathcal{M} \hookrightarrow \mathcal{N}' \equiv \mathcal{N} \text{ such that } \mathcal{M} \models T\}$
(2)   $T_\exists = \mathrm{Th}\{\mathcal{N} : \text{there are } \mathcal{M} \hookrightarrow \mathcal{N}' \equiv \mathcal{N} \text{ such that } \mathcal{M} \models T\}$

Exercise 8.4.5 is asking for an example of a theory $T$ and a model of $T_\exists$ that does not contain a model of $T$ (this showing that we cannot improve on the previous exercise).

## 6.3   Finitely generated structures and local properties

We are going to investigate properties of structures that are determined by their finitely generated substructures.

**Remark.** The intersection of an arbitrary collection of substructures of a given structure is again a substructure of that structure (possibly empty). More precisely, the intersection of the universes is closed under functions from the signature and is therefore the universe of a uniquely determined substructure.

This allows us to make the following definition. Suppose $\mathcal{M}$ is an $L$-structure and $X \subseteq M$. The **substructure generated** by $X$ in $\mathcal{M}$ is, by definition, the uniquely determined substructure with universe $\bigcap\{N : X \subseteq N, \mathcal{N} \subseteq \mathcal{M}\}$, which we denote by $\mathcal{M}_X$. (Thus $\mathcal{M}_X$ is the smallest substructure of $\mathcal{M}$ containing $X$.) $\mathcal{M}$ is said to be **generated** by $X$ if $\mathcal{M}_X = \mathcal{M}$.

A substructure of $\mathcal{M}$ is said to be **finitely generated** if it is generated by a finite set, i. e., if it is of the form $\mathcal{M}_X$ for some finite $X \subseteq M$. A structure is called **finitely generated** if it is a finitely generated substructure of itself.

Let $\wp$ be an arbitrary property of $L$-structures. We say that $\mathcal{M}$ **has locally** $\wp$ if every finitely generated substructure of $\mathcal{M}$ has the property $\wp$.

We often identify a property of structures with the class of structures it defines. So a property $\wp$ is said to be **elementary** (or **axiomatizable**) if the class of all $L$-structures with this property is elementary, i. e. axiomatizable. Similarly, a property is said to be **hereditary** if it is preserved in substructures. We often identify an axiomatizable property with the set of axioms it is axiomatized by. (One could avoid the usage of the term 'property' altogether, but it may serve as a shorter way of saying things sometimes.)

**Exercise 6.3.1.** Show that the substructure of an $L$-structure generated by the empty set is empty if and only if $L$ has no constant symbols.

**Exercise 6.3.2.** Let $\mathcal{M}$ be an $L$-structure and $X \subseteq M$. Verify that the universe, $M_X$, of $\mathcal{M}_X$ is the set $\{t^{\mathcal{M}}(\bar{a}) : t$ is an $L$-term, $\bar{a}$ is a matching tuple from $X\}$.

**Exercise 6.3.3.** Let $\mathcal{M}$ and $\mathcal{N}$ be $L$-structures and $X \subseteq M$. Suppose $f : X \to N$ satisfies $(\mathcal{M}, X) \equiv_{\mathbf{at}} (\mathcal{N}, f[X])$ (and is thus injective). Extend $f$ to an isomorphism $F : \mathcal{M}_X \cong \mathcal{N}_{f[X]}$ and prove $(\mathcal{M}, M_X) \equiv_{\mathbf{at}} (\mathcal{N}, F[M_X])$.

Examples of local properties are **locally finite**, **locally embeddable** in a structure $\mathcal{N}$ (i. e., every finitely generated substructure is embeddable in $\mathcal{N}$), **locally embeddable** in a structure of a given class $\mathbf{K}$ (i. e., every finitely generated substructure is embeddable in some structure from $\mathbf{K}$). For the latter we simply say **locally embeddable** in the class $\mathbf{K}$. A group-theoretic example is **locally cyclic**locally!cyclic (i. e., every finitely generated subgroup is cyclic).

Next we consider a few examples showing that a structure having a property $\wp$ locally need not itself have $\wp$.

**Examples.**

(1) The easiest one is an infinite set $M$ with no further structure, i. e. a model of $T^{\infty}$. Since the substructure generated by a subset is just this subset itself, $M$ is locally finite (but not finite).

(2) The same applies to just any purely relational structure, i. e. a structure $\mathcal{M}$ whose signature has no constant or function symbols. Also then the substructure generated by a subset $X \subseteq M$ is simply the restriction of $\mathcal{M}$ to that set $X$. Hence $\mathcal{M}$ is locally finite (but not necessarily finite).

(3)   Less trivial examples are the Prüfer groups of §11.1, which are locally finite, but infinite, and also locally cyclic, but not itself cyclic.

(4)   Consider the ordering of the natural numbers, $\mathcal{N} = (\mathbb{N}, <)$, as a structure in the language $L_<$. Choose pairwise disjoint chains $\mathcal{M}_n$ of length $n$ ($n \in \mathbb{N}$) in $\mathbb{N}$. Now consider the $L_<$-structure $\mathcal{M}$ on $M = \bigcup_{n \in \mathbb{N}} M_n$ defined as follows: given $a, b \in \mathbb{N}$, set $a < b$ if and only if there is $n \in N$ such that $a, b \in M_n$ and $\mathcal{M}_n \models a < b$.

Then $\mathcal{N}$ is locally embeddable, but not embeddable in $\mathcal{M}$, for $\mathcal{M}$ has no infinite chains. (However, $\mathcal{N}$ is embeddable in Mod Th$\mathcal{M}$, as the set of $L_<(\mathbb{N})$-sentences Th$\mathcal{M} \cup \{i < j \; : \; i, j \in \mathbb{N}, i < j\}$ is consistent and each of its models has an infinite chain.)

The argument of the last statement in parentheses can be generalized as follows.

**Proposition 6.3.1.** (Henkin's Criterion) *Let* $\mathcal{M}$ *be an L-structure and* **K** *an axiomatizable class of L-structures.*

*Then* $\mathcal{M}$ *is embeddable (in a member of)* **K** *if and only if* $\mathcal{M}$ *is locally embeddable in* **K**.

[Henkin, L. : Some interconnections between modern algebra and mathematical logic, Transactions of the American Mathematical Society **74** (1953) 410 – 427]

*Proof.* For the nontrivial direction, choose an axiomatization $T$ of **K** (i. e., **K** $= \text{Mod}_L T$) and consider $T \cup D(\mathcal{M})$. If $\mathcal{N}' \models T \cup D(\mathcal{M})$, then the $L$-reduct $\mathcal{N}$ of $\mathcal{N}'$ is a model of $T$, hence $\mathcal{N} \in$ **K**. Further, $\mathcal{N}' \models D(\mathcal{M})$ implies $\mathcal{M} \hookrightarrow \mathcal{N}$, by the diagram lemma.

Thus, all we have to do is show that $T \cup D(\mathcal{M})$ is consistent. By the finiteness theorem it suffices that every finite subset of $T \cup D(\mathcal{M})$ have a model. Every such finite subset is easily seen to be contained in a set of the form $T \cup D(\mathcal{M}_0)$ for an appropriate choice of a finitely generated substructure $\mathcal{M}_0 \subseteq \mathcal{M}$. By hypothesis, however, there is an embedding $f$ of $\mathcal{M}_0$ in some $\mathcal{N}_0 \in$ **K**. Hence $T \cup D(\mathcal{M}_0)$ has a model, namely $(\mathcal{N}_0, f[M_0])$.   □

Here we have a case where a local property implies the corresponding 'global' property—one also says that the *local theorem* holds for this property. We will see much deeper ones in due course.

**Exercise 6.3.4.** Prove: every linear ordering can be embedded in a dense linear ordering.

**Example.** As an application of Henkin's criterion, we prove the well-known theorem that every abelian group is embeddable in a **divisible** abelian group, i. e. a model of AG $\cup \{\forall x \exists y\,(x = ny)\ :\ n \in \mathbb{N} \smallsetminus \{0\}\}$. Henkin's criterion implies that an abelian group is embeddable in a divisible abelian group, whenever it is so locally. Recall the *fundamental theorem of abelian group theory* saying that every finitely generated abelian group is a direct sum of finitely many cyclic groups. So, in order to embed such a group into a divisible one, it is enough to be able to do this for the various cyclic summands (and then just patch those embeddings together). This now is not too hard: the finite cyclic groups are embeddable in a Prüfer group (cf. §11.1), while the infinite one (there's only $\mathbb{Z}$) is embeddable in $\mathbb{Q}$.

The proof of Henkin's criterion can be modified to yield the local theorem for every property that is both, hereditary and elementary (i. e. preserved under substructures and axiomatizable). We conclude this section with a proof of this. Some remarks are in order before that.

**Remarks.** (1) Since elementary properties are invariant under isomorphism (cf. Proposition 6.1.3), every hereditary and elementary property passes from any structure $\mathcal{M}$ down to all structures *embeddable* in $\mathcal{M}$, cf. the situation in the remark after Lemma 6.2.1.

(2) By the Łoś-Tarski preservation theorem (Corollary 6.2.5), the hereditary and elementary properties are exactly those which are expressible by sets of $\forall$-sentences, i. e. by $\forall$-theories.

**Examples.** Consider the language of signature $(1; \cdot, ^{-1})$ appropriate for group theory.

(5) The property of being abelian is hereditary and elementary (take $\Sigma = \{\forall xy\,(xy = yx)\}$).

(6) The property of being of **exponent** $k$ (i. e., satisfying $\forall x\,(x^k = 1)$, is hereditary and elementary.

(7) The property of being **periodic** (i. e., every element has finite order) clearly is hereditary, but not elementary (see Corollary 5.2.2).

(8) The property of being locally finite clearly is hereditary, but not elementary (exercise!).

(9) The property of being divisible is elementary, but not hereditary.

**Exercise 6.3.5.** Check the last two examples.

**Proposition 6.3.2.** (The Local Theorem for Hereditary, Elementary Properties) *Suppose $\mathcal{M}$ is an L-structure and $\wp$ is a hereditary and elementary property of L-structures.*

*Then $\mathcal{M}$ has $\wp$ if and only if $\mathcal{M}$ has $\wp$ locally.*

*Proof.* Let $\mathbf{K} = \mathrm{Mod}_L \Sigma$ be the class of $L$-structures having $\wp$.

By the diagram lemma, $\Sigma \cup D(\mathcal{M})$ is consistent if and only if $\mathcal{M}$ is embeddable in $\mathbf{K}$. As $\mathbf{K}$ is hereditary, the latter is the case if and only if $\mathcal{M}$ itself is in $\mathbf{K}$. Thus $\mathcal{M} \in \mathbf{K}$ iff $\Sigma \cup D(\mathcal{M})$ is consistent (for any $L$-structure $\mathcal{M}$). The finiteness argument from the proof of Henkin's criterion completes the proof. □

Using a clever expansion of the language, Malcev extended the applicability of this kind of model-theoretic argument to a much wider class of not necessarily elementary properties like the one introduced next. We will deal with this in §6.5*.

Let $\Sigma_1, \ldots, \Sigma_n$ be elementary properties of groups or just sets of sentences in the signature $(1; \cdot, ^{-1})$. Following Malcev we say that a group $\mathcal{G}$ has **type** $(\Sigma_1, \ldots, \Sigma_n)$ if there is a normal series $\mathcal{G} = \mathcal{G}_0 \rhd \mathcal{G}_1 \rhd \ldots \rhd \mathcal{G}_n = \{1\}$ such that $\mathcal{G}_k/\mathcal{G}_{k+1} \models \Sigma_{k+1}$ for all $k < n$.

In §6.5* we will prove the local theorem for this group-theoretic property. For this we need

# 6.4*   The interpretation lemma

The general technique of interpretation[1] is a tool to code structures in other structures—possibly (and often desirably) of different signature. We confine ourselves to the particular case of interpretation of factor structures of substructures in a given structure, since this suffices for the group-theoretic applications we are after. In particular, our interpretations here do not change signatures—a restriction one would not want to make in a general treatment of interpretation. (See also the first two exercises below for another direction of generalization.)

In order to code factor structures of substructures in $L$-structures we expand the language $L$ by a predicate (i. e. a unary relation symbol), $P$, for the substructure and a binary relation symbol, $E$, for the factorization. The resulting language is denoted by $L^*$. (Abusing the notation slightly, one often writes $L^* = L \cup \{P, E\}$ to express this.)

Consider $\Sigma_{PE}$, the union of the following (sets of) $L^*$-sentences.

(1)  $\forall xyz\, (P(x) \wedge P(y) \wedge P(z) \to E(x,x) \wedge (E(x,y) \leftrightarrow E(y,x)) \wedge (E(x,y) \wedge$
     $E(y,z) \to E(x,z)))$          ('$E$ is an equivalence relation on $P$'),

---

[1]This term has formally nothing to do with 'interpretations' of symbols of a signature in a structure, as introduced in Chapter 3.

(2) $\{\forall \bar{x} \, (\bigwedge_i P(x_i) \to P(f(\bar{x}))) : f \in \mathbf{F}\} \cup \{P(c) : c \in \mathbf{C}\}$   ('$P$ is a substructure'),[2]

(3) $\{\forall \bar{x}\bar{y} \, (\bigwedge_i (P(x_i) \wedge P(y_i) \wedge E(x_i, y_i)) \to (E(f(\bar{x}), f(\bar{y})) \wedge (R(\bar{x}) \leftrightarrow R(\bar{y})))) :$
$f \in \mathbf{F}, R \in \mathbf{R}\}$   ('$E$ is a **strict congruence relation** on $P$').

If $\mathcal{M} \models \Sigma_{PE}$, then, in the following canonical way, $P(\mathcal{M})/E$ can be regarded as an $L$-structure.

First some notation. Given $a \in P(\mathcal{M})$, set $a/E = \{a' \in P(\mathcal{M}) : E(a, a')\}$. If $A \subseteq P(\mathcal{M})$, let $A/E$ denote $\{a/E : a \in A\}$; and for a tuple $\bar{a} = (a_0, \dots, a_{n-1})$ from $P(\mathcal{M})$ use $\bar{a}/E$ to denote $(a_0/E, \dots, a_{n-1}/E)$.

Then $\mathcal{M}^* =_{\text{def}} P(\mathcal{M})/E$ becomes an $L$-structure by the following stipulations. Given $c \in \mathbf{C}$, set $c^{\mathcal{M}^*} = c^{\mathcal{M}}/E$; given $f \in \mathbf{F}$ and a matching tuple $\bar{a}$ from $P(\mathcal{M})$, set $f^{\mathcal{M}^*}(\bar{a}/E) = f^{\mathcal{M}}(\bar{a})/E$; and, given $R \in \mathbf{R}$ and a matching tuple $\bar{a}$ from $P(\mathcal{M})$, let $\mathcal{M}^* \models R(\bar{a}/E)$ if and only if $\mathcal{M} \models R(\bar{a})$. Since $E$ is a strict congruence relation (in the above sense), this definition does not depend on the choice of representatives.

We call $\mathcal{M}^*$ the **canonical factor structure** of $P(\mathcal{M})$ modulo $E$ and say that the structure $\mathcal{M}^*$ (and every structure isomorphic to it) is **interpreted** in $\mathcal{M}$.

**Example.** Let $L$ be the language of signature $(1; \cdot, ^{-1})$.
Then (2) above is TG-equivalent to

$$P(1) \wedge \forall x_0 x_1 \, (P(x_0) \wedge P(x_1) \to P(x_0^{-1}) \wedge P(x_0 x_1)),$$

and also to

$$P(1) \wedge \forall x_0 x_1 \, (P(x_0) \wedge P(x_1) \to P(x_0 x_1^{-1})).$$

Thus (2) says that $P(\mathcal{M})$ is closed under multiplication and inverses, i. e. that $P(\mathcal{M})$ is a subgroup. (3) is TG-equivalent to

$$\forall x_0 x_1 y_0 y_1 \, ((P(x_0) \wedge P(x_1) \wedge P(y_0) \wedge P(y_1) \wedge E(x_0, y_0) \wedge E(x_1, y_1)) \to$$
$$(E(x_0 x_1, y_0 y_1) \wedge E(x_0^{-1}, y_0^{-1}))).$$

Along with (1) and (2), this implies

$$\forall xyz \, ((P(x) \wedge P(y) \wedge P(z) \wedge E(x, 1) \wedge E(y, 1)) \to (E(xy^{-1}, 1) \wedge E(z^{-1}xz, 1))),$$

hence that $1/E = \{a \in P(\mathcal{M}) : E(a, 1)\}$ is a normal subgroup of $P(\mathcal{M})$. It further implies

$$\forall xy \, ((P(x) \wedge P(y)) \to (E(x, y) \leftrightarrow E(y^{-1}x, 1))),$$

---

[2]The $i$ run over the indices of the entries of $\bar{x}$ here. Similarly in (3).

i. e. that $b/E = \{a \in P(\mathcal{M}) : E(a,b)\}$ is a coset of $1/E$ for every $b \in P(\mathcal{M})$. One can show that, conversely, every relation $E$ is a strict congruence relation in the sense of (3) above, whenever $1/E$ is a normal subgroup of $P(\mathcal{M})$ and the $b/E$ are cosets of $1/E$.

**Lemma 6.4.1.** (The Interpretation Lemma)

*In the notation above, there is a map $^* : L \to L^*$ with $^* : L_n \to L_n^*$ for all $n \in \mathbb{N}$ and such that if $\mathcal{M} \models \Sigma_{PE}$, $\varphi \in L$, and $\bar{a}$ is a matching tuple from $P(\mathcal{M})$, then*

$\mathcal{M}^* \models \varphi(\bar{a}/E)$ *if and only if* $\mathcal{M} \models \varphi^*(\bar{a})$.

*More concretely,* $^* : L \to L^*$ *is given as follows.*

$(t_1 = t_2)^* = E(t_1, t_2)$ *for all unnested terms $t_1$ and $t_2$;*
$(R(x_0, \ldots, x_{n-1}))^* = R(x_0, \ldots, x_{n-1})$ *for all $R \in \mathbf{R}$;*
$\neg(\psi)^* = \neg\psi^*$; $(\psi_1 \wedge \psi_2)^* = \psi_1^* \wedge \psi_2^*$; *and* $(\exists x \, \theta)^* = \exists x \, (P(x) \wedge \theta^*)$
*for all $\psi, \psi_1, \psi_2$, and $\theta$ in $L$.*

*Proof.* By induction on the complexity of formulas.

In the initial step we have $\mathcal{M}^* \models f(\bar{a}/E) = g(\bar{a}/E)$ iff $\mathcal{M}^* \models f(\bar{a})/E = g(\bar{a})/E$ iff $\mathcal{M} \models E(f(\bar{a}), g(\bar{a}))$ iff $\mathcal{M} \models (f(\bar{a}) = g(\bar{a}))^*$, and analogously for relations and constants. Note that Exercise 3.3.6 justifies to consider only unnested atomic formulas in this.

The induction steps for $\neg$ and $\wedge$ are trivial. Hence we are left with the quantifier step. So let $\varphi(\bar{x})$ be the formula $\exists y \, \theta(\bar{x}, y)$, and assume $\theta$ already satisfies the assertion, i. e., $\mathcal{M}^* \models \theta(\bar{a}/E, b/E)$ iff $\mathcal{M} \models \theta^*(\bar{a}, b)$, whenever $\mathcal{M} \models \Sigma_{PE}$, $b \in P(\mathcal{M})$, and $\bar{a}$ is a matching tuple from $P(\mathcal{M})$.

Then $\mathcal{M}^* \models \varphi(\bar{a}/E)$ if and only if there is some $b \in \mathcal{M}^*$ such that $\mathcal{M}^* \models \theta(\bar{a}/E, b/E)$, if and only if there is some $b \in P(\mathcal{M})$ such that $\mathcal{M} \models \theta^*(\bar{a}, b)$, if and only if $\mathcal{M} \models \exists x \, (P(x) \wedge \theta^*(\bar{a}, x))$, if and only if $\mathcal{M} \models \varphi^*(\bar{a})$.    $\square$

An $n$-fold application of the lemma yields

**Corollary 6.4.2.** (The $n$-fold Interpretation Lemma) *Suppose $P_i$ and $E_i$ ($i < n$) are like $P$ and $E$ before; $L_i^*$ is like $L^*$ before; and $\mathcal{M}_i^*$ is defined as $\mathcal{M}^*$ before for $P = P_i$ and $E = E_i$.*

*Then there are maps $_i^* : L \to L_i^*$ ($i < n$) with $_i^* : L_m \to (L_i^*)_m$ for all $m \in \mathbb{N}$ such that if $\mathcal{M} \models \bigcup_{i<n} \Sigma_{P_iE_i}$, $\varphi \in L$, and $\bar{a}$ is a matching tuple from $P_i(\mathcal{M})$, then*

$\mathcal{M}_i^* \models \varphi(\bar{a}/E_i)$ *iff* $\mathcal{M} \models \varphi_i^*(\bar{a})$    $\square$

**Exercise 6.4.1.** Find and prove an analogous interpretation lemma for many-place $P$.

**Exercise 6.4.2.** Extend the concept of interpretation slightly as to be able to interpret the ring of complex numbers in that of the reals, as well as the ring of rational numbers in that of integers.

**Exercise 6.4.3.** Interpret the cyclic group with five elements in $(\mathbb{Z}; 0; +, -)$.

**Exercise 6.4.4.** Let $T$ be an $L$-theory, $P$ a new predicate, and $L'$ the language $L \cup \{P\}$.

Find an $L'$-theory $T'$ such that the models of $T'$ are exactly the $L'$-structures $\mathcal{M}'$ of the following kind: the $L$-reduct $\mathcal{M}$ of $\mathcal{M}'$ is a model of $T$, $P(\mathcal{M}')$ is closed under functions of $L$, and the therefore uniquely determined $L$-substructure of $\mathcal{M}$ on $P(\mathcal{M}')$ is a model of $T$.

# 6.5* Malcev's local theorems

Throughout this section, $L$ denotes the language of signature $(1; \cdot, ^{-1})$.

Using the interpretation lemma, in an appropriate expansion of a group we first elementarily express the property of being of type $(\Sigma_1, \ldots, \Sigma_n)$. We have seen that in a group $\mathcal{G}$ the formula $P'(x) \Longleftrightarrow_{\mathrm{def}} P(x) \wedge E(x, 1)$ defines the normal subgroup $1/E$ of $P(\mathcal{G})$ and that $P(\mathcal{G})/E$ is isomorphic to the factor group $P(\mathcal{G})/P'(\mathcal{G})$. If, conversely, a formula $P'(x)$ defines a normal subgroup $P'(\mathcal{G})$ of $P(\mathcal{G})$ in $\mathcal{G}$, then setting $E(x, y) \Longleftrightarrow_{\mathrm{def}} P'(xy^{-1})$ we obtain a congruence relation such that $P(\mathcal{G})/P'(\mathcal{G}) \cong P(\mathcal{G})/E$.

Consider an expansion of the language $L$ by two new predicates $P$ and $P'$. Let $\Sigma_{PP'}$ be the set obtained from $\Sigma_{PE}$ by replacing each subformula of the form $E(u, v)$ by $P'(uv^{-1})$. Then, given a group $\mathcal{G}$ and a corresponding expansion $\mathcal{G}'$, we have $\mathcal{G}' \models \Sigma_{PP'}$ iff $P'(\mathcal{G}) \lhd P(\mathcal{G})$.

Again we can do this $n$-foldly. Given new predicates $P_0, \ldots, P_n$, let $L' = L \cup \{P_0, \ldots, P_n\}$ and $\Gamma = \bigcup_{i<n} \Sigma_{P_i P_{i+1}} \cup \{\forall x \, P_0(x) \wedge \forall x \, (P_n(x) \to x = 1)\}$. Then $(\mathcal{G}, P_0, \ldots, P_n) \models \Gamma$ iff $\mathcal{G} = P_0(\mathcal{G}) \rhd P_1(\mathcal{G}) \rhd \ldots \rhd P_n(\mathcal{G}) = \{1\}$.

**Lemma 6.5.1.** *Suppose* $\Sigma_1, \ldots, \Sigma_n$ *are elementary properties. For all* $i < n$, *consider the set* $(\Sigma_{i+1})_i^* = \{\varphi_i^* : \varphi \in \Sigma_{i+1}\}$ *(where the* $_i^*$ *are the maps from Corollary 6.4.2) and the set* $\Sigma' = \mathrm{TG} \cup \Gamma \cup (\Sigma_1)_0^* \cup \ldots \cup (\Sigma_n)_{n-1}^*$ *(where* $\Gamma$ *is as above).*

*Then* $\mathcal{G}$ *is a group of type* $(\Sigma_1, \ldots, \Sigma_n)$ *if and only if* $\mathcal{G}$ *has an* $L'$-*expansion* $\mathcal{G}'$ *that is a model of* $\Sigma'$.

*Proof.* $\Longrightarrow$. If $\mathcal{G}' = (\mathcal{G}, P_0, \ldots, P_n) \models \Sigma'$, then the $\mathcal{G}_i =_{\mathrm{def}} P_i(\mathcal{G}')$ constitute a normal series satisfying $\mathcal{G}' \models (\Sigma_{i+1})_i^*$. Hence, by the $n$-fold interpretation lemma, $\mathcal{G}_i/\mathcal{G}_{i+1} \models \Sigma_{i+1}$, i. e., $\mathcal{G}$ is of type $(\Sigma_1, \ldots, \Sigma_n)$.

$\Longleftarrow$. If $\mathcal{G}_0 \rhd \mathcal{G}_1 \rhd \ldots \rhd \mathcal{G}_n$ is as required, $\mathcal{G}$ becomes an $L'$-structure $\mathcal{G}'$ on setting $P_i(\mathcal{G}') = \mathcal{G}_i$. $\qquad \square$

The property of being of type $(\Sigma_1, \ldots, \Sigma_n)$ is **pseudo-elementary** in the sense that it is axiomatizable in an expansion of the original language; more precisely, there is an axiomatizable class of structures in some expanded language whose $L$-reducts are exactly the groups of type $(\Sigma_1, \ldots, \Sigma_n)$. (More about this notion can be found e. g. in Hodges' *Model Theory*.)

Finally we are prepared to prove

**Theorem 6.5.2.** (Malcev's Main Local Theorem) *Suppose* $\Sigma_1, \ldots, \Sigma_n$ *are hereditary, elementary properties (in L).*

(1)   *Being of type* $(\Sigma_1, \ldots, \Sigma_n)$ *is a hereditary property (i. e. preserved in substructures).*

(2)   *A group* $\mathcal{G}$ *is of type* $(\Sigma_1, \ldots, \Sigma_n)$ *if and only if it is so locally.*

*Proof.* Ad (1). Let $\mathcal{G}$ be of type $(\Sigma_1, \ldots, \Sigma_n)$. Then there is a normal series $\mathcal{G} = \mathcal{G}_0 \rhd \mathcal{G}_1 \rhd \ldots \rhd \mathcal{G}_n = \{1\}$ such that $\mathcal{G}_i/\mathcal{G}_{i+1} \models \Sigma_{i+1}$ for all $i < n$. We are going to show that every subgroup $\mathcal{H}$ of $\mathcal{G}$ also is of type $(\Sigma_1, \ldots, \Sigma_n)$. Setting $\mathcal{H}_i = \mathcal{G}_i \cap \mathcal{H}$ we have $\mathcal{H} = \mathcal{H}_0 \rhd \mathcal{H}_1 \rhd \ldots \rhd \mathcal{H}_n = \{1\}$, for if $h' \in \mathcal{H}_{i+1}$ and $h \in \mathcal{H}_i$, then $h^{-1}h'h \in \mathcal{H} \cap \mathcal{G}_{i+1} = \mathcal{H}_{i+1}$ (hence $\mathcal{H}_{i+1} \lhd \mathcal{H}_i$).

It remains to prove $\mathcal{H}_i/\mathcal{H}_{i+1} \models \Sigma_{i+1}$. For this, in turn, by heredity of the $\Sigma_i$, it suffices to prove $\mathcal{H}_i/\mathcal{H}_{i+1} \hookrightarrow \mathcal{G}_i/\mathcal{G}_{i+1}$. However, the map from $\mathcal{H}_i/\mathcal{H}_{i+1}$ to $\mathcal{G}_i/\mathcal{G}_{i+1}$ given by $h\mathcal{H}_{i+1} \mapsto h\mathcal{G}_{i+1}$ is an embedding, as for all $h \in \mathcal{H}_i$, we have $h \in \mathcal{H}_{i+1}$ if and only if $h \in \mathcal{H} \cap \mathcal{G}_{i+1}$, hence if and only if $h \in \mathcal{G}_{i+1}$.

Ad (2). $\Longrightarrow$ is a special case of (1).

$\Longleftarrow$. In analogy to Propositions 6.3.1 and 6.3.2, the just proved assertion (1) and Lemma 6.5.1 imply $\mathcal{G}$ is a group of type $(\Sigma_1, \ldots, \Sigma_n)$ if and only if $TG \cup \Sigma' \cup D(\mathcal{G})$ is consistent. A finiteness argument as in the propositions mentioned completes the proof then.                                                    $\square$

There are numerous applications of this result, of which we consider only a few here (and in the exercises to §7.2*). Other applications can be found in the cited literature, especially in §§22 and 26 of the group theory text by Kargapolov and Merzljakov (two younger colleagues of Malcev at Novosibirsk), where a stronger version of Malcev's local theorem—for more general logics—can be found. See also Corollary 6.6.8 in Hodges' *Model Theory*, where a proof of the local theorem for the property of having a faithful $n$-dimensional linear representation is given. Following Hodges, this latter 1940 result of Malcev was the first application of the finiteness theorem in algebra!

A group $\mathcal{G}$ is said to be **abelian-by-finite** if it has an abelian normal subgroup of **finite index**, i. e. an abelian subgroup $\mathcal{A} \lhd \mathcal{G}$ such that $\mathcal{G}/\mathcal{A}$

is finite. The local theorem for this property can be stated somewhat more strongly.

**Corollary 6.5.3.** *A group is abelian-by-finite if and only if there is $n \in \mathbb{N}$ such that every finitely generated subgroup of $\mathcal{G}$ has an abelian normal subgroup of index $\leq n$.*

*Proof.* Consider the hereditary and elementary properties $\Sigma_n = \{\exists^{\leq n} x \, (x = x)\}$ and $\Sigma = \{\forall xy \, (xy = yx)\}$. A group $\mathcal{G}$ is of type $(\Sigma_n, \Sigma)$ if and only if it has an abelian normal subgroup $\mathcal{A}$ of **index** $\leq n$ (i. e. such that $|\mathcal{G}/\mathcal{A}| \leq n$).

If now $\mathcal{G}$ is abelian-by-finite, it has a normal subgroup $\mathcal{A} \lhd \mathcal{G}$ of a certain finite index $n = |\mathcal{G}/\mathcal{A}|$. Then $\mathcal{G}$ is of type $(\Sigma_n, \Sigma)$. Now this direction, as well as its converse, follow from Malcev's local theorem. $\quad\square$

**Example.** Consider a periodic group $\mathcal{G}$ of $m \times m$-matrices over the field $\mathbb{C}$ of complex numbers. Schur's so-called second theorem asserts that $\mathcal{G}$ is abelian-by-finite. This can be derived, using the preceding corollary, from Schur's so-called first theorem saying that $\mathcal{G}$ is locally finite and a theorem of Jordan saying that finite groups of $m \times m$-matrices over $\mathbb{C}$ have an abelian normal subgroup whose index depends only on $m$.

A group is said to be **solvable** of **degree** $\leq n$ if it has a normal series of length $\leq n$ with abelian factors. There is an easy proof of the local theorem for this property using a different definition (via some well-known commutator conditions). However, to practice this section's material, we include

**Exercise 6.5.1.** Derive the local theorem for solvability of degree at most $n$ from Malcev's theorem.

# Chapter 7

# Some theory of orderings

This chapter contains various fundamental order-theoretic topics. The first one is a modification of the previous chapter's diagram method.

## 7.1 Positive diagrams

The **positive diagram** $D_+(\mathcal{M})$ of an $L$-structure $\mathcal{M}$ is defined to be the set of all atomic $L(M)$-sentences satisfied in $(\mathcal{M}, M)$.

**Remark.**
$D(\mathcal{M}) = D_+(\mathcal{M}) \cup \{\neg\varphi(\bar{a}) : \varphi \in L \text{ is atomic}, \bar{a} \text{ in } M, \varphi(\bar{a}) \notin D_+(\mathcal{M})\}$.
As in Exercise 6.1.5, it suffices to consider unnested sentences $\varphi$ (exercise!).

**Example.** The positive diagram of a group is (up to TG-equivalence) its Cayley diagram (i. e. its multiplication table).

As an analogue of Lemma 6.1.2 we have

**Lemma 7.1.1.** (The Positive Diagram Lemma) *Suppose $\mathcal{M}$ and $\mathcal{N}$ are $L$-structures.*
(1) *If $h : M \to N$, then $h : \mathcal{M} \to \mathcal{N}$ if and only if $h : \mathcal{M} \xrightarrow{\text{at}} \mathcal{N}$, if and only if $(\mathcal{N}, h[M]) \models D_+(\mathcal{M})$.*
(2) *There is a homomorphism $\mathcal{M} \to \mathcal{N}$ if and only if $\mathcal{N}$ has an expansion which is a model of $D_+(\mathcal{M})$.*

*Proof.* Ad (1). The first equivalence is Remark (3) after Lemma 6.1.1. The second is obvious (for the given map $h$).
(2) follows from (1) as in Lemma 6.1.2. □

87

**Exercise 7.1.1.** Verify that the positive diagram of a structure is logically equivalent to the subset of all its *unnested* atomic formulas.

**Exercise 7.1.2.** Prove $\mathbf{qf} \cap + \cap \mathrm{Th}(\mathcal{M}, M) \subseteq \mathrm{D}_+(\mathcal{M})^\models$ (cf. notation in §2.4).

## 7.2*  The theorem of Marczewski-Szpilrajn

This theorem is a neat application of positive diagrams.

**Theorem 7.2.1.** *Any partial order on a set $A$ can be extended to a linear order on $A$.*

[Szpilrajn, E. : Sur l'extension de l'ordre partiel, Fundamenta Math. **16** (1930) 386 – 389] (The author later published under the name E. Marczewski.)

*Proof.* Let $\mathcal{A} = (A, \prec)$ be a partial ordering (regarded as an $L_<$-structure). As a first step we reduce the result to finite orderings.

If $(B, \{b_a : a \in A\}, <)$ is a model of

$$\Sigma = \mathrm{D}_+(\mathcal{A}) \cup \mathrm{LO} \cup \{\underline{a} \neq \underline{b} : a, b \in A, a \neq b\}$$

and if $\mathcal{B}$ is its $L_<$-reduct, then $h(a) = b_a$ defines, by Lemma 7.1.1, a homomorphism $h : \mathcal{A} \to \mathcal{B}$, which is obviously an embedding. In particular, $h$ respects the ordering, i. e.,

(*)  $a \prec b \Longrightarrow h(a) < h(b)$.

As $\mathcal{B}$ is a linear ordering, so is $h(\mathcal{A}) \subseteq \mathcal{B}$. Since $h$ is a bijection between $A$ and $h[A]$, we therefore obtain a linear ordering of $A$ on setting $a < b$ if $h(a) < h(b)$, for any $a, b \in A$. By (*), the order $<$ extends the partial order $\prec$ on $A$.

Thus it remains to prove that $\Sigma$ is consistent. So let $\Sigma_0$ be a finite subset of $\Sigma$ and let $a_0, \ldots, a_{n-1}$ be the new constants occurring in $\Sigma_0$. Then $\Sigma_0 \subseteq \mathrm{D}_+(\{a_0, \ldots, a_{n-1}\}, \prec) \cup \mathrm{LO} \cup \{a_i \neq a_j : i < j < n\}$. If now, on $\{a_0, \ldots, a_{n-1}\}$, the partial order $\prec$ can be extended to a linear order (on $\{a_0, \ldots, a_{n-1}\}$), we get a model of $\Sigma_0$, and are done.

Thus we have reduced the theorem to finite partial orderings and we are left with                                                                  $\square$

**Lemma 7.2.2.** *Every finite partial ordering can be extended to a linear ordering.*

*Proof.* By induction on the number of elements, $n$, the cases $n = 0, 1, 2$ being trivial.

Let $(\{a_0, \ldots, a_n\}, \prec)$ be a partial ordering enumerated in such a way that $a_n$ is a maximal element, and assume (by induction hypothesis) that the partial ordering $(\{a_0, \ldots, a_{n-1}\}, \prec)$ can be extended to a linear ordering $(\{a_0, \ldots, a_{n-1}\}, <)$. Then $< \cup \{(a_i, a_n) : i \le n\}$ is a linear ordering on $\{a_0, \ldots, a_n\}$ extending $\prec$. $\square$

Another proof of the preceding lemma is the following.[1]

**Exercise 7.2.1.** Define recursively a rank function $r$ from a finite partial ordering $\mathcal{A} = (A, \prec)$ to $\mathbb{N}$ by setting $r(a) = 0$ for every $\prec$-minimal element $a$ in $\mathcal{A}$, and setting $r(b) = 1$ for every $\prec$-minimal element $b$ in $A \smallsetminus \{a \in A : r(a) = 0\}$, and so forth.

Use this to extend $\prec$ to a linear order on $\mathcal{A}$.

A partially ordered group $(\mathcal{G}, <)$ is said to be **orderable** if $<$ can be extended to a linear ordering of the group $\mathcal{G}$ (in the sense of §5.5). $\mathcal{G}$ is called **orderable** if there is some linear ordering of the group $\mathcal{G}$, while $\mathcal{G}$ is called **freely orderable** if every partial order of the group $\mathcal{G}$ can be extended to a linear order of $\mathcal{G}$.

**Exercise 7.2.2.** Prove the following local theorems.
(a)  A partially ordered group $(\mathcal{G}, <)$ is orderable if and only if $(\mathcal{G}_0, <)$ is orderable for every finitely generated subgroup $\mathcal{G}_0$ of $\mathcal{G}$.
(b)  A group is orderable if and only if it is locally orderable.
(c)  A group is freely orderable if and only if it is so locally.

**Exercise 7.2.3.** An abelian group is orderable if and only if it is torsionfree.

# 7.3  Cantor's theorem

As early as at the end of the 19th century Georg Cantor discovered an important feature of countable dense linear orderings.

**Theorem 7.3.1.** *All countable dense linear orderings without endpoints are isomorphic.*

[Cantor, G. : Beiträge zur Begründung der transfiniten Mengenlehre I, Mathematische Annalen **46** (1895) 481 – 512]

---

[1]Thanks, Ulrich!

*Proof.* First recall that dense linear orderings are infinite (cf. Exercise 5.5.1).

Consider two countable dense linear orderings without endpoints, $\mathcal{A} = (\{a_i : i \in \mathbb{N}\}; <)$ and $\mathcal{B} = (\{b_i : i \in \mathbb{N}\}; <)$. (The same symbol $<$ is used in both cases, as no confusion will arise.) We successively choose sequences $(a'_i : i \in \mathbb{N})$ in $A$ and $(b'_i : i \in \mathbb{N})$ in $B$ such that, for all $i \in \mathbb{N}$, we have

(*)    $a'_i < a'_j$ iff $b'_i < b'_j$

(this making $f : a'_i \mapsto b'_i$ a homomorphism of orderings). Further we require $\{a'_i : i \in \mathbb{N}\} = A$ and $\{b'_i : i \in \mathbb{N}\} = B$, which then yields $f : \mathcal{A} \cong \mathcal{B}$. We accomplish this by the so-called *back-and-forth* method as follows.

Assume we have already found $a'_i \in A$ and $b'_i \in B$ satisfying (*) for all $i < n$, where $n$ is any given natural number.

If $n$ is even, let $a'_n$ be the $a_j$ of smallest index $j$ which is not contained in $\{a'_0, \ldots, a'_{n-1}\}$. Density and the lack of endpoints guarantees the existence of some $b'_n \in B$ such that

(**)    $(\{a'_0, \ldots, a'_n\}, <) \cong (\{b'_0, \ldots, b'_n\}, <)$

If $n$ is odd, let, conversely, $b'_n$ be the $b_j$ of smallest index $j$ which is not contained in $\{b'_0, \ldots, b'_{n-1}\}$. Again there is $a'_n \in A$ satisfying (**).

In both cases (**) implies (*), and, as we choose alternatingly from $A$ and $B$, we exhaust them both. Hence we eventually obtain an everywhere defined and surjective map $f : A \to B$, which, by (*), is a homomorphism of orderings.                                                    □

Incidentally, in contrast to a usual misconception, Cantor himself did *not* use back-and-forth (but went only 'forth'). The method seems to have first appeared in

[Huntington, E. V. : The continuum as a type of order: an exposition of the modern theory, Annals of Math., **6** (1904/05) §45]

(see Hodges' *Model theory* for other relevant references). Later it became a fundamental method of model theory (cf. the proof of Theorem 12.1.1(2)).

**Exercise 7.3.1.** Find a proof of Cantor's theorem that goes only 'forth'.

**Exercise 7.3.2.** Find the four isomorphism types of countable dense linear orderings (and justify your answer).

## 7.4  Well-orderings

In the remaining three sections of this chapter we introduce some set-theoretic tools—also mostly due to Cantor—that are used in many parts of mathematics.[2] We present this material as part of our model-theoretic treatment of orderings; we will add it, however, to our set-theoretic tools on the metalevel, i. e. as part of our methods of proof. As usual in most parts of mathematics, we haven't formally outlined or precisely fixed the latter, but rather made use of some sort of naive set theory (including the axiom of choice). However, in order to justify the method of transfinite induction, which we will need later, we are forced to be somewhat more precise in our set-theoretic setting. (Nevertheless, we will remain on a quite naive level and refer the interested reader to the cited set-theoretic literature for a more mature treatment.)

Usually one works in a realm of *pure sets*, i. e. sets whose elements are sets again (as opposed to so-called *urelements*). This realm can be thought of as a huge 'structure' $\mathbb{V}$ whose signature has only one symbol, the binary relation symbol $\in$. This is not a structure in the formal sense of Chapter 1, as its universe is not a set anymore (it is too big). However, besides that it looks just like a usual structure. Further, one assumes that $\mathbb{V}$ satisfy the so-called *Zermelo-Fraenkel axioms* ZF and, with the exception of a few philosophically concerned mathematicians, also the *axiom of choice*, AC, which, altogether, form the axioms ZFC. The elements of $\mathbb{V}$ are the **sets**, and the axioms of ZFC control which 'composites' of these are to be sets (i. e. elements of $\mathbb{V}$) again. In general, those 'composites' are **classes**. Thus some of the classes are sets. But not all of them are: an example of such a **proper class**, as one says, is $\mathbb{V}$ (which is not an element of itself). Other examples are isomorphism types of arbitrary nonempty structures. The underlying idea for the distinction of sets among the classes is that sets are 'small' classes. Thus it is natural to assume that finite classes are sets and that a subclass of a set is a set again. These are essentially two of the Zermelo-Fraenkel axioms. We will mention more when we need them in the course of our arguments.

To make our model theory part of such a set-theoretic setting, as the rest of mathematics, we included in the concept of 'structure' that the collection of all its elements be a set (cf. §1.2), i. e. an element of $\mathbb{V}$. As mentioned before, $(\mathbb{V}, \in)$ is not a structure in this sense. It is no problem, however, to extend the notion of structure to arbitrary classes by simply dropping this requirement. In this way one obtains 'big structures'—like $(\mathbb{V}, \in)$. In order to have the previous material at our disposal we allow such structures in the remainder of this chapter. For instance, by a **partial ordering** we

---

[2]Those who are more or less familiar with the rudiments of the theory of cardinal and ordinal numbers may skip this and pass on immediately to Part III.

here simply mean a *class* equipped with an irreflexive and transitive binary relation.

Suppose $(X, <)$ is a linear ordering. By a(n **initial**) **segment** of $(X, <)$ we mean a subclass of $X$ which contains, along with every $a \in X$, also all $b \in X$ smaller than $a$. A(n **initial**) **section** of $(X, <)$ is a class of the form $\{x \in X : x < a\}$ for some $a \in X$. This section is **given** by the element $a$; and we denote it by $X_a$. A linear ordering is said to be **narrow** if each of its sections is a set.

Clearly, any section is a segment. Further, any proper segment of a narrow linear ordering $(X, <)$ is contained in a section and is thus a set. Consequently, if $(X, <)$ is narrow, but not itself a set, then $X$ is the only segment of $(X, <)$ that is not a set. In particular, $X$ cannot be a section, which means that $(X, <)$ has no greatest element.

A **well-ordering** is, by definition, a narrow linear ordering in which every nonempty subclass has a **least** element, i. e. an element that is smaller than or equal to *all* elements in that subclass. A well-ordering which itself is a set is simply called a **well-ordered set**. A well-ordered set obviously is narrow. Note further that any well-ordering is a linear ordering (consider any two-element subset).

Recall that our orderings are strict (i. e. irreflexive), hence homomorphisms of linear orderings are automatically monomorphisms.

**Examples.** (1)  Every finite linear ordering is a well-ordered set.
(2)  The ordering of the natural numbers (i. e. the usual ordering on N ) is a well-ordered set.

**Remarks.** (1)  Every subordering of a well-ordering is a well-ordering, and every proper segment of a well-ordering is a well-ordered set.
(2)  A narrow linear ordering is a well-ordering if and only if it is **well-founded** in the sense that it contain no infinite descending chain of elements $a_0 > a_1 > a_2 > \ldots > a_n > \ldots$.
(3)  The empty set is a segment of every linear ordering. Since every well-ordering has a smallest element, the empty set even is a section in every well-ordering.
(4)  Any segment of a well-ordering $(X, <)$ is a section or the entire class $X$.
(5)  Every well-ordering is isomorphic (as an ordering) to the class of all of its sections ordered by set inclusion.

*Proof.* Ad (4). Let $I$ be a segment of $X$, but not $X$ itself. Let further $a$ be the least element of $X \smallsetminus I$. We show that $I = X_a$. Pick any $b \in X$. If

$b < a$, then $b \notin X \smallsetminus I$ by the choice of $a$, hence $b \in I$. If, on the other hand, $b = a$ or $b > a$, then $b \notin I$, for otherwise $a$ would be in $I$. Consequently, $I = \{x \in X : x < a\} = X_a$.

For the proof of (5) consider the map given by $a \mapsto X_a$. $\qquad\qquad$ □

**Lemma 7.4.1.**
(1) *If $f$ is a homomorphism of a well-ordering $(X, <)$ into itself, then $f(x) \geq x$ for all $x \in X$.*
(2) *No well-ordering is isomorphic to any of its sections.*

*Proof.* Ad (1). Iterated application of $f$ to the inequality $x > f(x)$ would yield $x > f(x) > f^2(x) > \ldots > f^n(x) > \ldots$, contradicting the well-foundedness of $(X, <)$.

(2) follows from (1). $\qquad\qquad$ □

Every well-ordering satisfies the

**Induction principle.** *Let $(X, <)$ be a well-ordering and $\wp$ a property of elements of $X$.*

*Suppose every element $a \in X$ satisfies the* **induction hypothesis**
(a) *if all $b < a$ in $X$ have the property $\wp$, then so does $a$.*
*Then all elements of $X$ have the property $\wp$.*

*Proof.* Otherwise there would be a least element $a \in X$ not having the property $\wp$, hence the condition (a) would be violated. $\qquad\qquad$ □

While the induction principle constitutes the fundamental method of **proof by transfinite induction**, there is a similar fundamental tool for the *construction* of functions on well-orderings, the so-called recursion principle. Applications of the recursion principle are also known as **definitions by transfinite induction**. They typically look as follows.

We are given a well-ordering $(X, <)$, an arbitrary class W, and a **recursive condition** $F$ by which we are to successively define a function $f : X \to W$, that is, this condition says how to extend an already constructed $f_a : X_a \to W$ to $a \in X$. Finally one defines $f$ to be the union of all the $f_a$. The recursive condition $F$ thus has to be a function assigning, to every function from a section $X_a$ (of $X$) to W, a value $f(a) \in W$ of $a$. That is, if $f_a : X_a \to W$ has been already constructed, we set $f(a) = F(f_a)$ (and thus obtain a 'larger' function $f_b : X_b \to W$, say, where $b$ is the least element of $X \smallsetminus X_a$).

**Recursion principle.** *Let $(X, <)$ be a well-ordering and $W$ an arbitrary class. Given $a \in X$, consider the class $\mathcal{F}_a$ of all functions of the section $X_a$ to $W$.*

*For every function $F : \bigcup_{a \in X} \mathcal{F}_a \to W$ there is a uniquely determined function $f : X \to W$ satisfying the recursive condition $f(a) = F(f \restriction X_a)$ for all $a \in X$.*

*Proof.* Let $\mathcal{F}$ be the class $\bigcup_{a \in X} \{ g \in \mathcal{F}_a \ : \ g(b) = F(g \restriction X_b) \text{ for all } b < a \}$. We claim,

(*) for all $a \in X$, the set $\mathcal{F} \cap \mathcal{F}_a$ contains at most one function; moreover, for all $g, h \in \mathcal{F}$, we have $g \subseteq h$ or $h \subseteq g$ (where, as usual in set theory, we identify a function with its graph).

For a proof, pick $g \in \mathcal{F} \cap \mathcal{F}_a$ and $h \in \mathcal{F} \cap \mathcal{F}_c$, where $a \leq c$. We claim that $g = h \restriction X_a$ (hence $g \subseteq h$). If not, there would be a least $b < a$ in $X$ with $g(b) \neq h(b)$. Then we would have $g \restriction X_b = h \restriction X_b$ and hence, by the choice of $\mathcal{F}$, also $g(b) = F(g \restriction X_b) = F(h \restriction X_b) = h(b)$. This contradiction proves (*).

Thus $\mathcal{F}$ forms a chain with respect to set inclusion. Hence, its union $\bigcup \mathcal{F}$ is a function, $f'$, whose domain is either all of $X$ or a section thereof.

In the former case we are done, for the function $f'$ clearly satisfies the recursive condition on its domain. Assume therefore that dom $f'$ is a section, $X_a$ say, for some $a \in X$. Then $f'$ is in $\mathcal{F} \cap \mathcal{F}_a$. Let $f$ be the function $f' \cup \{ (a, F(f')) \}$ (i. e., we set $f \restriction X_a = f'$ and $f(a) = F(f')$). Obviously, this latter function satisfies the recursive condition on dom $f = X_a \cup \{a\}$. If there were some $b > a$ in $X$, we would have $f \in \mathcal{F}_b \cap \mathcal{F}$ for the smallest such $b$, hence also $f \subseteq f'$. This would imply $a \in$ dom $f'$, contradicting the assumption. Thus $a$ is the greatest element of $X$, and therefore dom $f = X_a \cup \{a\} = X$.

In any case, $f : X \to W$ has the desired properties. The uniqueness of $f$ one finally derives as in the proof of (*). $\qquad\qquad\square$

In the next section we will specify these principles to ordinal numbers.

**Remark.** The *axiom of choice* is equivalent to the **well-ordering principle**, which says that *every* set can be well-ordered. Another equivalent is the following statement, known as **Zorn's lemma**.[3]

*If every chain in a partially ordered set $X$ has a least upper bound in $X$, then every element in $X$ is smaller than or equal to a **maximal** element,*

---

[3]It is said that Max Zorn himself rejected this name for the simple reason that it is neither Zorn's nor a lemma.

*i. e. an element for which there is no bigger element.* (Note, if $X$ is not linearly ordered, it can have several maximal elements.)

We assume the reader familiar with these equivalents. They can be found in any of the cited set-theoretic (as well as in in most of the cited algebraic) literature.

**Exercise 7.4.1.** Given orderings $\mathcal{X}$ and $\mathcal{Y}$, one defines their **sum**, $\mathcal{X} + \mathcal{Y}$, to be the **disjoint union** of $X$ and $Y$ equipped with the order extending the orders of $\mathcal{X}$ and $\mathcal{Y}$ defined by $x < y$ for all $x \in X$ and $y \in Y$.

Prove that the sum of two well-orderings again is a well-ordering.

**Exercise 7.4.2.** Show that the class of well-orderings is not axiomatizable (as a class of $L_<$-structures).

## 7.5 Ordinal numbers and transfinite induction

We now investigate sets partially ordered by the element relation $\in$. In particular, on such sets the relation $\in$ is transitive. In set theory there is another important concept of transitivity: a class $X$ is said to be **transitive** if $x \in y \in X$ implies $x \in X$.

**Remarks.**
(1)   The following are equivalent for any class $X$.
    (i)   $X$ is transitive.
    (ii)   $x \in X$ implies $x \subseteq X$.
    (iii)   $Y \subseteq X$ implies $\bigcup Y \subseteq X$.

*Proof.* (i) $\Longleftrightarrow$ (ii) is clear.

(iii) means that $y \in Y$ implies $y \subseteq X$, provided $Y \subseteq X$. This, however, is immediate from (ii).

Conversely, (ii) is the special case $Y = \{x\}$ of (iii).     □

Trivially, $X$ is the greatest element with respect to $\in$ in $X \cup \{X\}$ (for $x \in X$ implies $x \in X$). The transition from $X$ to $X \cup \{X\}$ plays a particular role in what follows.

**Lemma 7.5.1.** *Suppose $X$ is a transitive set on which the element relation $\in$ forms a partial (resp. a well-) ordering.*

*Then the same is true for $X \cup \{X\}$.*

*Proof.* Since every element of the transitive set $X$ is a subset of $X$, the same is true for $X \cup \{X\}$. Hence the latter is transitive.

If $\in$ is irreflexive on $X$, we have $X \notin X$ (for otherwise $X \in X \in X$ would contradict the irreflexivity of $\in$ on $X$). Hence $\in$ is irreflexive also on $X \cup \{X\}$.

Let $\in$ be, in addition, transitive on $X$. We show it is so also on $X \cup \{X\}$. For this, let $x, y, z \in X \cup \{X\}$ with $x \in y \in z$. If $x, y, z$ are all in $X$, then $x \in z$ follows from the transitivity of $\in$ on $X$. It is easily seen from $X \notin X$ and the transitivity of $X$ that otherwise $x \in y \in z = X$, hence $x \in z$.    □

An **ordinal number** (or just an **ordinal**) is, by definition, a transitive *set* on which the element relation $\in$ forms a well-ordering (and is, in particular, irreflexive). We denote the class of all ordinal numbers by **On**.

(Their name stems from the fact that the ordinals represent the order types of well-ordered sets, see Theorem 7.5.6 below.)

**Remarks.**
(2)   There is no infinite descending chain of ordinal numbers $\alpha_0 \ni \alpha_1 \ni \alpha_2 \ni \ldots$ (cf. Remark (2) in §7.4).
(3)   Intersections of ordinal numbers are ordinal numbers.
(4)   The previous lemma shows that if $\alpha$ is an ordinal, so is $\alpha \cup \{\alpha\}$. (Here we used that by the Zermelo-Fraenkel axioms $\alpha \cup \{\alpha\}$ is a set again.)
(5)   Every transitive subset of an ordinal is an ordinal.
(6)   Every element of an ordinal is an ordinal.
(7)   The union of a class (resp. of a set) of transitive sets is a transitive class (resp. a transitive set).

*Proof.* Ad (6). Let $\alpha \in$ **On** and $z \in \alpha$, hence also $z \subseteq \alpha$. By (5) we need only show that $z$ is a transitive set. So let $x \in y \in z$. By the transitivity of $\alpha$, we have $x \in y \in \alpha$ and hence $x \in \alpha$. Consequently, $x, y, z \in \alpha$ and the transitivity of $\in$ on $\alpha$ implies $x \in z$, as desired.

Ad (7). Let each $x \in X$ be transitive. Using (1)(ii) above we show that $\bigcup X$ is transitive. So let $y \in \bigcup X$. Then $y \in x$ for some $x \in X$. Now the transitivity of $x$ yields $y \subseteq x$, hence also $y \subseteq \bigcup X$.

Finally, if $X$ is a set all of whose elements are sets then, by the axioms of set theory, also $\bigcup X$ is a set.    □

We will see shortly that also unions of arbitrary sets of ordinals as well as transitive sets of ordinals are themselves ordinals.

The simplest ordinal is the empty set $\emptyset$. Starting from this and using (4) we obtain infinitely many ordinals. Given an ordinal $\alpha$, we denote by $\alpha + 1$ the ordinal $\alpha \cup \{\alpha\}$. Further, we write

0   for   $\emptyset$,
1   for   $0 + 1$         $(= \{\emptyset\})$,
2   for   $1 + 1$         $(= \{\emptyset, \{\emptyset\}\})$,
3   for   $2 + 1$         $(= \{\emptyset, \{\emptyset\}, \{\emptyset, \{\emptyset\}\}\})$,
etc., and
$\omega$   for   $\{0, 1, 2, 3, \dots\}$.
(In Lemma 7.5.3(4) below we will see that $\omega \in \mathbf{On}$.)
Further, we write

$\alpha + 2$   for   $(\alpha + 1) + 1$,
$\alpha + 3$   for   $(\alpha + 2) + 1$,
etc.

From now on we will regard natural numbers as elements of $\omega$, i. e. as ordinal numbers (and identify N with $\omega$).

**Lemma 7.5.2.** *Let* $\alpha, \beta \in \mathbf{On}$.
(1)   $\alpha \in \beta$ *iff* $\alpha \subset \beta$.
(2)   *Either* $\alpha \in \beta$ *or* $\alpha = \beta$ *or else* $\beta \in \alpha$ *(i. e.,* $(\mathbf{On}, \in)$ *is a linear ordering).*

*Proof.* Ad (1). $\Longrightarrow$. Let $\alpha \in \beta$. The transitivity of $\beta$ implies $\alpha \subseteq \beta$. The irreflexivity of $\in$ on $\alpha$ implies $\alpha \neq \beta$.
$\Longleftarrow$. If $\alpha \subset \beta$, the set $\beta \smallsetminus \alpha$ is a nonempty subset of $\beta$, and has thus an $\in$-least element $\gamma \in \beta \smallsetminus \alpha$. We are going to prove $\alpha = \gamma$ (and hence $\alpha \in \beta$).
If $\delta \in \gamma$, the transitivity of $\beta$ yields $\delta \in \beta$. Together with the minimality of $\gamma$ this implies $\delta \in \alpha$. Hence $\gamma \subseteq \alpha$.
To show that also $\alpha \subseteq \gamma$ let $\delta \in \alpha$ ($\subseteq \beta$). Since $\in$ orders the set $\beta$ linearly, $\delta$ and $\gamma$ are comparable with respect to $\in$. Thus $\gamma \in \delta$, $\gamma = \delta$, or $\delta \in \gamma$. In the first two cases we would derive $\gamma \in \alpha$ from $\delta \in \alpha$ (and the transitivity of $\alpha$). As this would contradict the choice of $\gamma$, we must have $\delta \in \gamma$, as desired.
Ad (2). (1) implies that $\alpha \cap \beta = \beta$ or $\alpha \cap \beta \in \beta$, for $\alpha \cap \beta \subseteq \beta$. Analogously we have $\alpha \cap \beta = \alpha$ or $\alpha \cap \beta \in \alpha$. This yields four cases to consider, $\alpha = \beta$, $\beta \in \alpha$, $\alpha \in \beta$, or, as the fourth case, $\alpha \cap \beta \in \beta$ *and* $\alpha \cap \beta \in \alpha$. But the fourth case is impossible, for $\alpha \cap \beta \in \alpha$ and $\alpha \cap \beta \in \alpha \cap \beta$ contradict the irreflexivity of $\in$ on $\alpha \in \mathbf{On}$.
Finally, irreflexivity and antisymmetry (of $\in$ on $\mathbf{On}$) guarantee that no two of the remaining cases can take place simultaneously.   $\square$

From now on, given ordinals $\alpha$ and $\beta$, we write $\alpha < \beta$ instead of $\alpha \in \beta$ (which is the same as $\alpha \subset \beta$), and we write $\alpha \leq \beta$ instead of '$\alpha < \beta$ or $\alpha = \beta$' (which is the same as $\alpha \subseteq \beta$).

Thus the ordinal numbers form an infinite ascending chain $(\mathbf{On}, <)$ with respect to $<$ (which, for ordinals, is the same as the element relation $\in$ or the inclusion $\subseteq$). This chain $(\mathbf{On}, <)$ contains no infinite descending chains. We will see shortly that it even is a well-ordering.

As suggested by the notation, $\alpha + 1$ is the immediate successor of $\alpha$: if $\alpha < \beta$ (hence $\alpha \in \beta$ and $\alpha \subseteq \beta$), then $\alpha \cup \{\alpha\} \subseteq \beta$, i. e., $\alpha + 1 \leq \beta$. Starting from 0, which obviously is the smallest ordinal (for $\emptyset$ is contained in every ordinal), we thus obtain the ascending chain $0 < 1 < 2 < 3 < \ldots < \omega < \omega + 1 < \omega + 2 < \ldots$, which constitutes an initial segment of the chain of all ordinals, $(\mathbf{On}, <)$. More generally we have

**Lemma 7.5.3.**

(1)  *An ordinal is precisely the set of all its sections with respect to the order*
     $<$. *Further,* $\alpha_\beta = \mathbf{On}_\beta = \beta$ *for all sections* $\alpha_\beta$ *of* $\alpha \in \mathbf{On}$ *($\beta \in \alpha$).*
(2)  $(\mathbf{On}, <)$ *is a well-ordering.*
(3)  *Every transitive subset of* $\mathbf{On}$ *is an ordinal.*
(4)  *The union of any subset of* $\mathbf{On}$ *is an ordinal.*

*Proof.* Ad (1). Suppose $\alpha \in \mathbf{On}$ and $\beta \in \alpha$.

As $\beta \subseteq \alpha$, we have $\alpha_\beta = \{\gamma \in \alpha : \gamma < \beta\} = \{\gamma \in \alpha : \gamma \in \beta\} = \alpha \cap \beta = \beta$. As every element of $\beta$ is an ordinal, we further have $\mathbf{On}_\beta = \{\gamma \in \mathbf{On} : \gamma < \beta\} = \{\gamma \in \mathbf{On} : \gamma \in \beta\} = \beta$.

Ad (2). By (1), $(\mathbf{On}, <)$ is narrow. Since $\mathbf{On}$ is also well-founded, Remark (2) of §7.4 shows that $\mathbf{On}$ is well-ordered by $<$.

(3) now is immediate from the definition of ordinal number, while (4) follows on invoking the above Remark (7). □

Not only is $(\mathbf{On}, <)$ a well-ordering, it is even transitive, for every $\beta \in \mathbf{On}$ itself is a set of ordinals, hence $\beta \subseteq \mathbf{On}$. Nevertheless $\mathbf{On}$ is *not* an ordinal: otherwise it would be an element of itself, i. e., $\mathbf{On} \in \mathbf{On}$, contradicting the irreflexivity of $\in$ on $\mathbf{On}$. What is wrong here? Nothing: $\mathbf{On}$ is only a *class* and not a set, while ordinal numbers are sets.[4] Its not being a set is the only obstacle for $\mathbf{On}$ to be an ordinal.

Now that we know that $\mathbf{On}$ is a well-ordering, we may specify the principles of induction and recursion to ordinal numbers.

**Transfinite Induction.** *Suppose* $X \in \mathbf{On}$ *or* $X = \mathbf{On}$, *and* $\wp$ *is a property of ordinals.*

*Assume that the induction hypothesis*
($\alpha$)  *if every* $\beta < \alpha$ *has* $\wp$, *then so does* $\alpha$

---

[4]The assumption that $\mathbf{On}$ be a set is known as Burali-Forti's antinomy.

*holds for every $\alpha \in X$.*

*Then all $\alpha \in X$ have the property $\wp$.*     □

**Transfinite Recursion.** *Suppose $X \in \mathbf{On}$ or $X = \mathbf{On}$, and $W$ is an arbitrary class. Suppose further that $F$ is a function assigning, to each function from some $\alpha \in X$ to $W$, a value in $W$.*

*Then there is a uniquely determined function $f : X \to W$ such that $f(\alpha) = F(f \restriction \alpha)$ for all $\alpha \in X$.*     □

A **limit ordinal** is, by definition, an ordinal that is the union of all smaller ordinals. All other ordinals are called **successor ordinals**.

**Lemma 7.5.4.**
(1)  $\alpha \in \mathbf{On}$ *is a successor ordinal if and only if $\alpha = \beta+1$ for some $\beta \in \mathbf{On}$.*
(2)  *0 is (the smallest) limit ordinal.*
(3)  *$\omega$ is the next (and the smallest infinite) limit ordinal.*

*Proof.* Ad (1). If $\alpha = \beta \cup \{\beta\}$, then $\bigcup_{\gamma < \alpha} \gamma = \bigcup_{\gamma \in \alpha} \gamma = \beta \cup \bigcup_{\gamma \in \beta} \gamma = \beta \neq \alpha$, hence $\alpha = \beta + 1$ is a successor ordinal.

Let, conversely, $\alpha$ be a successor ordinal, i. e., $\alpha \neq \bigcup_{\gamma < \alpha} \gamma$. Pick $\beta \in \alpha \smallsetminus \bigcup_{\gamma < \alpha} \gamma$. Then, for all occurring $\gamma$, we have $\beta \notin \gamma$, hence $\beta \not< \gamma$. But then $\beta \geq \gamma$, hence $\beta$ is the greatest element in $\alpha$. Finally, for the initial section $\alpha_\beta = \{\gamma \in \alpha : \gamma < \beta\}$ we have $\alpha = \alpha_\beta \cup \{\beta\} = \beta \cup \{\beta\} = \beta + 1$.

Ad (2). Remember, $0 = \emptyset$.

Ad (3). $\omega = \mathbb{N} = \{0\} \cup \{n + 1 : n \in \mathbb{N}\}$.     □

We now reformulate the principle of **transfinite induction** for ordinals. According to their division into two classes, the limit ordinals and the successor ordinals, one obtains two essentially different inductive steps, the **successor step** and the **limit step**.

*Suppose $X \in \mathbf{On}$ or $X = \mathbf{On}$, and $\wp$ is a property of ordinals. Assume that, for all $\alpha \in X$, we have*
*$(\alpha + 1)$   if $\alpha$ has the property $\wp$, then so does $\alpha + 1$,*
*and for all limit ordinals $\delta \in X$ we have*
*$(\delta)$   if all $\beta < \delta$ have the property $\wp$, so does $\delta$.*
*Then all ordinals from $X$ have the property $\wp$.*

One can formulate $(\delta)$ separately for $\delta = 0$ and all infinite limit ordinals $\delta$. The former is then also called the **initial step** or the **basis** of the induction:
*$(0)$   0 has the property $\wp$.*

The reformulation of the recursion principle for ordinals is left to the reader. Instead we want to derive the aforementioned correspondence between ordinals and isomorphism types of well-orderings (also called **well-order types**). More precisely, we are going to show that every well-order type contains exactly one ordinal. This enables us to transfer the transfinite induction as specified to ordinals to arbitrary sets: by the well-ordering principle every set can be well-ordered and is then, by the theorem below, isomorphic to an ordinal. This means nothing more than that every set $X$ can be indexed by an ordinal number $\alpha$, i. e., $X = \{x_\beta : \beta < \alpha\}$. Then statements about the elements of $X$ can be proved by induction on $\beta < \alpha$.

Any two ordinals are comparable. Hence, if they are different, by Lemma 7.5.3(1), the smaller one of them must be a section of the other. Then Lemma 7.4.1(2) implies that they cannot be isomorphic. Thus every well-order type contains at most one ordinal. In order to show that every well-order type does contain an ordinal we need the following

**Lemma 7.5.5.** *Suppose $(X, <)$ is a well-ordering and $f : X \to \textbf{On}$ is such that the ordering $(X_a, <)$ is isomorphic to the ordinal $f(a)$ for every $a \in X$.*

*Then $f$ is an order isomorphism from $(X, <)$ onto an initial segment of $(\textbf{On}, <)$. Further, $f[X] = \textbf{On}$ if and only if $X$ is not a set.*

*Proof.* The map $f$ is an embedding of $(X, <)$ in $(\textbf{On}, <)$, for $c < b$ iff $X_c \subset X_b$ iff $f(c)$ is isomorphic (hence equal) to a section of $f(b)$.

We claim that $f[X]$ is an initial segment of $(\textbf{On}, <)$. This is trivially true in case $f[X] = \textbf{On}$. If this is not the case, there is a least element $\beta$ in $\textbf{On} \smallsetminus f[X]$. We show that $\beta = f[X]$ (and hence $(X, <)$ is isomorphic to $\beta$). By the minimal choice of $\beta$, we have $\beta \subseteq f[X]$. If there were $\alpha > \beta$ in $f[X]$, there would be also $a \in X$ with $f(a) = \alpha$. The isomorphism $\alpha \cong (X_a, <)$ would then yield $b \in X_a$ with $\beta \cong (X_b, <)$ (for $\beta$ is a section of $\alpha$). Then we would obtain $\beta \cong f(b)$, hence also $\beta = f(b)$, contradicting the choice of $\beta$. Thus, in any case, $f[X]$ is a segment of $(\textbf{On}, <)$, as required for the first assertion of the lemma.

The set-theoretic *axiom of replacement* says that, given a function $f$ and a set $X$, also $f[X]$ is a set. Since $f$ is a bijection onto $f[X]$, it has an inverse, and so the replacement axiom yields that $X$ is a set if and only if so is $f[X]$. As $\textbf{On}$ is not a set while each of its sections is, the second assertion follows as well.                                                                                            $\square$

**Theorem 7.5.6.** *Every well-ordered set is isomorphic to a unique ordinal number. Every well-ordering that is not a set is isomorphic to $(\textbf{On}, <)$.*

*Proof.* First we show that every well-ordered set $(X, <)$ is isomorphic to a proper initial segment of $(\mathbf{On}, <)$, i. e. to an ordinal. If this were not the case, the preceding lemma would yield $a$ in $X$ such that $X_a$ is not isomorphic to an ordinal. But then every section of $X_a$ would be isomorphic to an ordinal, contradicting this lemma.

Thus every section of a well-ordering is isomorphic to a unique ordinal. If now $(X, <)$ is an arbitrary well-ordering, we may assign to every $a \in X$ a (uniquely determined) ordinal $f(a) \cong X_a$. The resulting function $f : X \to \mathbf{On}$ then satisfies the above lemma, which implies the assertions. $\square$

**Corollary 7.5.7.** *Given well-orders $\mathcal{X}$ and $\mathcal{Y}$, the well-ordering $\mathcal{X}$ is isomorphic to an initial segment of $\mathcal{Y}$, or $\mathcal{Y}$ is isomorphic to an initial segment of $\mathcal{X}$. (In any case, one of them is embeddable in the other.)* $\square$

Cantor, who introduced the ordinal numbers, defined them as well-order types (and he didn't bother to give a precise set-theoretic justification for their existence). In contrast, John von Neumann defined them as particular well-ordered sets. We followed the latter's apparently more artificial way, as it is more elegant and set-theoretically precise. The theorem just proved, however, shows that the two well-ordered classes, $(\mathbf{On}, <)$ on the one hand and the class of well-order types with the order given by inclusion of corresponding representatives on the other, are isomorphic. In this sense the two definitions—von Neumann's and Cantor's original one—are equivalent.

[Cantor, G. : Beiträge zur Begründung der transfiniten Mengenlehre II, Mathematische Annalen **49** (1897) 207 – 246]

[v. Neumann, J. : Zur Einführung der transfiniten Zahlen, Acta Litt. Scient. Univ. Szeged., Sectio scient. math. 1 (1923) 199 – 208]

For more on the history of ordinals see

[Bachmann, H. : **Transfinite Zahlen**, Ergebnisse der Mathematik und ihrer Grenzgebiete 1, Springer, Berlin 1955, 204 pp.].

**Exercise 7.5.1.** Given ordinals $\alpha$ and $\beta$, their sum $\alpha + \beta$ is defined to be the uniquely determined ordinal which is isomorphic (as an ordering) to the sum of the ordering $\alpha$ and $\beta$ (in the sense of Exercise 7.4.1).
(a)   Show that the addition on $\omega$ coincides with the usual addition of natural numbers.
(b)   Prove that the successor of $\alpha$ is the sum of $\alpha$ and 1 (this justifying the notation $\alpha + 1$ for the successor of $\alpha$).
(c)   Verify $1 + \alpha = \alpha$ for all infinite $\alpha \in \mathbf{On}$ (and that addition of ordinals is not commutative).

**Exercise 7.5.2.** Let $X$ be a subset of $\mathbf{On}$. Prove that $\bigcup X$ is the supremum of $X$ in $(\mathbf{On}, <)$.

**Exercise 7.5.3.** Show that the complete $L_<$-theory of any infinite ordinal number has a non-well-founded model, i. e. a model that is not well-ordered.

## 7.6 Cardinal numbers

Recall that two sets have the same *power* (or *cardinality*) if there is a bijection between them. The relation of having the same power is easily seen to be an equivalence relation on the universe of sets. Similarly as well-order types (i. e. equivalence classes of well-orderings modulo isomorphism) contain as a particular representative an ordinal number, we see next that also the equivalence classes of sets with respect to having the same power contain as a representative a particular kind of number, a so-called cardinal number. This will allow us to speak of *the* cardinality of a set —as a set-theoretic object (as opposed to just having around an equivalence relation of having the same power).

Namely, we define a **cardinal number** (or just a **cardinal**) to be an ordinal that has not the same power as any smaller ordinal (i. e. there is no bijection onto a smaller ordinal). In other words, a cardinal is the smallest ordinal in the class of all sets of the same cardinality. For this reason cardinals are also known as **initial** (ordinal) **numbers**. We denote the class of all cardinal numbers by **Cn**.

**Example.** Obviously, finite ordinals are cardinals (since no finite set can be bijectively mapped onto a proper subset). Thus all natural numbers are cardinals. Clearly, also $\omega$ is a cardinal. However, $\omega + 1$ is not a cardinal, for $\omega + 1$ and $\omega$ have the same power.

Recall the well-ordering principle (from the remark at the end of §7.4) that says that every set $A$ has an ordering $<$ making $(A, <)$ a well-ordered set $A$ and hence isomorphic to an ordinal. The smallest thus occurring ordinal is defined to be the **cardinality** of $A$, in symbols $|A|$. Clearly, $|A|$ is a cardinal.

Being a subclass, **Cn** inherits various features from **On**, first of all the ordering, which is a well-ordering also on **Cn**. Then, being an ordinal, every cardinal is equal to the set of all smaller ordinals, i. e., to the set of all ordinals of smaller cardinality. Further, by Theorem 7.5.6, every set of cardinals is, with respect to the inherited order, isomorphic to a unique ordinal.

As mentioned, 'finite ordinal', 'finite cardinal' (and 'natural number') are synonymous. The situation is different in the infinite. Let us look at the infinite cardinals a little closer. First of all, denote the class of all such by $\mathbf{Cn}^\infty$. The class of (infinite) cardinals is a proper subclass of that of (infinite) ordinals. However, even though there seem to be 'much more' ordinals than cardinals, $\mathbf{Cn}^\infty$ is a proper class, for it is unbounded within $\mathbf{On}$ and hence, by Theorem 7.5.6, cannot be a set. In particular, as orderings, $(\mathbf{Cn}^\infty, <)$ and $(\mathbf{On}, <)$ are isomorphic. This isomorphism is of great importance. It provides every infinite cardinal with a uniquely determined ordinal index, and conversely assigns to every ordinal $\alpha$ a unique infinite cardinal, which, following Cantor, is denoted by $\aleph_\alpha$.[5] Thus every infinite cardinal is of the form $\aleph_\alpha$ (and all these are different). Therefore infinite cardinals are also called **alephs**.

For instance, being the first infinite cardinal, the ordinal $\omega$ is equal to the cardinal $\aleph_0$. The smallest uncountable cardinal is $\aleph_1$. Then comes $\aleph_2$, and so on. In general, given $n < \omega$, $\aleph_n$ is the $n + 1$st infinite cardinal. The smallest aleph with an infinite index is $\aleph_\omega$. In general, given any ordinals $\alpha$ and $\beta$, we have $\aleph_\alpha \leq \aleph_\beta$ iff $\alpha \leq \beta$. Further, within $(\mathbf{Cn}^\infty, <)$, the cardinal $\aleph_{\alpha+1}$ is the successor of $\aleph_\alpha$. Given a cardinal number $\kappa$, not necessarily in aleph notation, its successor (in $(\mathbf{Cn}^\infty, <)$) is denoted by $\kappa^+$. Thus, if $\kappa = \aleph_\alpha$, then $\kappa^+ = \aleph_{\alpha+1}$ (while $\kappa^+ = n + 1$ in case $\kappa = n < \omega$).

Since cardinals behave quite differently from arbitrary ordinals, it is very useful to have this separate notation. E. g., the successor of the ordinal $\omega$ is $\omega + 1$, i. e. the second countably infinite ordinal (which is not a cardinal, as it has the same power as its predecessor), while the successor of the cardinal $\omega$, that is the successor of $\aleph_0$, is $\aleph_1$ (an ordinal much bigger than $\omega + 1$ or even $\omega + n$ for all $n$, for it is uncountable).

Another advantage of this indexing of infinite cardinals by ordinals is that it allows us to apply the principles of transfinite induction and recursion not only to $\mathbf{Cn}$, but also to $\mathbf{Cn}^\infty$ (directly via the ordinal indices).

Since the successor of an infinite ordinal has the same cardinality, no cardinal can be a successor ordinal. In other words, every cardinal is a limit *ordinal*. However, the reader might have already guessed that it is reasonable to call a cardinal of the form $\kappa^+$ (where $\kappa$ is an arbitrary cardinal) a **successor cardinal**. Correspondingly, a **limit cardinal** is defined to be a cardinal of the form $\aleph_\delta$, where $\delta$ is a limit *ordinal*. Thus $\aleph_0$ is the smallest limit cardinal, and $\aleph_\omega$ is the second biggest limit cardinal.

We defined the sum of ordinals in one of the exercises of the preceding section. Also the other arithmetical operations can be defined in a canonical way so that their arithmetic, restricted to natural numbers, coincides

---

[5] $\aleph$ [read: 'aleph'] is the first letter of the Hebrew alphabet.

with the usual one. In the infinite this arithmetic proves useful for the investigation of well-orderings (due to Theorem 7.5.6). However, we have to refer the reader to one of the set-theoretic texts for more details, and also to Rosenstein's book cited in Appendix C, which contains some interesting model theory of ordinal numbers.

The arithmetic of cardinal numbers is completely different. While that of ordinals is designed to reflect certain compositions of well-orderings, the arithmetic of cardinals is designed to reflect cardinalities of compound sets (and we will need the rudiments of it in the sequel to calculate certain cardinalities). It is therefore natural to define the arithmetical operations as follows.

Given $\kappa, \lambda \in \mathbf{Cn}$, set $\kappa + \lambda =_{\text{def}} |\kappa \sqcup \lambda|$, $\kappa \cdot \lambda =_{\text{def}} |\kappa \times \lambda|$, and $\kappa^\lambda =_{\text{def}} |{}^\lambda \kappa|$.

As in the case of ordinal arithmetic, these operations coincide with the usual ones (for natural numbers) on the set $\omega$. But we will see in due course that in the infinite things are different. Nevertheless, many common arithmetical rules hold true for arbitrary cardinals. For instance, using appropriate bijections the reader may verify the ones listed below.

**Remarks.** Let $\kappa, \lambda, \mu \in \mathbf{Cn}$.
(1) $(\mathbf{Cn}; 0; +)$ is a commutative monoid with neutral element 0 (cf. §5.2).
(2) If $\kappa \leq \lambda$, then $\kappa + \mu \leq \lambda + \mu$.
(3) $(\mathbf{Cn}; 1; \cdot)$ is a commutative monoid with neutral element 1.
(4) $\kappa \cdot (\lambda + \mu) = \kappa \cdot \lambda + \kappa \cdot \mu$.
(5) If $\kappa \leq \lambda$, then $\kappa \cdot \mu \leq \lambda \cdot \mu$.
(6) $0 \cdot \kappa = \kappa \cdot 0 = 0$.
(7) $\kappa^{\lambda + \mu} = \kappa^\lambda \cdot \kappa^\mu$.
(8) $\kappa^{\lambda \cdot \mu} = (\kappa^\lambda)^\mu$.
(9) $(\kappa \cdot \lambda)^\mu = \kappa^\mu \cdot \lambda^\mu$.
(10) If $\kappa \leq \lambda$ and $\mu > 0$, then $\mu^\kappa \leq \mu^\lambda$ and $\kappa^\mu \leq \lambda^\mu$.
(11) $\kappa^0 = 1$, in particular, $0^0 = 1$.
(12) $\kappa^1 = \kappa$.
(13) If $\kappa \neq 0$, then $0^\kappa = 0$.
(14) If $\kappa \geq 2$, then $\kappa + \kappa \leq \kappa \cdot \kappa$.

These are the basic laws of cardinal arithmetic which look (and are verified) the same on $\mathbf{Cn}$ as on $\omega$. Next we turn to some properties that are essentially different from those of natural numbers. Most of them follow more or less directly from the following.

**Theorem 7.6.1.** (Hessenberg)
$\lambda \cdot \lambda = \lambda$ for all $\lambda \in \mathbf{Cn}^\infty$.

*Proof.* By induction on $\lambda \in \mathbf{Cn}^\infty$.

First we order the pairs of ordinals from the set $\lambda \times \lambda$ by the following rule.

$(\alpha, \beta) \prec (\gamma, \delta)$, provided

either $\quad \max\{\alpha, \beta\} < \max\{\gamma, \delta\}$,

or else $\quad \max\{\alpha, \beta\} = \max\{\gamma, \delta\} \quad$ *and* $\quad$ either $\quad \alpha < \gamma$,

$\qquad\qquad\qquad\qquad\qquad\qquad\qquad\qquad$ or else $\quad \alpha = \gamma$ and $\beta < \delta$.

Obviously, $(\lambda \times \lambda, \prec)$ is a (strict) linear ordering. In the case $\lambda = \aleph_0 = \omega$ this ordering is the well-known ordering of $\omega \times \omega$ found by Cantor in order to prove the countability of $\mathbb{Q}$. This also constitutes the initial step of our induction (the case $\aleph_0 \cdot \aleph_0 = \aleph_0$).

Let now $\lambda \in \mathbf{Cn}^\infty$ be arbitrary, and assume that $\kappa \cdot \kappa = \kappa$ for all $\kappa < \lambda$ (induction hypothesis). All we have to do in order to complete the induction is derive $\lambda \cdot \lambda = \lambda$ from this.

Let $(\gamma, \delta) \in \lambda \times \lambda$ and $\varepsilon = \max\{\gamma, \delta\} + 1$. Being an infinite cardinal, $\lambda$ is a limit *ordinal*. Hence $\varepsilon \in \lambda$ and $(\gamma, \delta) \prec (\varepsilon, \varepsilon) \in \lambda \times \lambda$. Moreover, the entire section $(\lambda \times \lambda)_{(\gamma,\delta)}$ given by $(\gamma, \delta)$, is contained in $\varepsilon \times \varepsilon$ and therefore has a cardinality $\leq |\varepsilon \times \varepsilon| = |\varepsilon| \cdot |\varepsilon|$. The latter is, by induction hypothesis, equal to $|\varepsilon|$, hence smaller than $\lambda$.

Consequently, every section of $\lambda \times \lambda$ has cardinality $< \lambda$. If we could prove that $(\lambda \times \lambda, \prec)$ is a well-ordered set, hence isomorphic to some ordinal $\alpha$ (by Theorem 7.5.6), this would imply that $\alpha$ cannot be bigger than $\lambda$ (for otherwise it would contain a section of power $\lambda$). In particular, we would obtain $\lambda \cdot \lambda = |\lambda \times \lambda| \leq |\lambda| = \lambda$.

On the other hand, Remark (5) above says that always $\lambda \leq \lambda \cdot \lambda$. Thus, indeed, it remains to verify that $(\lambda \times \lambda, \prec)$ is a well-ordered set.

First of all, the Zermelo-Fraenkel axioms imply that $\lambda \times \lambda$ is a set. Let now $\emptyset \neq X \subseteq \lambda \times \lambda$. Pick the least ordinal $\eta$ in $\{\max\{\alpha, \beta\} : (\alpha, \beta) \in X\}$ and then the least ordinal $\gamma$ in the set $Y =_{\text{def}} \{\alpha \in \mathbf{On} :$ there is $\beta$ with $(\alpha, \beta) \in X$ and $\max\{\alpha, \beta\} = \eta\}$. If now $\delta$ is the least ordinal such that $(\gamma, \delta) \in X$, then obviously $(\gamma, \delta)$ is the least element in $X$, as desired. $\qquad\qquad\square$

**Corollary 7.6.2.** *If one of the nonzero cardinals $\kappa$ and $\lambda$ is infinite, then $\kappa + \lambda = \kappa \cdot \lambda = \max\{\kappa, \lambda\}$.*

*Proof.* Without loss of generality, let $\lambda$ be infinite and $\kappa \leq \lambda$.

Clearly $\max\{\kappa, \lambda\} = \lambda \leq \kappa + \lambda \leq \lambda + \lambda$. The latter sum is, by Remark (14), not greater than $\lambda \cdot \lambda$, hence, by Hessenberg's theorem, not greater than $\lambda = \max\{\kappa, \lambda\}$. This shows $\kappa + \lambda = \max\{\kappa, \lambda\}$.

If $\kappa \neq 0$, an analogous argument can be used to prove also $\kappa \cdot \lambda = \max\{\kappa, \lambda\}$. $\square$

**Corollary 7.6.3.** *If $\lambda \in \mathbf{Cn}^\infty$ and $0 < n < \omega$, then $\lambda + n = \lambda \cdot n = \lambda^n = \lambda$.* $\square$

Next we see why sets of the form $\mathfrak{P}(X)$ are called *power*sets.

**Lemma 7.6.4.**
(1)  $|\mathfrak{P}(X)| = 2^{|X|}$ *for every set $X$.*
(2)  *Let $\kappa \in \mathbf{Cn}$ and $\lambda \in \mathbf{Cn}^\infty$.*
     *If $2 \leq \kappa \leq \lambda$, then $\kappa^\lambda = 2^\lambda$.*

*Proof.* Ad (1). The subsets of $X$ are in 1-1 correspondence with the characteristic functions on $X$, hence with functions from $^X\{0,1\}$, i. e. $^X 2$.

Ad (2). Invoking Remark (10) above, $2 \leq \kappa \leq \lambda \leq |\mathfrak{P}(\lambda)| \leq 2^\lambda$ implies $2^\lambda \leq \kappa^\lambda \leq (2^\lambda)^\lambda$. But Remark (8) and Hessenberg's theorem yield $(2^\lambda)^\lambda = 2^{\lambda \cdot \lambda} = 2^\lambda$. $\square$

The inequality $|\mathfrak{P}(X)| > |X|$ is obvious for finite sets $X$ (for $2^n > n$ for all $n < \omega$). Cantor found an elegant, so-called *diagonal* argument to prove this for arbitrary sets $X$. Variants of this argument occurred later in all parts of mathematical logic.

**Theorem 7.6.5.** (Cantor)
   $2^\kappa > \kappa$ *for all cardinal numbers $\kappa$.*

*Proof.* We prove $|\mathfrak{P}(X)| > |X|$. Consider singletons to see $|\mathfrak{P}(X)| \geq |X|$. If we had equality here, there would be a function $f$ from $X$ onto $\mathfrak{P}(X)$. Then consider the set $Y =_{\mathrm{def}} \{x \in X : x \notin f(x)\}$. The surjectivity of $f$ would yield $y \in X$ with $f(y) = Y$. Then, for all $x \in X$, we would have $x \in f(y)$ iff $x \notin f(x)$. But in the case $x = y$ this is a contradiction. $\square$

**Remark.** (About the cardinalities of number systems)
   Clearly, $|\mathbb{N}| = \aleph_0$, hence, by Hessenberg's theorem, $|\mathbb{Q}| = |\mathbb{N} \times \mathbb{N}| = \aleph_0 \cdot \aleph_0 = \aleph_0$. Corollary 7.6.2 yields $|\mathbb{Z}| = |\mathbb{N} \setminus \{0\} \sqcup \mathbb{N}| = \aleph_0 + \aleph_0 = \aleph_0$. Lemma 7.6.4(2) implies that the **continuum** $\mathbb{R}$ has power $|\mathbb{R}| = |^\mathbb{N}\mathbb{Z}| = \aleph_0^{\aleph_0} = 2^{\aleph_0}$. Invoking Hessenberg's theorem again, we then obtain $|\mathbb{C}| = |\mathbb{R} \times \mathbb{R}| = 2^{\aleph_0} \cdot 2^{\aleph_0} = 2^{\aleph_0}$.

As a special case of Cantor's theorem we have $2^{\aleph_0} \geq \aleph_1$ or, equivalently, $\mathbb{R} \geq \aleph_1$. So the question arrises if this inequality is proper, i. e., if there is an uncountable set having smaller cardinality than the continuum. Cantor was convinced that this was not the case. He tried to prove what later became the **continuum hypothesis** (and Hilbert's first problem):

(CH) $2^{\aleph_0} = \aleph_1$.

Even though he was completely convinced of the truth of his conjecture, Cantor did not succeed in proving it, for a good reason. For, although in 1938 Kurt Gödel was able to construct a model of the set-theoretic axioms ZFC satisfying a much stronger hypothesis, the so-called **general continuum hypothesis**

(GCH) $2^{\aleph_\alpha} = \aleph_{\alpha+1}$ for all $\alpha \in \mathbf{On}$,

in 1963 Paul Cohen found a model of ZFC that satisfies the negation of (CH). Hence, the statements (CH) and (GCH) are, as one says, **independent** of ZFC, and the (deductive closure of) ZFC is thus an incomplete theory.

Finally we want to determine the cardinality of languages. The following bound turns out to be of use in this.

**Lemma 7.6.6.** *Suppose $\kappa$ and $\lambda$ are cardinals and $|X_i| \leq \lambda$ for all $i < \kappa$. Then $|\bigcup_{i<\kappa} X_i| \leq \kappa \cdot \lambda$.*

*Proof.* $|X_i| \leq \lambda = |\{i\} \times \lambda|$ implies $|\bigcup_{i<\kappa} X_i| \leq |\bigcup_{i<\kappa}(\{i\} \times \lambda)| = |\kappa \times \lambda| = \kappa \cdot \lambda$. □

We need some more set-theoretic notation. Given a set $X$ and a cardinal $\lambda$, we use $^{<\lambda}X$ to denote the union of the sets $^\mu X$, where $\mu$ runs over all cardinals $< \lambda$. Since no confusion can arise, one often writes $^{<\omega}X$ instead of $^{<\aleph_0}X$. Further, given a cardinal $\kappa$, the cardinality of the set $^{<\lambda}\kappa$ is denoted by $\kappa^{<\lambda}$.

Of particular interest is the case $\lambda = \aleph_0$: the set $^{<\omega}X$ is the set of all finite sequences of elements of (i. e. of all tuples from) $X$. Its cardinality, for infinite $X$, can be computed as follows.

**Proposition 7.6.7.** $\kappa^{<\aleph_0} = \kappa$ for all infinite cardinals $\kappa$.

*Proof.* Corollary 7.6.3 yields $|^n\kappa| = \kappa^n = \kappa$, hence, together with the preceding lemma, also $\kappa^{<\aleph_0} = |\bigcup_{n<\omega}{}^n\kappa| = \aleph_0 \cdot \kappa$. But the latter is, by Corollary 7.6.2, equal to $\kappa$. □

Thus the set of all tuples from an infinite set has the same power as that set. This leads to the computation of the following important cardinalities.

**Corollary 7.6.8.** *Let $L$ be a language of signature $\sigma$.*

(1) $|L| = |L_n| = |L_0| = \max\{\aleph_0, |\sigma|\} = |\sigma| + \aleph_0$ *(here $|\sigma|$ is the cardinality of the set of non-logical symbols of $\sigma$, cf. §1.1).*

(2) *If $C$ is an arbitrary set of new constants, then $|L(C)| = |L| + |C|$.*

*Proof.* Ad (1). Every $L$-formula is a finite sequence of symbols of the alphabet of $L$. The latter consists of finitely many logical symbols, countably infinitely many variables, two parentheses, and $|\sigma|$ non-logical symbols. Thus the alphabet has power $\kappa =_{\text{def}} \max\{\aleph_0, |\sigma|\}$. By the preceeding proposition, there are $\kappa$ such finite sequences, whence $|L_0| \leq |L_n| \leq |L| \leq \kappa$. On the other hand, using $\aleph_0$ variables and $|\sigma|$ non-logical constants it is easy to build $\max\{\aleph_0, |\sigma|\}$ different $L$-sentences, whence also $\kappa \leq |L_0|$.

(2) follows from (1), for the signature of $L(C)$ has power $|\sigma| + |C|$.  $\square$

Also the cardinal numbers were introduced by Cantor, albeit as abstract powers, i. e. equivalence classes of sets modulo the equivalence relation of having the same power (and also here he didn't care much about a set-theoretic justification of their existence). Again it was von Neumann who turned them into concrete objects of the set-theoretic universe, namely into ordinal numbers.

[Cantor, G. : Beiträge zur Begründung der transfiniten Mengenlehre I, Mathematische Annalen **46** (1895) 481 – 512]

[v. Neumann, J. : Die Axiomatisierung der Mengenlehre, Math. Zeitschrift **27** (1928) 669 – 752]

**Exercise 7.6.1.** Prove Remarks (1)–(14) above.

**Exercise 7.6.2.** Prove that a vector space of dimension $\lambda > 0$ over a division ring $\mathcal{K}$ has power $\max\{|\mathcal{K}|, \lambda\}$.

**Exercise 7.6.3.** Show that every set of cardinal numbers has a supremum in **Cn**.

A cardinal $\lambda$ is said to be **regular** if it is not a union of less than $\lambda$ lesser cardinals, i. e., if there are no $\kappa < \lambda$ and $\kappa_i < \lambda$ $(i < \kappa)$ such that $\bigcup_{i<\kappa} \kappa_i = \lambda$.

**Exercise 7.6.4.** Show that $\aleph_0$ and $\aleph_1$ (and, moreover, all successor cardinals) are regular.

**Exercise 7.6.5.** Show that $\aleph_\omega$ is not regular.

# Part III

# Basic properties of theories

The material presented in the previous part was based mostly on some more or less clever applications of the finiteness theorem in order to produce models of certain theories. To ensure that these theories expressed what they were to express we often formulated them using diagrams or interpretations.

We now turn to general properties of theories and more advanced methods of investigation of the relationship between their models. Fundamental here is the notion of elementary embedding between structures of the same signature (Chapter 8). Cases where this concept simplifies are dealt with in Chapter 9 about quantifier elimination and related issues (where we prove, among other things, Hilbert's Nullstellensatz). Chapter 10 finally deals with iterated elementary embeddings, that is with so-called elementary chains, a basic tool of model theory.

*As a general assumption, let $\sigma$ be an arbitrary signature and $L = L(\sigma)$ the corresponding language, unless we say otherwise.*

# Chapter 8

# Elementary maps

## 8.1 Elementary equivalence

Recall from §6.1 that the $L$-structures $\mathcal{M}$ and $\mathcal{N}$ are said to be **elementarily equivalent** if $\mathcal{M} \equiv \mathcal{N}$, which is the same as $\mathrm{Th}\,\mathcal{M} = \mathrm{Th}\,\mathcal{N}$.

Proposition 6.1.3 says, among other things, that isomorphic structures are elementarily equivalent. We prove first that the converse is true if and only if the structures under consideration are finite.

**Proposition 8.1.1.** *The following are equivalent for any $L$-structure $\mathcal{M}$.*
(i)   $\mathcal{N} \equiv \mathcal{M}$ *implies* $\mathcal{N} \cong \mathcal{M}$ *for any $L$-structure $\mathcal{N}$.*
(ii)  $\mathcal{M}$ *is finite.*

*Proof.* The isomorphism type of a structure is uniquely determined by its diagram.

If $\mathcal{M}$ is a finite structure with universe $M = \{a_0, \dots, a_{n-1}\}$ in a language of finite signature, then the diagram of $\mathcal{M}$ is essentially finite (see Exercise 6.1.6). In this case we can axiomatize the diagram by a single sentence $\varphi(a_0, \dots, a_{n-1})$. Then $\mathcal{M}$ is, up to isomorphism, the only model of the sentence $\exists^{=n} x(x = x) \wedge \exists \bar{x}\, \varphi(\bar{x})$, and (i) follows.

If the signature of $\mathcal{M}$ is infinite, the diagram of $\mathcal{M}$ need not be finite anymore (in the simplest such case, there could be infinitely many constant symbols that are interpreted in $\mathcal{M}$ by the same element). Then, for any finite subset $\Delta$ of $\mathrm{D}(\mathcal{M})$, we form the conjunction $\varphi_\Delta(a_0, \dots, a_{n-1})$ of all sentences in $\Delta$ and see that $\mathcal{M}$ is, up to isomorphism, the only model of the set $\{\exists^{=n} x(x = x)\} \cup \{\exists \bar{x}\, \varphi_\Delta(\bar{x}) : \Delta \Subset \mathrm{D}(\mathcal{M})\}$. This implies (i) again.

If, for the converse, $\mathcal{M}$ is infinite, then, by Löwenheim-Skolem upward (Theorem 5.1.1), its complete theory has arbitrarily large models. As we will see in the next proposition, these are elementarily equivalent to $\mathcal{M}$. To violate (i) it therefore remains to consider such a model of a cardinality bigger than that of $\mathcal{M}$ (for isomorphic structures certainly have the same cardinality). □

We just used that all models of a complete theory are elementarily equivalent. Now we verify this, along with a certain converse.

**Proposition 8.1.2.** *A theory is complete if and only if all of its models are elementarily equivalent.*

*Proof.* Obviously, $\mathcal{M} \models T$ implies $T \subseteq \text{Th}\,\mathcal{M}$ and, provided $T$ is complete, also $T = \text{Th}\,\mathcal{M}$. Hence, if $T$ is complete, then $\text{Th}\,\mathcal{M} = \text{Th}\,\mathcal{N}$ (i. e. $\mathcal{M} \equiv \mathcal{N}$) for all $\mathcal{M}, \mathcal{N} \models T$.

If now $T$ is not complete, then $T \subset \text{Th}\,\mathcal{M}$ for any model $\mathcal{M} \models T$. Pick $\varphi \in \text{Th}\,\mathcal{M} \setminus T$. Since then $T \not\models \varphi$, the set $T \cup \{\neg\varphi\}$ has a model $\mathcal{N}$. Then $\mathcal{M}$ and $\mathcal{N}$ are two models of $T$ that are not elementarily equivalent. □

**Corollary 8.1.3.** *A complete theory has a finite model if and only if it has, up to isomorphism, only one model.*

*Proof.* If the complete theory $T$ has a model of power $n < \omega$, then $\exists^{=n} x \,(x = x) \in T$, hence *every* model has power $n$. Since $T$ is complete, all models of $T$ are elementarily equivalent. But for finite structures elementary equivalence is the same as isomorphism.

If, conversely, $T$ has only one model, Löwenheim-Skolem implies that $T$ cannot have an infinite model. □

**Exercise 8.1.1.** Let $\varphi \in L$ define a finite set $X$ in the $L$-structure $\mathcal{M}$. Show that in every $\mathcal{N}$ elementarily equivalent to $\mathcal{M}$, the set defined by $\varphi$ has the same power as $X$. Formulate and prove a converse of this.

**Exercise 8.1.2.** Show that $T_{\cong}^{\infty}$ (from §3.4) is complete.

**Exercise 8.1.3.** Prove that every structure elementarily equivalent to the group $\mathbb{Q}$ of rational numbers (in the signature $(0; +, -)$) is a direct sum of copies of $\mathbb{Q}$ (hence a vector space over $\mathbb{Q}$).

## 8.2   Elementary maps

As already seen in §6.1, we obtain a stronger relationship between structures if we consider the transfer not only of sentences, but also of arbitrary formulas.

Let $\mathcal{M}$ and $\mathcal{N}$ be $L$-structures. A map $f : M \to N$ is called **elementary** if $f : \mathcal{M} \overset{\equiv}{\hookrightarrow} \mathcal{N}$ (cf. §6.1). The structure $\mathcal{M}$ is said to be **elementarily embeddable** in $\mathcal{N}$, in symbols $\mathcal{M} \overset{\equiv}{\hookrightarrow} \mathcal{N}$, if there is an elementary map $f : \mathcal{M} \overset{\equiv}{\hookrightarrow} \mathcal{N}$ from $\mathcal{M}$ to $\mathcal{N}$.

**Remarks.**
(1) If $\mathcal{M} \overset{\equiv}{\hookrightarrow} \mathcal{N}$, then $\mathcal{M} \equiv \mathcal{N}$.
(2) While elementary equivalence is weaker than isomorphism, every elementary map is automatically an isomorphic embedding (cf. Lemma 6.1.2(1)). Therefore elementary maps are also called **elementary embeddings**, and the notation $\mathcal{M} \overset{\equiv}{\hookrightarrow} \mathcal{N}$ is justified.
(3) The converse is not true in general, unless the embedding is surjective (i. e. an isomorphism), since
(4) Proposition 6.1.3 says that every isomorphism $f : \mathcal{M} \cong \mathcal{N}$ is an elementary map.

Similarly as in the diagram lemmas 6.1.2 and 7.1.1 we can describe elementary maps and elementary embeddability by theories in an expanded language. For this we define the **elementary diagram** of an $L$-structure $\mathcal{M}$ to be the complete $L(M)$-theory $\text{Th}(\mathcal{M}, M)$.

**Lemma 8.2.1.** (The Elementary Diagram Lemma) *Let $\mathcal{M}$ and $\mathcal{N}$ be $L$-structures.*
(1) *A map $f : M \to N$ is elementary if and only if $(\mathcal{M}, M) \equiv (\mathcal{N}, f[M])$ if and only if $(\mathcal{N}, f[M]) \models \text{Th}(\mathcal{M}, M)$.*
(2) *$\mathcal{M} \overset{\equiv}{\hookrightarrow} \mathcal{N}$ iff $\mathcal{N}$ has an expansion that is a model of $\text{Th}(\mathcal{M}, M)$.*

*Proof.* Ad (1). $f : \mathcal{M} \overset{\equiv}{\hookrightarrow} \mathcal{N}$ is the same as $f : \mathcal{M} \overset{L}{\to} \mathcal{N}$, hence also the same as $(\mathcal{M}, M) \equiv (\mathcal{N}, f[M])$. By the completeness of $\text{Th}(\mathcal{M}, M)$, however, the latter is equivalent to $(\mathcal{N}, f[M]) \models \text{Th}(\mathcal{M}, M)$.
(2) follows from (1) as in 6.1.2 or 7.1.1. $\square$

It is handy to use the diagram[1] notation

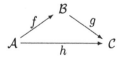

---
[1]This term has nothing to do with the model-theoretic diagrams discussed before.

known from homological algebra, which is a shorthand for the following statement.

$\mathcal{A}$, $\mathcal{B}$, $\mathcal{C}$ *are structures of the same signature;* $f : \mathcal{A} \to \mathcal{B}$, $g : \mathcal{B} \to \mathcal{C}$, $h : \mathcal{A} \to \mathcal{C}$ *(are homomorphisms), and* $gf = h$, *i. e.,* $g(f(a)) = h(a)$ *for all* $a \in A$ *(for the latter, one also says, the diagram* **commutes***).*

Similarly one can form more complicated diagrams, where we indicate elementary maps by putting '$\equiv$' next to the corresponding arrow.

**Lemma 8.2.2.** *Suppose*

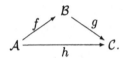

(1)  *If $f$ and $g$ are elementary, then so is $h$.*
(2)  *If $g$ and $h$ are elementary, then so is $f$.*
(3)  *$g$ need not be elementary, if $f$ and $h$ are.*

*Proof.* $f$ is elementary iff $(\mathcal{A}, A) \equiv (\mathcal{B}, f[A])$, and analogously for $g$ and $h$. Now the transitivity of $\equiv$ and $(\mathcal{C}, g[f[A]]) = (\mathcal{C}, h[A])$ imply (1).

Ad (2).  Since $(\mathcal{B}, B) \equiv (\mathcal{C}, g[B])$ implies $(\mathcal{B}, X) \equiv (\mathcal{C}, g[X])$ for every $X \subseteq B$, we have $(\mathcal{A}, A) \equiv (\mathcal{C}, h[A]) = (\mathcal{C}, g[f[A]]) \equiv (\mathcal{B}, f[A])$.

Ad (3).  We construct an example. Let $\mathcal{A}$ be an infinite $L$-structure. By Löwenheim-Skolem, $\mathrm{Th}(\mathcal{A}, A)$ has a model $\mathcal{B}^*$ of a power bigger than $\mathcal{A}$. If $\mathcal{B}$ denotes its $L$-reduct, Lemma 8.2.1 yields $f : \mathcal{A} \overset{\equiv}{\hookrightarrow} \mathcal{B}$. Setting $\mathcal{C} = \mathcal{A}$ and $h = \mathrm{id}_A$ we obtain $h : \mathcal{A} \overset{\equiv}{\hookrightarrow} \mathcal{C}$. For cardinality reasons we cannot have $\mathcal{B} \overset{\equiv}{\hookrightarrow} \mathcal{C}$ (for elementary maps are injective).

It remains to find $g : \mathcal{B} \to \mathcal{C}$ such that $gf = h$. For this consider $L = L_=$. Then every map is a homomorphism. Hence we may define $g$ by $g(b) = a$, in case $f(a) = b$ for some $a \in A$, and $g(b) = c$ for an arbitrary $c \in A$ otherwise. Then the diagram commutes.                                                        □

**Exercise 8.2.1.** Relativise Lemma 8.2.1 to an arbitrary set $\Delta \subseteq L$, that is to $f : \mathcal{M} \overset{\Delta}{\to} \mathcal{N}$.

**Exercise 8.2.2.** Given a structure $\mathcal{M}$ and an ultrafilter $U$ on a nonempty set $I$, prove that the canonical embedding (from Exercise 4.1.2) of $\mathcal{M}$ in its ultrapower $\mathcal{M}^I/U$ is elementary.

## 8.3 Elementary substructures and extensions

Next we deal with the elementariness of identical inclusions of structures.

Suppose $\mathcal{M}$ and $\mathcal{N}$ are $L$-structures and $\Delta$ is an arbitrary set of formulas. We use the notation $\mathcal{M} \preccurlyeq_\Delta \mathcal{N}$ (or also $\mathcal{N} \succcurlyeq_\Delta \mathcal{M}$) to mean that $M \subseteq N$ and the following implication holds for all $\varphi \in \Delta \cap L$ and matching (possibly empty) tuples $\bar{a}$ from $M$: if $\mathcal{N} \models \varphi(\bar{a})$, then $\mathcal{M} \models \varphi(\bar{a})$.

In case $\Delta$ contains all of $L$, we write $\mathcal{M} \preccurlyeq \mathcal{N}$ and $\mathcal{N} \succcurlyeq \mathcal{M}$ and say, $\mathcal{M}$ is an **elementary substructure** of $\mathcal{N}$, or $\mathcal{N}$ is an **elementary extension** of $\mathcal{M}$.

[Tarski, A. and Vaught, R. : Arithmetical extensions of relational systems, Compositio Math. **13** (1957) 81 – 102]

Trivially, $\mathcal{M} \preccurlyeq \mathcal{M}$. Further, $\mathcal{M} \preccurlyeq_\Delta \mathcal{N}$ is equivalent to $(\mathcal{N}, M) \Rightarrow_{\Delta(M)} (\mathcal{M}, M)$ (cf. the notation in §6.1).

If $\Gamma$ is the class of all negations of formulas from $\Delta$, then we have $\mathcal{M} \preccurlyeq_\Delta \mathcal{N}$ if and only if $M \subseteq N$ and $\mathrm{id}_M : \mathcal{M} \xrightarrow{\Gamma} \mathcal{N}$. In particular, $\mathcal{M} \preccurlyeq \mathcal{N}$ if and only if $M \subseteq N$ and $\mathrm{id}_M : \mathcal{M} \xrightarrow{\equiv} \mathcal{N}$. Further, $\mathcal{M} \preccurlyeq \mathcal{N}$ if and only if $M \subseteq N$ and $\varphi(\mathcal{N}) \cap M^n = \varphi(\mathcal{M})$ for all $n \in \mathbb{N}$ and $\varphi \in L_n$, i. e., the definable sets of $\mathcal{M}$ are just the intersections with $M$ of the definable sets of $\mathcal{N}$. For sets defined with parameters from $M$, this yields $\varphi(\mathcal{N}, \bar{c}) \cap M^n = \varphi(\mathcal{M}, \bar{c})$ for all $\varphi \in L_{n+m}$ and $\bar{c} \in M^m$.

As every elementary embedding is an (isomorphic) embedding, every elementary substructure is a substructure (and the use of the term is justified).

**Remarks.** The following are true for any $L$-structures $\mathcal{A}$, $\mathcal{B}$, and $\mathcal{C}$.
(1) $\mathcal{A} \preccurlyeq \mathcal{B}$ if and only if $A \subseteq B$ and $(\mathcal{A}, A) \equiv (\mathcal{B}, A)$ (special case of Lemma 8.2.1(1)).
(2) $f : \mathcal{A} \xrightarrow{\equiv} \mathcal{B}$ if and only if there is $\mathcal{A}' \preccurlyeq \mathcal{B}$ such that $f : \mathcal{A} \cong \mathcal{A}'$.
(3) If $\mathcal{A} \preccurlyeq \mathcal{B}$, then $\mathcal{A} \equiv \mathcal{B}$.
(4) $\mathcal{A} \preccurlyeq \mathcal{A}$.
(5) If $\mathcal{A} \preccurlyeq \mathcal{B} \preccurlyeq \mathcal{C}$, then $\mathcal{A} \preccurlyeq \mathcal{C}$ (special case of Lemma 8.2.2(1)).
(6) If $\mathcal{A} \subseteq \mathcal{B} \preccurlyeq \mathcal{C}$ and $\mathcal{A} \preccurlyeq \mathcal{C}$, then $\mathcal{A} \preccurlyeq \mathcal{B}$ (special case of Lemma 8.2.2(2)).

The following simple, but useful trick allows us to reduce the relation of (elementary) embeddability to the existence of an (elementary) extension.

**Lemma 8.3.1.** (Isomorphic Correction) *Let $\mathcal{M}$ and $\mathcal{N}$ be $L$-structures, and $\Delta \subseteq L$.*
(1) *If $\mathcal{M} \xrightarrow{\Delta} \mathcal{N}$, then there is $\mathcal{N}' \cong \mathcal{N}$ such that $M \subseteq N'$ and $\mathrm{id} : \mathcal{M} \xrightarrow{\Delta} \mathcal{N}'$.*
(2) *If $\mathcal{M} \hookrightarrow \mathcal{N}$, then there is $\mathcal{N}' \cong \mathcal{N}$ such that $M \subseteq N'$.*

(3)   *If $M \xrightarrow{\equiv} N$, then there is $N' \cong N$ such that $M \preccurlyeq N'$.*

*Proof.* (2) and (3) are special cases of (1). For the proof of the latter, suppose $f : M \xrightarrow{\Delta} N$. Take the union of the map $f^{-1}$ (defined on $f[M]$) with an arbitrary bijection from $N \smallsetminus f[M]$ onto a set disjoint from $M$. Call the resulting map $g$ and the image of this map $N'$. Then $N'$ contains $M$ and $g$ is a bijection from $N$ onto $N'$ extending $f^{-1}$. Hence $gf$ is an identical inclusion of $M$ into $N'$. Make $N'$ an $L$-structure $N'$ in such a way that $g$ becomes an isomorphism from $N$ onto $N'$ (i. e., interpret the non-logical symbols in $N'$ by the $g$-images of the interpretations in $N$). Being an isomorphism, $g$ is an elementary map. By an obvious $\Delta$-version of Lemma 8.2.2(1), so is $gf$, and we are done.                                                     $\square$

The next result shows that, when verifying by induction on the complexity of formulas that an embedding is elementary, all but the existential quantifier step are trivial.

**Proposition 8.3.2.** (The Tarski-Vaught Test) *Let $M$, $N$ be $L$-structures. $M \preccurlyeq N$ if and only if $M \subseteq N$ and the following condition holds.*

(TV)    *For all $n < \omega$, all $\varphi \in L_{n+1}$, and all $n$-tuples $\bar{a}$ from $M$, if $N \models \exists x\, \varphi(x, \bar{a})$, then there exists $b \in M$ such that $N \models \varphi(b, \bar{a})$.*

*Proof.* If $M \preccurlyeq N$, then $M \subseteq N$ and for all $\varphi$ and $\bar{a}$ as above we have both, $N \models \exists x\, \varphi(x, \bar{a})$ iff $M \models \exists x\, \varphi(x, \bar{a})$, and $N \models \varphi(b, \bar{a})$ iff $M \models \varphi(b, \bar{a})$ (whenever $b \in M$). Therefore, $N \models \exists x\, \varphi(x, \bar{a})$ yields $b \in M$ such that $M \models \varphi(b, \bar{a})$, hence also $N \models \varphi(b, \bar{a})$. This proves the easy direction.

For the harder direction, assuming $M \subseteq N$ and the condition (TV), we show by induction on the complexity of $\psi \in L$ that

(*)   $M \models \psi(\bar{a})$ iff $N \models \psi(\bar{a})$

for all matching tuples $\bar{a}$ from $M$.

As $M \subseteq N$, this is true for all atomic $\psi$. The induction step for conjunction and negation being trivial, we are left with verifying (*) for $\psi(\bar{a})$ of the form $\exists x\, \varphi(x, \bar{a})$. Now, $M \models \exists x\, \varphi(x, \bar{a})$ if and only if there is $b \in M$ with $M \models \varphi(b, \bar{a})$. By induction hypothesis, this is equivalent to the existence of $b \in M$ with $N \models \varphi(b, \bar{a})$. This in turn is, by condition (TV), equivalent with $N \models \exists x\, \varphi(x, \bar{a})$.                                         $\square$

Here's a reformulation of the Tarski-Vaught test: $M \preccurlyeq N$ if and only if $M \subseteq N$ and for all $n < \omega$, all $\psi(x, \bar{x}) \in L_{n+1}$, and all $n$-tuples $\bar{a}$ from $M$, if $\psi(N, \bar{a}) \neq \emptyset$, then $M \cap \psi(N, \bar{a}) \neq \emptyset$.

**Corollary 8.3.3.** (The Tarski-Vaught Test) *Suppose $\mathcal{N}$ is an L-structure and $M \subseteq N$ is such that condition (TV) holds.*

*Then $M$ is closed under functions, and the restriction $\mathcal{M}$ of $\mathcal{N}$ onto $M$ is an elementary substructure of $\mathcal{N}$.*

*Proof.* We need only check the aforementioned closedness. For this consider (TV) for the formulas $x = c$ $(c \in \mathbf{C})$ and $x = f(\bar{a})$ $(f \in \mathbf{F}$ and $\bar{a}$ from $M)$. □

**Corollary 8.3.4.** *Let $\mathcal{M}$ and $\mathcal{N}$ be L-structures with $\mathcal{M} \subseteq \mathcal{N}$.*

*If for all finite sets $A \Subset M$ and all $c \in N$ there is an automorphism $f \in \mathrm{Aut}_A \mathcal{N} = \{f \in \mathrm{Aut}\,\mathcal{N} : f \upharpoonright A = \mathrm{id}_A\}$ such that $f(c) \in M$, then $\mathcal{M} \preccurlyeq \mathcal{N}$.*

*Proof.* Suppose $n < \omega$, $\varphi \in L_{n+1}$, and $\bar{a}$ is an $n$-tuple from $M$ with $\mathcal{N} \models \exists x\,\varphi(x, \bar{a})$. We have to find $b \in M$ with $\mathcal{N} \models \varphi(b, \bar{a})$.

$\mathcal{N} \models \exists x\,\varphi(x, \bar{a})$ yields $c \in N$ with $\mathcal{N} \models \varphi(c, \bar{a})$. Choose $f \in \mathrm{Aut}_{\bar{a}} \mathcal{N}$ such that $f(c) \in M$. Since $f$ is an elementary map, $\mathcal{N} \models \varphi(c, \bar{a})$ implies $\mathcal{N} \models \varphi(f(c), f[\bar{a}])$, hence $\mathcal{N} \models \varphi(f(c), \bar{a})$. □

Let us apply this to two examples.

**Proposition 8.3.5.**
(1) *Every infinite subset $\mathcal{M}$ of a set $\mathcal{N}$ is, as an $L_=$-structure, an elementary substructure.*
(2) *Let $\eta$ denote the (usual) ordering of the rational numbers, and $\rho$ that of the reals (as $L_<$-structures).*
   *Then $\eta \preccurlyeq \rho$.*

*Proof.* Ad (1). Let $\mathcal{M}$ and $\mathcal{N}$ be infinite sets and $M \subseteq N$. Let further $A \Subset M$ and $c \in N$. We will find an automorphism of $\mathcal{N}$ (i. e. a bijection of $\mathcal{N}$ onto itself) which pointwise fixes $A$ and brings $c$ to some $d \in M$. Without loss of generality, $c \notin A$ (otherwise set $d = c$ and $f = \mathrm{id}_N$).

As $\mathcal{M}$ is infinite and $A$ is finite, we can pick $d \in M \smallsetminus A$. Set $f \upharpoonright (N \smallsetminus \{c, d\}) = \mathrm{id}_{N \smallsetminus \{c,d\}}$, $f(d) = c$, and $f(c) = d$. Then Corollary 8.3.4 yields the assertion.

Ad (2). Let $A \Subset \mathbb{Q}$ and $c \in \mathbb{R}$. We will find $d \in \mathbb{Q}$ and an automorphism of the ordering $\rho$ sending $c$ to $d$ and pointwise fixing $A$. As in (1), let $c \notin A$. The set $A \cup \{c\}$ partitions $\mathbb{R}$ into finitely many pairwise disjoint left-open intervals. Consider the two intervals that have $c$ as a joint endpoint, say $(a, c]$ and $(c, b]$ for some $a, b \in A$. Pick $d \in \mathbb{Q} \cap (a, b)$ arbitrarily. Then there is an order automorphism of $(a, b]$ onto itself (i. e. an order respecting bijection from $(a, b]$ onto itself) which sends $c$ to $d$. (Take, for example, $f(x) = a + (x - a)\frac{d-a}{c-a}$ if $x \in (a, c]$, and $f(x) = b - (b - x)\frac{b-d}{b-c}$ if $x \in (c, b]$). Extending this bijection by the identity on $\mathbb{R} \setminus (a, b]$, we obtain the desired automorphism. Again the preceding corollary yields the assertion.    □

Now that we have learned how to test given maps for elementariness, we should go on and 'construct' elementary maps, which we will do in the next section.

**Exercise 8.3.1.** Let $\mathcal{M}$ be an $L$-structure. Prove that the following are equivalent for any formula $\varphi \in L_n(M)$.
(i)   $\varphi(\mathcal{M})$ is finite.
(ii)  $\varphi(\mathcal{M}) = \varphi(\mathcal{N})$ for every $\mathcal{N} \succcurlyeq \mathcal{M}$.
(iii) $\varphi(\mathcal{N}) \subseteq M^n$ for every $\mathcal{N} \succcurlyeq \mathcal{M}$.

**Exercise 8.3.2.** Show that an elementary substructure $\mathcal{H}$ of an abelian group $\mathcal{G}$ is a **pure subgroup** in the sense that if $h \in H$ is **divisible** by $n$ in $\mathcal{G}$ (i. e. $h = ng$ for some $g \in G$), then $h$ is divisible by $n$ in

**Exercise 8.3.3.** Prove that embeddings of infinite-dimensional vector spaces are elementary (in the language of §5.4).

**Exercise 8.3.4.** Suppose $T$ is an $L$-theory, $P$ is a new predicate, and $L'$ is the language obtained from $L$ by joining $P$.
   Find an $L'$-theory $T'$ such that the models of $T'$ are the $L'$-structures $\mathcal{M}'$ of the following kind: the $L$-reduct $\mathcal{M}$ of $\mathcal{M}'$ is a model of $T$; the set $P(\mathcal{M}')$ is closed under functions from $L$; the thus determined $L$-substructure of $\mathcal{M}$ on the universe $P(\mathcal{M}')$ is an elementary substructure of $\mathcal{M}$.

**Exercise 8.3.5.** Show by counterexample that the criterion 8.3.4 is not necessary for being an elementary substructure.

**Exercise 8.3.6.** Prove that, given an arbitrary subordering $\mathcal{M} = (M, <) \subseteq \rho$, this inclusion is elementary if and only if $\mathcal{M} \models \text{DLO}_{--}$.

# 8.4  Existence of elementary substructures and extensions

First we construct 'small' elementary substructures. Note that only in the infinite case can we hope for something nontrivial, for a finite structure has, by Proposition 8.1.1, only itself as an elementary substructure or an elementary extension.

**Theorem 8.4.1.** (Löwenheim-Skolem downward for $\preccurlyeq$)
*Every infinite $L$-structure $\mathcal{M}$ has an elementary substructure of a power $\leq |L|$.*

*Moreover, every subset $A$ of $M$ is contained in an elementary substructure of $\mathcal{M}$ of a power $\leq |L| + |A|$. Consequently, $\mathcal{M}$ has elementary substructures of every cardinality $\kappa$ with $|L| \leq \kappa \leq |M|$.*

*Proof.* The third assertion follows from the second if $A$ is taken to have power $\kappa$. The first assertion follows from the second if $A$ is taken empty. Further, all of them are redundant in case $|M| \leq |L|$, for then we may take $\mathcal{M} \preccurlyeq \mathcal{M}$.

So let $A \subseteq M$ and $|M| > |L|$, and set $\kappa = |L| + |A|$. Successively we choose an ascending chain $A \subseteq A_0 \subseteq A_1 \subseteq \ldots \subseteq A_i \subseteq \ldots$ $(i < \omega)$ of sets $A_i \subseteq M$ of power $\kappa$ satisfying

(*)   if $\varphi \in L_1(A_i)$ and $\mathcal{M} \models \exists x\, \varphi$, then there is $a \in A_{i+1}$ such that $\mathcal{M} \models \varphi(a)$,

for all $i < \omega$.

We choose the $A_i$ as follows. Start with an arbitrary subset $A_0$ of $M$ of power $\kappa$ containing $A$. Having chosen $A_i$, we add to it, for every $\varphi(x)$ from $L(A_i)$ with $\mathcal{M} \models \exists x\, \varphi(x)$, an arbitrary $b_\varphi \in M$ with $\mathcal{M} \models \varphi(b_\varphi)$, and call the resulting set $A_{i+1}$.

As $|L(A_i)| = |L| + |A_i| = \kappa$ (Corollary 7.6.8), we add at most $\kappa$ such elements $b_\varphi$, hence $\kappa = |A_i| \leq |A_{i+1}| \leq \kappa + |A_i| = \kappa$.

Set $N = \bigcup_{i < \omega} A_i$. Clearly, $N$ has power $\kappa$. We are going to show, using the Tarski-Vaught test (Corollary 8.3.3), that this set is the universe of an elementary substructure $\mathcal{N}$ of $\mathcal{M}$.

So let $n < \omega$, $\varphi \in L_{n+1}$, and $\bar{a}$ an $n$-tuple from $N$ with $\mathcal{M} \models \exists x\, \varphi(x, \bar{a})$. We want $b \in N$ satisfying $\mathcal{M} \models \varphi(b, \bar{a})$. But $\bar{a}$ is contained already in some $A_k$ $(k < \omega)$, as $N = \bigcup_{i < \omega} A_i$. Since then $\varphi(x, \bar{a}) \in L_1(A_k)$, condition (*) provides us with some $b \in A_{k+1} \subseteq N$ satisfying $\mathcal{M} \models \varphi(b, \bar{a})$. $\qquad\square$

Note that this says nothing about the precise power of elementary substructures of a power $< |L|$, and the same applies to the next corollary. Models whose cardinality is smaller than the language are also known as **tiny**. It is, in general, a hard and largely open question to find the possible cardinalities of tiny models.

**Corollary 8.4.2.** (Löwenheim-Skolem)
(1)   *Every $L$-theory has a model of a power $\leq |L|$.*
(2)   *Every $L$-theory with an infinite model has a model of every power $\geq |L|$.*

[Löwenheim, L. :  Über Möglichkeiten im Relativkalkül, Mathematische Annalen **76** (1915) 447 – 470]

[Skolem, Th.  :  Logisch-kombinatorische Untersuchungen über die Erfüllbarkeit oder Beweisbarkeit mathematischer Sätze nebst einem Theorem über dichte Mengen, Skrifter, Videnskabsakademie i Kristiania I. Mat.-Nat. Kl. no. 4 (1920) 1 – 36]

*Proof.* Let $\mathcal{M}$ be a model of the $L$-theory $T$. Every finite model of $T$ has power even $< |L|$ (i. e. is tiny). So let $T$ have an infinite model. Then, for every $\kappa \geq |L|$, the upward Löwenheim-Skolem yields a model $\mathcal{M}$ of $T$ of power $\geq \kappa$. Then the preceding theorem provides us with an elementary substructure of $\mathcal{M}$ of power $\kappa$. Since the latter is elementarily equivalent to $\mathcal{M}$, it is a model of $T$ too. $\qquad\qquad\square$

**Skolem's Paradox.** Consider the Zermelo-Fraenkel set theory ZFC (in the language $L_\in$, cf. §7.4). If the theory ZFC is consistent (which is unknown but largely assumed to be true), then it has a countable model. By Cantor's theorem 7.6.5, given any set $A$, its power set has power $> |A|$. Consequently, this countable model contains uncountable sets. This seeming contradiction is Skolem's paradox. (For Skolem it was a reason for serious doubts about axiomatic set theory.) However, the contradiction is only apparent and can be resolved as follows. Uncountability in a model of set theory is not 'real' (that is, not metatheoretical uncountability). All it means for a set in a model of set theory to be uncountable is that there is no bijection between that set and the set of natural numbers *in that model*. (Remember, set-theoretically functions are just sets, namely graphs.) From this point of view, Skolem's paradox becomes, on the contrary, very plausible: the less sets (and hence functions) a model contains, the less bijections will there be with the set of natural numbers.

[Skolem, Th. :  Einige Bemerkungen zur axiomatischen Begründung der Mengenlehre, in **Proc. 5th Scand. Math. Congress**, Helsinki 1922, 217 – 232]

Applying the upward Löwenheim-Skolem to elementary diagrams, we obtain arbitrarily large elementary extensions. Letting a downward application (of Theorem 8.4.1) follow, as in the preceding proof, we can exactly prescribe the cardinality (but, as before, only in cardinalities that are not 'tiny').

**Theorem 8.4.3.** (Löwenheim-Skolem upward for $\preccurlyeq$) *Suppose $\mathcal{M}$ is an infinite $L$-structure and $\kappa \geq |L| + |M|$.*
   *Then $\mathcal{M}$ has an elementary extension of power $\kappa$.*

*Proof.* The elementary diagram of $\mathcal{M}$ is the $L(M)$-theory $\mathrm{Th}(\mathcal{M}, M)$. By the preceding corollary, it has a model $\mathcal{N}^*$ of power $\kappa$, for $|L(M)| \leq |L| + |M| \leq \kappa$. Denote its $L$-reduct by $\mathcal{N}$. By Lemma 8.2.1(2), $\mathcal{M} \stackrel{\equiv}{\hookrightarrow} \mathcal{N}$. Using isomorphic correction (Lemma 8.3.1), we then find $\mathcal{N}' \cong \mathcal{N}$ with $\mathcal{N}' \succcurlyeq \mathcal{M}$. □

We can strengthen this as follows. (Remember from Exercise 7.6.3: every *set* of cardinals has a supremum in **Cn**.)

**Theorem 8.4.4.** *Let* **K** *be an arbitrary set of elementarily equivalent $L$-structures.*

*For every cardinal $\kappa \geq |\mathbf{K}| + |L| + \sup_{\mathcal{M} \in \mathbf{K}}(|\mathcal{M}|)$ there is an $L$-structure $\mathcal{N}$ of power $\kappa$ such that $\mathcal{M} \stackrel{\equiv}{\hookrightarrow} \mathcal{N}$ for all $\mathcal{M} \in \mathbf{K}$. Moreover, for any given $\mathcal{M}_0 \in \mathbf{K}$, we can find $\mathcal{N}$ so that $\mathcal{M}_0 \preccurlyeq \mathcal{N}$.*

*Proof.* Consider $T = \bigcup \{\mathrm{Th}(\mathcal{M}, M) : \mathcal{M} \in \mathbf{K}\}$, where, without loss of generality, for any two different $\mathcal{M}, \mathcal{M}' \in \mathbf{K}$ we may assume that the new constants $\{\underline{a} : a \in M\}$ and $\{\underline{a}' : a' \in M'\}$ are chosen to be disjoint too (this can be accomplished by renaming them). It suffices to show that $T$ is consistent. So let $\Delta$ be a finite subset of $T$. The various occurring elementary diagrams are complete theories, hence closed under finite conjunction. Therefore we may assume that $\Delta = \{\varphi_i : i < n\}$, where $\varphi_i \in \mathrm{Th}(\mathcal{M}_i, M_i)$ for some $\mathcal{M}_i \in \mathbf{K}$ $(i < n)$ such that $\mathcal{M}_i \neq \mathcal{M}_j$ for all $i < j < n$. Then, for all $i < n$, there are a number $m_i < \omega$, a formula $\varphi_i' \in L_{m_i}$, and an $m_i$-tuple $\bar{a}_i$ from $M_i$ such that $\varphi_i$ is of the form $\varphi_i'(\bar{a}_i)$. By the disjoint choice of the constants, also the $\bar{a}_i$ (as sets) are disjoint. Thus we may assume the tuples of variables, $\bar{x}_i$, to be pairwise disjoint as well. But then we have

$$\models (\exists \bar{x}_0 \ldots \bar{x}_{n-1} \bigwedge_{i<n} \varphi_i'(\bar{x}_i)) \leftrightarrow \bigwedge_{i<n} \exists \bar{x}_i \, \varphi_i'(\bar{x}_i),$$

hence every model of the right-hand conjunction is, after an appropriate expansion by constants, already a model of $\Delta$.

So it remains to find a model of that conjunction. For all $i < n$ we have $\varphi_i \in \mathrm{Th}(\mathcal{M}_i, M_i)$, hence also $\mathcal{M}_i \models \exists \bar{x}_i \, \varphi_i'(\bar{x}_i)$. But the latter is an $L$-sentence, which, by $\mathcal{M}_0 \equiv \mathcal{M}_i$, must be true also in $\mathcal{M}_0$. Thus $\mathcal{M}_0$ is a model of $\Delta$, and $T$ is consistent, as desired.

The moreover clause can be derived using isomorphic correction (Lemma 8.3.1). □

**Corollary 8.4.5.** *Given a set* **K** *of models of a complete theory $T$, there is a model of $T$ in which every model from* **K** *is elementarily embeddable.*

*In particular, any two models $\mathcal{M}_0$ and $\mathcal{M}_1$ of a complete theory $T$ can be elementarily embedded in a joint model $\mathcal{N} \models T$. Moreover, we can choose $\mathcal{N}$ so that $\mathcal{M}_0 \preccurlyeq \mathcal{N}$.*

*Proof.* Proposition 8.1.2 shows that all models are elementarily equivalent. Then we may apply the preceding theorem, for if $\mathbf{K}$ is a *set*, $\{|M| : \mathcal{M} \in \mathbf{K}\}$ has a supremum in $\mathbf{Cn}$.                                                                $\square$

**Exercise 8.4.1.** Suppose $\mathcal{M}$ is an $L$-structure and $\varphi \in L_n$ $(n > 0)$ is such that $\varphi(\mathcal{M})$ is infinite.

Then for every $\kappa \in \mathbf{Cn}$ with $\kappa \geq |L|$ there is an $L$-structure $\mathcal{N} \equiv \mathcal{M}$ of power $\kappa$ and such that $|\varphi(\mathcal{N})| = \kappa$.

**Exercise 8.4.2.** Let $\mathcal{K}$ be a division ring.

Show that the theory $T_{\mathcal{K}}^{\infty}$ (of the infinite $\mathcal{K}$-vector spaces, cf. §5.4) is complete; moreover, every embedding of infinite $\mathcal{K}$-vector spaces is elementary.

**Exercise 8.4.3.** Verify that ultrapowers (along with isomorphic correction) provide us with another tool to construct elementary extensions.

**Exercise 8.4.4.** Consider $\mathcal{M}$ and $\mathcal{N}$ as in Example (4) of §6.3 and the disjoint union of $\mathcal{M}$ and $\mathcal{N}$, regarded as an $L_<$-structure $\mathcal{N}'$.

Prove that $\mathcal{M}$ and $\mathcal{N}'$ are elementarily equivalent.

**Exercise 8.4.5.** Find an example of a theory $T$ and a model of $T_{\exists}$ that does not contain a model of $T$ (this showing that we cannot improve on Exercise 6.2.6).

# 8.5   Categoricity and prime models

A theory is said to be **categorical** in a given power $\lambda \in \mathbf{Cn}$ (or just $\lambda$-**categorical**) if it has, up to isomorphism, exactly one model of power $\lambda$. An $L$-theory is said to be **categorical** if it is categorical in some power $\geq |L|$. We call an $L$-theory **totally categorical** if it has infinite models and every two models of the same power (finite or infinite) are isomorphic.

Notice, the definition of $\lambda$-categoricity has two parts: the existence of a model of power $\lambda$ *and* the isomorphism of any two models of power $\lambda$, i. e., the number of nonisomorphic models of power $\lambda$ is 1. There is a subtilty in the definition of total categoricity: by Löwenheim-Skolem (Corollary 8.4.2), an $L$-theory is totally categorical if and only if it is categorical in all $\lambda \geq |L|$ and also in all $\lambda < |L|$ in which it has a model; in other words, if and only if it has an infinite model and the number of nonisomorphic models in any given cardinality is 1 or 0 (and for no $\lambda \geq |L|$ is this number 0.) In particular, a totally categorical theory is categorical.

**Theorem 8.5.1.** (The Łoś-Vaught Test)
*A categorical theory is complete if and only if it has no finite models.*

[Łoś, J. : On the categoricity in power of elementary deductive systems and some related problems, Colloqium Math. **3** (1954) 58 – 62]

[Vaught, R. : Applications of the Löwenheim-Skolem-Tarski theorem to problems of completeness and decidability, Koninkl. Ned. Akad. Wettensch. Proc. Ser. A **57** (1954) 467 – 472]

*Proof.* Since a complete categorical theory has infinite models (which can't be elementarily equivalent to a finite one), no complete categorical theory has a finite model (cf. Propositions 8.1.1 and 8.1.2).

For the converse, let $T$ be a $\lambda$-categorical $L$-theory (where $\lambda \geq |L|$) having no finite models. By Proposition 8.1.2, it suffices to show that, given $\mathcal{M}, \mathcal{N} \models T$, we have $\mathcal{M} \equiv \mathcal{N}$.

Using Löwenheim-Skolem (8.4.2), choose structures $\mathcal{M}' \equiv \mathcal{M}$ and $\mathcal{N}' \equiv \mathcal{N}$ such that $|\mathcal{M}'| = |\mathcal{N}'| = \lambda$. The $\lambda$-categoricity implies that $\mathcal{M}'$ and $\mathcal{N}'$ are isomorphic, hence, by Proposition 6.1.3, elementarily equivalent. This implies also $\mathcal{M} \equiv \mathcal{N}$. □

**Examples.**
(1) The theory $T_=$ of pure identity is totally categorical, but not complete (it obviously has in every nonzero cardinality up to isomorphism exactly one model). Also $T_=^\infty$ is totally categorical; in particular, it is complete.
(2) The theory DLO__ is, by Cantor's theorem, $\aleph_0$-categorical, hence complete (since it has no finite models). In one of the exercises below it is asked for a proof that this theory is not categorical in any uncountable power.
(3) It is proved in Corollary 14.2.3 that every complete theory of algebraically closed fields is categorical in all uncountable cardinalities.
(4) More examples of categorical theories will be discussed in the exercises.

In his 1954 paper cited above, Łoś conjectured that a theory in a countable language is categorical in *all* uncountable cardinalities, once it is categorical in *some* uncountable cardinality. Michael Morley was able to confirm this in 1963. Thus, for countable complete theories (with infinite models) there are only three possible kinds of categoricity: countably categorical (=$\aleph_0$-categorical), but not uncountably categorical; uncountably categorical (categorical in all $\lambda > \aleph_0$), but not countably categorical; and totally categorical (categorical in all infinite cardinalities).

[Morley, M. : Categoricity in power, Transactions of the American Mathematical Society **114** (1965) 514 – 538]

The proof of this result is beyond the scope of this text. The reader may consult the quoted literature instead, for instance Hodges' *shorter model theory*. It should be noted, though, that Morley's deep result was the starting point for stability (or classification) theory—as developed by Saharon Shelah—which constitutes the heart of contemporary model theory (see also §15.8).

In §5.3 we noted that every field contains a prime field, which depends only on the given characteristic. This leads to the following definitions.

Let **K** be a class of $L$-structures. An $L$-structure $\mathcal{M}$ is said to be a **prime structure** for **K**, if $\mathcal{M} \hookrightarrow \mathcal{N}$ for all $\mathcal{N} \in$ **K**. Let $T$ be an $L$-theory. An $L$-structure $\mathcal{M}$ is said to be an **algebraically prime model** of $T$, if $\mathcal{M}$ is both, a model of $T$ and a prime structure for Mod $T$. An $L$-structure $\mathcal{M}$ is called an **elementarily prime model**[2] of $T$, if $\mathcal{M} \overset{\equiv}{\hookrightarrow} \mathcal{N}$ for all $\mathcal{N} \models T$. Clearly, an elementarily prime model of $T$ is itself a model of $T$.

**Examples.**

(5)  The field $\mathbb{Q}$ is a prime structure (a prime field) for the class of fields of characteristic 0. The field $F_p$ is a prime structure (a prime field) for the class of fields of characteristic $p$.

(6)  The field $\mathbb{Q}$ is an algebraically prime model of the theory $\mathrm{TF}_0$ of all fields of characteristic 0. Similarly, $F_p$ is an algebraically prime model of the theory $\mathrm{TF}_p$ of all fields of characteristic $p$.

(7)  Every countably infinite set is, by Proposition 8.3.5(1), an elementarily prime model of $T_{\cong}^{\infty}$. Every finite set is a prime structure for the model class of $T_{\cong}^{\infty}$. (Somewhat pathologically, even the empty set is a prime structure for Mod $T_=$, for $L_=$ has no constant symbols.) Since the empty set is not a *model*, 'the' one-element set (a singleton) is the algebraically prime model of $T_=$.

(8)  The standard model N of Peano arithmetic is an elementarily prime model of true arithmetic Th N , see Exercise 13.2.3 below (and cf. §3.5).

(9)  If $L$ is a language without constant symbols, then the empty $L$-structure is a prime structure for every class of $L$-structures.

**Remarks.**

(1)  Prime models have cardinality $\leq |L|$.

(2)  Every elementarily prime model is an algebraically prime model.

(3)  Any theory with an elementarily prime model is complete.

(4)  Every countable $\aleph_0$-categorical theory without finite models has an elementarily prime model, namely 'the' countable model.

---

[2]The attributes *algebraically* and *elementarily* are to indicate with respect to which kind of maps we require embeddability, just (algebraic) embeddings or elementary embeddings.

(5) The ordering $\eta$ of the rationals is an elementarily prime model of DLO$_{--}$.

[Robinson, A. : **Complete Theories**, Studies in Logic and the Foundations of Mathematics, North-Holland, Amsterdam, 1956]

*Proof.* Ad (3). If a theory has an elementarily prime model, then the latter is elementarily embeddable in, hence equivalent to every model of that theory. Consequently, all of its models are elementarily equivalent, and the theory is complete.

Ad (4). By Löwenheim-Skolem, every model of a countable theory has a countably infinite elementary substructure. If the theory under consideration is in addition $\aleph_0$-categorical, this latter structure is unique up to isomorphism.

(5) follows from (4). □

It should be noted that there are complete theories without elementarily prime models, see §15.3.

**Exercise 8.5.1.** Show that DLO$_{-+}$, DLO$_{+-}$, and DLO$_{++}$ are complete theories.

**Exercise 8.5.2.** Prove that DLO$_{--}$ is in no uncountable power categorical.

**Exercise 8.5.3.** Prove the following statements for every division ring $\mathcal{K}$.
The theory $T_{\mathcal{K}}$ (of all $\mathcal{K}$-vector spaces, cf. Exercise 7.6.2) is categorical in all $\kappa > |\mathcal{K}| + \aleph_0$.
If $\mathcal{K}$ is infinite, $T_{\mathcal{K}}$ is not $|\mathcal{K}|$-categorical.
If $\mathcal{K}$ is finite, $T_{\mathcal{K}}$ is totally categorical. In which finite cardinalities does $T_{\mathcal{K}}$ have models?

**Exercise 8.5.4.** Prove the following three statements.
The theory of the abelian group $\mathbb{Q}$ is $\kappa$-categorical if and only if $\kappa$ is uncountable.
The theory of all infinite abelian groups of prime **exponent** $p$ (i. e. satisfying $\forall x \, (px = 0)$) is totally categorical . Both theories are complete.

**Exercise 8.5.5.** Show that the theory of the structure $\mathcal{M}$ from Exercise 8.4.4 is categorical in all uncountable powers.

**Exercise 8.5.6.** Find for every signature a totally categorical theory with a model in every cardinality $> 0$.

# Chapter 9

# Elimination

In the preservation theorems of §6.2 we found conditions for formulas to be equivalent (logically or modulo a given theory) to formulas of a particular form. In this chapter we investigate what it means for a theory that *all* formulas be, modulo that theory, equivalent to formulas of a particular form.

## 9.1  Elimination in general

Given a class of formulas, $\Delta$, and a theory $T$, we say that $T$ admits (or has) **elimination** up to formulas from $\Delta$ (or simply $\Delta$-**elimination**), if every formula $\varphi \in L_n$ is $T$-equivalent to a formula $\delta \in \Delta \cap L_n$ (i. e. $T \models \forall \bar{x}\,(\varphi \leftrightarrow \delta)$).

**Remark.** Suppose $T$ and $T'$ are theories in the same language and $T \subseteq T'$. If $T$ has $\Delta$-elimination, so does $T'$.

Of particular interest is the case where $\Delta$ is closed under boolean combinations. If it is not, we therefore pass to its *boolean closure* $\widetilde{\Delta}$, i. e. the class of all boolean combinations of formulas from $\Delta$. We first investigate $\widetilde{\Delta}$-elimination of *sentences*.

**Lemma 9.1.1.** *Suppose $\Sigma \subseteq L_0$, $\varphi$ is a sentence in $L_0$, $\Delta$ is an arbitrary class of formulas, and $\widetilde{\Delta}$ its boolean closure.*
  *Then $\varphi$ is $\Sigma$-equivalent to $\top$, to $\bot$, or to a formula from $\widetilde{\Delta} \cap L_0$ if and only if $\mathcal{M} \equiv_\Delta \mathcal{N}$ implies $\mathcal{M} \equiv_\varphi \mathcal{N}$, for all $\mathcal{M}, \mathcal{N} \models \Sigma$.*

*Proof.* Since $\mathcal{M} \equiv_\Delta \mathcal{N}$ implies $\mathcal{M} \equiv_{\widetilde{\Delta}} \mathcal{N}$ (even in the case $\Delta = \emptyset$), the direction from left to right is clear.

For the converse, consider the space $S_L$ of all complete $L$-theories (cf. §5.7) and the closed (hence compact) sets $S = \langle \varphi \rangle \cap \bigcap_{\sigma \in \Sigma} \langle \sigma \rangle$ and $S' = \langle \neg\varphi \rangle \cap \bigcap_{\sigma \in \Sigma} \langle \sigma \rangle$ therein. In case $S$ is empty, $\Sigma \cup \{\varphi\}$ is inconsistent, hence $\varphi$ is $\Sigma$-equivalent to $\bot$. Similarly, if $S'$ is empty, then $\varphi \sim_\Sigma \top$. Assume therefore that neither $S$ nor $S'$ are empty.

For every pair $T \in S$ and $T' \in S'$ and all $\mathcal{M} \models T$ and $\mathcal{M}' \models T'$, we have, by hypothesis, $\mathcal{M} \not\equiv_\Delta \mathcal{M}'$ (in particular, $\Delta \neq \emptyset$). Hence there must be a sentence $\gamma_{TT'} \in T$ with $\neg\gamma_{TT'} \in T'$ which is contained in $\Delta$ or whose negation is contained in $\Delta$ (in any case it is in $\widetilde{\Delta}$).

Then $\{\langle \neg\gamma_{TT'} \rangle : T' \in S'\}$ forms an open covering of $S'$. Compactness yields a finite subcovering, i. e. $T_i' \in S'$ ($i < n$) such that $\Sigma \models \neg\varphi \rightarrow \bigvee_{i<n} \neg\gamma_{TT_i'}$. Letting $\gamma_T$ be the sentence $\bigwedge_{i<n} \gamma_{TT_i'}$ (which is clearly in $\widetilde{\Delta} \cap L$), we obtain $\Sigma \models \gamma_T \rightarrow \varphi$.

But $\{\gamma_T : T \in S\}$ forms an open covering of $S$, which also has a finite subcovering. Hence there are $T_i \in S$ ($i < m$) such that $\Sigma \models \varphi \rightarrow \bigvee_{i<m} \gamma_{T_i}$. Together with the implication above this shows that $\varphi$ is $\Sigma$-equivalent to $\bigvee_{i<m} \gamma_{T_i}$, a sentence contained in $\widetilde{\Delta} \cap L$.                                                    □

We extend this, using expansions by constants, to arbitrary formulas.

**Proposition 9.1.2.** *Suppose $\Sigma \subseteq L_0$, $\varphi$ is a formula in $L_n$, $\Delta$ is an arbitrary set of formulas, and $\widetilde{\Delta}$ is its boolean closure.*

*$\varphi$ is $\Sigma$-equivalent to $\top$, to $\bot$, or to a formula from $\widetilde{\Delta} \cap L_n$ if and only if the following implication holds for all $\mathcal{M}, \mathcal{N} \models \Sigma$ and all $\bar{a} \in M^n$ and $\bar{b} \in N^n$: if, for all $\delta \in \Delta \cap L_n$, $\mathcal{M} \models \delta(\bar{a})$ iff $\mathcal{N} \models \delta(\bar{b})$, then $\mathcal{M} \models \varphi(\bar{a})$ iff $\mathcal{N} \models \varphi(\bar{b})$.*

*Proof.* Choose an $n$-tuple $\bar{c}$ of new constants (that is disjoint from $L \cup \Delta$). Since every $L(\bar{c})$-structure has the form $(\mathcal{M}, \bar{a})$, where $\mathcal{M}$ is an $L$-structure and $\bar{a} \in M^n$, the implication from the statement of the proposition reads as follows.

For all $L(\bar{c})$-structures $\mathcal{M}^*, \mathcal{N}^* \models \Sigma$, if $\mathcal{M}^* \equiv_{\Delta(\bar{c})} \mathcal{N}^*$, then $\mathcal{M}^* \equiv_{\varphi(\bar{c})} \mathcal{N}^*$.

By Lemma 9.1.1, this is the same as saying that $\varphi(\bar{c})$ is $\Sigma$-equivalent to a sentence $\delta(\bar{c})$ from $\widetilde{\Delta}(\bar{c}) \cap L(\bar{c})$ (as the boolean closure of $\Delta(\bar{c})$ is $\widetilde{\Delta}(\bar{c})$). Thus $\Sigma \models_{L(\bar{c})} \varphi(\bar{c}) \leftrightarrow \delta(\bar{c})$, hence Lemma 3.3.2 (about new constants) yields $\Sigma \models_L \forall \bar{x} (\varphi(\bar{x}) \leftrightarrow \delta(\bar{x}))$. Consequently, $\varphi$ is $\Sigma$-equivalent to $\delta$ (which obviously lies in $\widetilde{\Delta} \cap L$).                                                    □

**Remark.** Given any formula $\delta$, $\top$ is logically equivalent to $\delta \vee \neg\delta$, while $\bot$ is logically equivalent to $\delta \wedge \neg\delta$. Therefore, if $\widetilde{\Delta} \cap L_0$ is not empty, we do not need $\top$ and $\bot$ in the above lemma. Similarly, we do not need $\top$ and $\bot$ in the preceding proposition if $\tilde{\Delta} \cap L_n$ is not empty.

Letting $\varphi$ run over $L$, we obtain

**Corollary 9.1.3.** *Suppose $T$ is an $L$-theory and $\Delta$ is a class of formulas containing $\top$ and $\bot$ (and $\tilde{\Delta}$ is its boolean closure).*
*$T$ has $\tilde{\Delta}$-elimination if and only if, for all $\mathcal{M}, \mathcal{N} \models T$, all $n < \omega$, all $\bar{a} \in M^n$ and $\bar{b} \in N^n$, the following implication holds: if $\mathcal{M} \models \delta(\bar{a})$ iff $\mathcal{N} \models \delta(\bar{b})$ for all $\delta \in \Delta \cap L_n$, then $(\mathcal{M}, \bar{a}) \equiv (\mathcal{N}, \bar{b})$.* $\qquad \square$

We will see in §11.3 a more elegant way of formulating the preceding two results using the concept of type.

**Proposition 9.1.4.** *Suppose $T$ is an $L$-theory and $\Delta$ is a class of formulas that is closed under boolean combinations and contains $\top$ and $\bot$.*
*Then the following conditions are equivalent.*
(i)   *$T$ admits $\Delta$-elimination.*
(ii)  *For any $\mathcal{M} \models T$ and $A \subseteq M$, the deductive closure $\Sigma(\mathcal{M}, A)$ of $T \cup \mathrm{Th}_{\Delta(A)}(\mathcal{M}, A)$ is a complete $L(A)$-theory.*
(iii) *Like (ii), but with finite $A$ only.*
*If, in addition, $\Delta$ is closed under substitution by $L$-terms, another equivalent condition is*
(iv)  *Like (ii), but only with $A \subseteq M$ that are universes of (finitely generated) substructures of $\mathcal{M}$.*

*Proof.* (i) $\Longrightarrow$ (ii). Let $T$ have $\Delta$-elimination. Then, for all $\mathcal{M} \models T$ and $A \subseteq M$, the (complete) $L(A)$-theory $\mathrm{Th}(\mathcal{M}, A)$ is contained in, and hence equal to $\Sigma(\mathcal{M}, A)$. Then the latter is complete too.

(ii) $\Longrightarrow$ (iii) and (ii) $\Longrightarrow$ (iv) are trivial.

(iii) $\Longrightarrow$ (i). Assume $\mathcal{M}, \mathcal{N} \models T$, $\bar{a} \in M^n$, $\bar{b} \in N^n$, and that, for all $\delta \in \Delta \cap L_n$, we have $\mathcal{M} \models \delta(\bar{a})$ iff $\mathcal{N} \models \delta(\bar{b})$. In view of the preceding corollary we need only show that $(\mathcal{M}, \bar{a}) \equiv (\mathcal{N}, \bar{b})$. For this note that $(\mathcal{M}, \bar{a})$ and $(\mathcal{N}, \bar{b})$ are models of $\Sigma(\mathcal{M}, \bar{a})$ and that condition (iii) says that this latter is a complete $L(\bar{a})$-theory. Then Proposition 8.1.2 implies that $(\mathcal{M}, \bar{a})$ and $(\mathcal{N}, \bar{b})$ are indeed elementarily equivalent.

(iv) $\Longrightarrow$ (iii) (under the proviso that $\Delta$ be closed under substitution by $L$-terms). Suppose $\mathcal{M} \models T$ and $M_{\bar{a}}$ is the universe of the substructure of $\mathcal{M}$ generated by the finite sequence $\bar{a}$ of elements from $M$. All we have to do is show that condition (ii) for $A = M_{\bar{a}}$ implies condition (ii) for $A = \bar{a}$.

For this, let $\varphi$ be an $L(\bar{a})$-sentence that follows logically from $\Sigma(\mathcal{M}, M_{\bar{a}})$. We prove that $\varphi$ follows already from $\Sigma(\mathcal{M}, \bar{a})$. By the finiteness of $\models$ and the $\bigwedge$-closedness of $\Delta$, there is a single such sentence $\delta(\bar{c})$ as above implying $\varphi$, i. e. $T \models \delta(\bar{c}) \rightarrow \varphi$. Write $\delta(\bar{c})$ as $\delta(\bar{a}, \bar{b})$, where $\delta(\bar{x}, \bar{y}) \in \Delta$ and $\bar{a} \cap \bar{b} = \emptyset$. For notational simplicity we assume $l(\bar{y}) = l(\bar{b}) \leq 1$.

If $l(\bar{y}) = 0$, then $\delta(\bar{c})$ lies already in $\Sigma(\mathcal{M}, \bar{a})$ and there's nothing to prove.

So let $T \models \delta(\bar{a}, b) \rightarrow \varphi$, where $\delta(\bar{x}, y) \in \Delta$ and $b \in M_{\bar{a}} \smallsetminus \bar{a}$. Then there is an $L$-term $t(\bar{x})$ such that $b = t^{\mathcal{M}}(\bar{a})$ (cf. Exercise 6.3.1), and we have

$$T \models \delta(\bar{a}, t(\bar{a})) \wedge t(\bar{a}) = b \rightarrow \varphi.$$

The lemma about new constants (3.3.2) (together with Remark (8) from §3.3) now yields

$$T \models \exists y \, (\delta(\bar{a}, t(\bar{a})) \wedge t(\bar{a}) = y) \rightarrow \varphi$$

(for $b$ occurs neither in $\bar{a}$ nor in $\varphi$). However, $T \models \forall \bar{x} \, \exists y \, t(\bar{x}) = y$, whence $T \models \delta(\bar{a}, t(\bar{a})) \rightarrow \varphi$. The additional hypothesis on $\Delta$ says that $\delta(\bar{x}, t(\bar{x})) \in \Delta$, hence $\delta(\bar{a}, t(\bar{a})) \in \Sigma(\mathcal{M}, \bar{a})$. Consequently, $\Sigma(\mathcal{M}, \bar{a}) \models \varphi$.                        $\square$

**Exercise 9.1.1.** Find a nontopological proof of Lemma 9.1.1.

**Exercise 9.1.2.** Find a proof of Proposition 9.1.4 (iv) $\Longrightarrow$ (i) for the case $\Delta = \mathbf{qf}$ which uses Corollary 9.1.3 and Exercise 6.3.3.

## 9.2   Quantifier elimination

Elimination up to quantifier-free formulas, i. e. **qf**-elimination, is not surprisingly called **quantifier elimination**. Modulo theories with quantifier elimination one can do without quantifiers. In particular, all sentences are equivalent to quantifier-free ones. As by Lemma 6.1.1, quantifier-free sentences are preserved both ways, 'going up' and 'going down', we see that two models of a theory with quantifier elimination are elementarily equivalent, provided there is a structure (of the same signature) embeddable in both of them. We can derive somewhat more from Lemma 6.1.1.

**Remarks.**
(1)   A theory with quantifier elimination whose model class has a prime structure is complete. In particular, any theory with quantifier elimination in a language without constants is complete, cf. Example (9) in §8.5.

(2) Every embedding between models of theories with quantifier elimination is elementary. In particular, every algebraically prime model of such a theory is an elementarily prime model. For this reason, in theories with quantifier elimination we simply speak of **prime models**.

In Theorem 9.4.2 we will meet an incomplete theory admitting quantifier elimination.

Whether a theory admits quantifier elimination depends on the language it is formulated in. One can always achieve it by artificially 'adding' all formulas to the signature. More precisely, given an $L$-theory $T$, consider the extension $L^\star$ of $L$ by a new $n$-ary relation $R_\varphi$ for *every* $\varphi \in L_n$ $(n < \omega)$. Following Hodges we call the set of $L^\star$-sentences $T^\star = T \cup \{\forall \bar{x} (R_\varphi(\bar{x}) \leftrightarrow \varphi(\bar{x})) : \varphi = \varphi(\bar{x}) \in L\}$ an **atomization**[1] of $T$.

The atomization $T^\star$ of an $L$-theory $T$ is a theory with quantifier elimination (exercise!). Further, the reduction map $\upharpoonright L$ constitutes a bijection between $\operatorname{Mod} T^\star$ and $\operatorname{Mod} T$. This bijection preserves isomorphism types in both directions, i. e., it maps isomorphic $L^\star$-structures to isomorphic $L$-structures, and nonisomorphic ones to nonisomorphic ones. Further, $\mathcal{M}^\star \models T^\star$ and $\mathcal{M} = \mathcal{M}^\star \upharpoonright L$ have the same definable sets. Moreover, $\mathcal{M}^\star \subseteq \mathcal{N}^\star$ iff $\mathcal{M} \preccurlyeq \mathcal{N}$ (where $\mathcal{N}^\star \models T^\star$ and $\mathcal{N} = \mathcal{N}^\star \upharpoonright L$). That this map is surjective is immediate, for each $\mathcal{M} \models T$ can be expanded to $\mathcal{M}^\star \models T^\star$ by simply interpreting $R_\varphi$ in $\mathcal{M}$ by $\varphi$. Thus $T^\star$ is consistent. The rest follows from the fact that the map defined by $\varphi \mapsto R_\varphi$, from the Lindenbaum-Tarski algebra of $T$ to that of $T^\star$, is an embedding (of boolean algebras) satisfying $T^\star \models \exists x \, R_{\varphi(x,\bar{y})}(x, \bar{y}) \leftrightarrow R_{\exists x \, \varphi(x,\bar{y})}(\bar{y})$. We leave the verification of all these assertions as an exercise.

The upshot of this discussion is that if one is interested only in the investigation of the models of $T$ in terms of their definable sets (no matter *how* these are defined) and elementary maps between them, one can as well pass to the atomization $T^\star$, a theory with quantifier elimination.

Atomization yields an abundance of theories having quantifier elimination. In due course we will investigate some less trivial examples. Before we can do so we need a criterion for quantifier elimination, and for that, in turn, some more terminology.

Let $\mathbf{K}$ be an arbitrary class of structures. An $L$-theory $T$ is said to be **K-complete** if, for every model $\mathcal{M} \models T$ and every substructure $\mathcal{A} \subseteq \mathcal{M}$ which is a member of $\mathbf{K}$, the deductive closure of $T \cup D(\mathcal{A})$ is a complete $L(A)$-theory. In case $\mathbf{K}$ is the class of all structures, this is called **substructure-completeness**.

---

[1] Atomizations are also known as **Morleyizations** or **Morley expansions**.

Again we built in some redundancy for notational reasons: only those
structures from **K** count which are themselves $L$-structures (e. g., also a
$\mathrm{Mod}_L\emptyset$-complete theory is substructure-complete).

**Remarks.**
(3)   An $L$-theory $T$ is substructure-complete iff, for all $L$-structures $\mathcal{A}$, the
      consistency of $T \cup \mathrm{D}(\mathcal{A})$ implies its completeness.
(4)   By Proposition 8.1.2 (and isomorphic correction), $T$ is **K**-complete if
      and only if, for any two models $\mathcal{M}, \mathcal{N} \models T$ having a joint substructure
      $\mathcal{A} \in \mathbf{K}$, we have $(\mathcal{M}, A) \equiv (\mathcal{N}, A)$.
(5)   A **K**-complete theory $T$ has the following **elementary amalgamation**
      property over **K**.
      If $\mathcal{A} \in \mathbf{K}$ is a joint substructure of the models $\mathcal{M}, \mathcal{N} \models T$, then there
      are $\mathcal{M}' \succcurlyeq \mathcal{M}$ and $g : \mathcal{N} \overset{\equiv}{\hookrightarrow} \mathcal{M}'$ making the diagram

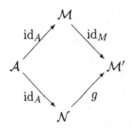

      commute, i. e. $g \restriction A = \mathrm{id}_A$. (Using Theorem 8.4.4, choose an ele-
      mentary extension $(\mathcal{M}', A) \succcurlyeq (\mathcal{M}, A)$ and an elementary embedding
      $g : (\mathcal{N}, A) \overset{\equiv}{\hookrightarrow} (\mathcal{M}', A)$. Similarly, one can verify a somewhat more
      general property, where the identical inclusions of $\mathcal{A}$ are replaced by
      arbitrary embeddings of $\mathcal{A}$.)
(6)   Suppose all substructures of $L$-structures generated by the empty set
      are isomorphic to a single $L$-structure $\mathcal{A}$. (This is the case in languages
      without constants, where $\mathcal{A}$ is the empty structure.)
      If $\mathcal{A} \in \mathbf{K}$, then, by (1), the **K**-completeness of an $L$-theory entails
      its (usual) completeness (for all models $\mathcal{M}$ and $\mathcal{N}$ contain $\mathcal{A}$, hence
      $(\mathcal{M}, A) \equiv (\mathcal{N}, A)$ yields $\mathcal{M} \equiv \mathcal{N}$).
(7)   For languages without constants, $\{\emptyset\}$-completeness is the same as com-
      pleteness. In particular, substructure-complete theories in a language
      without constants are complete.
(8)   More generally, if the model class of a substructure-complete theory
      $T$ (in any language) has a prime structure (in the sense of §8.5), then

$T$ is complete (for if $\mathcal{A}$ is this prime structure, then, after isomorphic correction, any two models $\mathcal{M}, \mathcal{N} \models T$ contain $\mathcal{A}$ as a substructure; the rest of the argument is as in (6)).

A **primitive** formula is, by definition, a formula of the form $\exists \bar{x}\, \psi$, where $\psi$ is a conjunction of literals; here $\bar{x}$ may be empty. (**Positive primitive** means that no *negated* atomic formulas may occur in $\psi$, cf. 4.2.1.) We call a formula **simply primitive** if it is a primitive formula with only one existential quantifier, i. e. a formula of the form $\exists x\, \psi$, where $\psi$ is a conjunction of literals. We denote the class of simply primitive formulas by $\exists^*$.

It is easy to see that $\exists \bar{x}\, (\varphi \vee \psi)$ and $\exists \bar{x}\, \varphi \vee \exists \bar{x}\, \psi$ are logically equivalent. Therefore, a formula is (equivalent to) an $\exists$-formula if and only if it is equivalent to a disjunction of simply primitive formulas.

**Lemma 9.2.1.** *An $L$-theory $T$ admits quantifier elimination if and only if every simply primitive $L$-formula is $T$-equivalent to a quantifier free $L$-formula.*

*Proof.* For the nontrivial direction, assume that every simply primitive $L$-formula is $T$-equivalent to a quantifier-free $L$-formula. Inductively on the complexity of $\varphi \in L$ we show that $\varphi$ is $T$-equivalent to a quantifier-free $L$-formula. The initial step $\varphi \in$ **at** is trivial. The inductive steps involving conjunction and negation are trivial as well, as conjunctions and negations of quantifier-free formulas are quantifier-free , and replacing subformulas by equivalent ones does not affect the equivalence.

We are left with the case that $\varphi$ is of the form $\exists y\, \theta(\bar{x}, y)$ and $\theta(\bar{x}, y)$ is, by induction hypothesis, $T$-equivalent to a quantifier-free $L$-formula $\psi(\bar{x}, y) \in$ **qf**. As noted above, $\exists y \bigvee_i \psi_i(\bar{x}, y)$ is logically equivalent to $\bigvee_i \exists y\, \psi_i(\bar{x}, y)$. Since **qf** is closed under disjunction, we may therefore assume that $\psi(\bar{x}, y)$ is a conjunction of atomic and negated atomic formulas, i. e. that $\exists y\, \psi(\bar{x}, y)$ is a simply primitive formula. By hypothesis, these formulas are $T$-equivalent to a quantifier-free $L$-formula, and the same is true for $\varphi$. $\square$

We are now able to prove the main result of this section.

**Theorem 9.2.2.** *The following are equivalent for any $L$-theory $T$.*
(i) *$T$ admits quantifier elimination.*
(ii) *$T$ is substructure-complete.*
(iii) *If $\mathcal{M}, \mathcal{N} \models T$ and $\mathcal{A}$ is a finitely generated substructure of $\mathcal{M}$ as well as of $\mathcal{N}$, then $(\mathcal{M}, A) \equiv_{\exists^*} (\mathcal{N}, A)$ (i. e., for every simply primitive $\varphi \in L_n$ ($n < \omega$) and every $\bar{a} \in A^n$, we have $\mathcal{M} \models \varphi(\bar{a})$ iff $\mathcal{N} \models \varphi(\bar{a})$).*

*Proof.* For the case $\Delta = \mathbf{qf}$ Proposition 9.1.4 implies that (i) is equivalent
to the condition

(*)   The deductive closure of $T \cup \mathrm{Th}_{\mathbf{qf}}(\mathcal{M}, A)$ is complete for every $\mathcal{M} \models T$
      and every subset $A \subseteq M$ that is the universe of a substructure $\mathcal{A}$ of
      $\mathcal{M}$.

As the truth of a boolean combination depends only on the truth of
its constituents, the theory $\mathrm{Th}_{\mathbf{qf}}(\mathcal{A}, A)$ is contained in $\mathrm{D}(\mathcal{A})^\models$. Hence the
deductive closures of $T \cup \mathrm{Th}_{\mathbf{qf}}(\mathcal{A}, A)$ and $T \cup \mathrm{D}(\mathcal{A})$ are the same. But,
by Lemma 6.1.1, also $\mathrm{Th}_{\mathbf{qf}}(\mathcal{A}, A)$ and $\mathrm{Th}_{\mathbf{qf}}(\mathcal{M}, A)$ are the same. Thus (*)
says nothing more than the completeness of $T \cup \mathrm{D}(\mathcal{A})$ for all $\mathcal{A} \subseteq \mathcal{M} \models T$.
Consequently, (ii) is equivalent to (*), and hence also to (i).

That (iii) follows from (ii) is clear, since the completeness of $T \cup \mathrm{D}(\mathcal{A})$
even implies $(\mathcal{M}, A) \equiv (\mathcal{N}, A)$, for all models $\mathcal{M}$ and $\mathcal{N}$ of $T$ such that
$\mathcal{A} \subseteq \mathcal{M}$ and $\mathcal{A} \subseteq \mathcal{N}$.

We are left with (iii) $\implies$ (i). The preceding lemma reduces the elimination to the case of simply primitive formulas, hence, by Proposition 9.1.2, it
suffices to prove the following claim.

If $n < \omega$, $\mathcal{M}, \mathcal{N} \models T$, $\bar{a} \in M^n$, $\bar{b} \in N^n$, and $(\mathcal{M}, \bar{a}) \equiv_{\mathbf{qf}} (\mathcal{N}, \bar{b})$, then
$(\mathcal{M}, \bar{a}) \equiv_{\exists^*} (\mathcal{N}, \bar{b})$.

So suppose $\mathcal{M}$, $\mathcal{N}$, $\bar{a}$, and $\bar{b}$ are as above, and $(\mathcal{M}, \bar{a}) \equiv_{\mathbf{qf}} (\mathcal{N}, \bar{b})$. Write
$\bar{b} = f[\bar{a}]$ for an appropriate bijection $f : \bar{a} \to \bar{b}$ such that $(\mathcal{N}, \bar{b}) = (\mathcal{N}, f[\bar{a}])$.
As $(\mathcal{M}, \bar{a}) \equiv_{\mathbf{qf}} (\mathcal{N}, f[\bar{a}])$, using Exercise 6.3.3 we can extend $f$ to an isomorphism $F : \mathcal{M}_{\bar{a}} \cong \mathcal{N}_{f[\bar{a}]} = \mathcal{N}_{\bar{b}}$. After an isomorphic correction of $\mathcal{N}$ (using an
extension of $F^{-1}$) we can easily achieve that $\bar{a} = \bar{b}$ (more exactly, $f = \mathrm{id}_{\bar{a}}$)
and even $\mathcal{M}_{\bar{a}} = \mathcal{N}_{\bar{a}}$. Now we are left with verifying $(\mathcal{M}, \bar{a}) \equiv_{\exists^*} (\mathcal{N}, \bar{a})$. However, if $A$ is the universe of $\mathcal{M}_{\bar{a}} = \mathcal{N}_{\bar{a}}$, then (iii) implies $(\mathcal{M}, A) \equiv_{\exists^*} (\mathcal{N}, A)$,
hence also $(\mathcal{M}, \bar{a}) \equiv_{\exists^*} (\mathcal{N}, \bar{a})$.                                   $\Box$

**Remark.**

(9)   (About languages without constants.)   Suppose $T$ is a theory with
      quantifier elimination in a language $L$ without constants. We know
      from Remark (1) above that $T$ is complete. One can also derive this
      from Remark (7)—invoking the previous theorem. Here is yet another
      proof. By Remark (3) of §2.5 all quantifier-free sentences are logically
      equivalent to $\top$ or $\bot$. Since $\top$ is true in every structure, while $\bot$ is
      true in none, we infer that all models of $T$ are elementarily equivalent,
      whence $T$ is complete.

The next example shows that the preceding theorem becomes false if in
the concept of substructure-completeness empty (sub-) structures are excluded.

**Example.** Let $L$ be the language (without constants) whose only non-logical symbol is a predicate $P$. Let further $T_P$ be the $L$-theory axiomatized by the sentence $\forall x\, P(x) \vee \forall x\, \neg P(x)$. The models of $T_P^\infty$ are exactly those infinite sets, all of whose elements satisfy $P$ or all of whose elements satisfy $\neg P$. Thus $T_P^\infty$ is not complete, hence, by the preceding remark, cannot have quantifier elimination. (One can see this directly: the sentences $\forall x\, P(x)$ and $\exists x\, P(x)$ cannot be $T_P^\infty$-equivalent to a quantifier-free formula.) However, if $\mathcal{A}$ is a *nonempty* substructure of a model of $T_P^\infty$ (e. g. a singleton) then $T_P^\infty \cup \mathrm{D}(\mathcal{A})$ is complete, since depending on whether $\forall x\, P(x)$ or $\forall x\, \neg P(x)$ holds true in $\mathcal{A}$, one of these sentences already *follows* from $T_P^\infty \cup \mathrm{D}(\mathcal{A})$. If one would not allow empty structures, $T_P^\infty$ would be substructure-complete without having quantifier elimination.

We conclude this section with a characterization of complete theories of *finite* structures having quantifier elimination in terms of extendability of isomorphisms. To this end, a structure $\mathcal{M}$ is said to be **ultrahomogeneous** if every isomorphism between finitely generated substructures of $\mathcal{M}$ can be extended to an automorphism of $\mathcal{M}$.

**Corollary 9.2.3.** *Let $\mathcal{M}$ be a finite, but nonempty structure.*

Th $\mathcal{M}$ *admits quantifier elimination if and only if $\mathcal{M}$ is ultrahomogeneous.*

*Proof.* Assume first that $T = \mathrm{Th}\,\mathcal{M}$ has quantifier elimination. Let $\mathcal{A}$ and $\mathcal{B}$ be isomorphic substructures of $\mathcal{M}$. This means nothing more than that $(\mathcal{M}, A)$ and $(\mathcal{M}, B)$ are models of $T \cup \mathrm{D}(\mathcal{A})$. Substructure-completeness now implies $(\mathcal{M}, A) \equiv (\mathcal{M}, B)$. As both structures are finite, this yields an isomorphism $h : (\mathcal{M}, A) \cong (\mathcal{M}, B)$. But this just means that $h \in \mathrm{Aut}\,\mathcal{M}$ and $h[\mathcal{A}] = \mathcal{B}$. This argument can be easily modified to yield the full ultrahomogeneity of $\mathcal{M}$ (which is left as an exercise).

Assume now that $\mathcal{M}$ is ultrahomogeneous. Given $\mathcal{A} \subseteq \mathcal{M}$, we will prove $(\mathcal{N}, B) \equiv (\mathcal{M}, A)$ for every model $(\mathcal{N}, B)$ of $T \cup \mathrm{D}(\mathcal{A})$. Such a set $B$ is the universe of a substructure $\mathcal{B}$ that is isomorphic to $\mathcal{A}$. Since $\mathcal{M}$ is finite, $\mathcal{N}$ is isomorphic to $\mathcal{M}$. Now it remains to extend the isomorphism between $\mathcal{B}$ and $\mathcal{A}$ to an automorphism of $\mathcal{M}$ to infer that $(\mathcal{M}, B) \cong (\mathcal{M}, A)$, hence also $(\mathcal{M}, B) \equiv (\mathcal{M}, A)$. $\qquad\square$

**Example.** By Exercise 15.1.2 below, all finite cyclic groups are ultrahomogeneous.

In the next two sections we illustrate the method of quantifier elimination by two prominent examples, DLO and ACF. As another application, the reader may pass to §§15.1–3, where the example of the (theory of the) integers is treated in detail.

**Exercise 9.2.1.** Fill in the missing details in the proof of the above corollary.

**Exercise 9.2.2.** Verify the assertions about atomizations, in particular that they admit quantifier elimination.

**Exercise 9.2.3.** Prove that conjunctions of finitely many primitive (resp., positive primitive) formulas are again primitive (resp., positive primitive). Refute an analogous statement for simply primitive formulas.

**Exercise 9.2.4.** Let $T$ be a $\lambda$-categorical and complete $L$-theory and $\lambda \geq |L|$. Show that $T$ has quantifier elimination, provided 'the' model of $T$ of power $\lambda$ is ultrahomogeneous.

**Exercise 9.2.5.** Let $\Delta$ be a class of formulas containing **at**, and let $\widetilde{\Delta}$ be its boolean closure.
Find generalizations of Lemma 9.2.1 and Theorem 9.2.2 for $\widetilde{\Delta}$-elimination.

**Exercise 9.2.6.** Show that $T_=$ has no quantifier elimination, but all completions of $T_=$ do.

**Exercise 9.2.7.** Prove that, for any division ring $\mathcal{K}$, the theory $T_{\mathcal{K}}^\infty$ of the infinite $\mathcal{K}$-vector spaces admits quantifier elimination.

**Exercise 9.2.8.** Let $A$ be an infinite set and $G$ a group of permutations of $A$ (that is a subgroup of the symmetric group of $A$). Consider a language $L_G$ whose nonlogical symbols are unary function symbols, one for each $g \in G$, which we again denote by $g$. This should not cause any confusion, since we interpret it on $A$ by $g$. Denote the resulting $L_G$-structure by $\mathcal{A}$ (so that it makes sense to say that $\mathcal{A}$ is the structure $(A, g)_{g \in G}$.
Prove that Th $\mathcal{A}$ admits quantifier elimination.

## 9.3  Dense linear orderings

First we consider the theory of dense linear orderings without endpoints, DLO$_{--}$, in the language $L_<$ (as in §5.5).

**Lemma 9.3.1.** *If $\mathcal{M}$ and $\mathcal{N}$ are models of* DLO$_{--}$ *and $\mathcal{A}$ is a substructure of $\mathcal{M}$ and of $\mathcal{N}$, then $(\mathcal{M}, A) \equiv (\mathcal{N}, A)$.*

*Proof.* Obviously, $(\mathcal{M}, A) \equiv (\mathcal{N}, A)$ if and only if $(\mathcal{M}, A_0) \equiv (\mathcal{N}, A_0)$ holds for every $A_0 \Subset A$. So let $A_0 = \{a_0, \ldots, a_{n-1}\} \subseteq A$. By Löwenheim-Skolem downward there are countable $\mathcal{M}' \preccurlyeq \mathcal{M}$ and $\mathcal{N}' \preccurlyeq \mathcal{N}$ such that $A_0 \subseteq M' \cap N'$ (remember, $L_<$ is countable). Since then $(\mathcal{M}, A_0) \equiv (\mathcal{M}', A_0)$

and $(\mathcal{N}, A_0) \equiv (\mathcal{N}', A_0)$, this reduces the assertion to the countable models $\mathcal{M}'$ and $\mathcal{N}'$ (and finite $A_0$).

If now $A_0 = \emptyset$, the assertion follows from Cantor's theorem (which even yields $\mathcal{M}' \cong \mathcal{N}'$). Any nonempty $A_0$ partitions $\mathcal{M}'$ and $\mathcal{N}'$ into the same finite number of intervals, all of which are (countable) dense linear orderings and therefore, by Cantor again, isomorphic. Patching corresponding isomorphisms together, we obtain $(\mathcal{M}', A_0) \cong (\mathcal{N}', A_0)$, hence also $(\mathcal{M}', A_0) \equiv (\mathcal{N}', A_0)$.                                    □

So DLO$_{--}$ is substructure-complete and thus admits quantifier elimination. Remark (1) of the preceding section implies that DLO$_{--}$ is complete (as $L_<$ has no constants). We summarize this together with what we can say about the other dense linear orderings.

**Theorem 9.3.2.** DLO *has exactly four different $L_<$-completions, namely* DLO$_{--}$, DLO$_{+-}$, DLO$_{-+}$, *and* DLO$_{++}$. *Each of these is $\aleph_0$-categorical (and thus has an elementarily prime model). Only* DLO$_{--}$ *of the above five theories has quantifier elimination.*

*Proof.* All of the four mentioned extensions of DLO are $\aleph_0$-categorical (which easily follows from the $\aleph_0$-categoricity of DLO$_{--}$ on adding the corresponding endpoints). Hence, by the Łoś-Vaught test (Theorem 8.5.1), they are all complete, for DLO has infinite models only. By Remark (4) of §8.5, all of them have an elementarily prime model.

Further, every completion of DLO has to 'decide' which endpoints exist. Hence each completion must contain—thus be equal to—one of these four complete theories.

That DLO$_{--}$ has quantifier elimination we saw already from the previous lemma. So we are left with showing that none of the other have quantifier elimination (then, being a subtheory, neither does DLO).

Suppose, for instance, the countable model $\mathcal{M} \models$ DLO has a left endpoint $c$. Given any point $d \in M$, let $\mathcal{A}_d$ denote the induced ordering on (i. e. the substructure generated by) the set $[d, \infty) = \{a \in M : d \leq a\}$. By $\aleph_0$-categoricity, $\mathcal{A}_d$ is isomorphic to $\mathcal{M}$, hence a model of Th $\mathcal{M}$. However, $(\mathcal{M}, [d, \infty))$ and $(\mathcal{A}_d, [d, \infty))$ are not elementarily equivalent. for $\exists x \, (x < d)$ holds in the former, but not in the latter. Consequently, Th $\mathcal{M}$ is not substructure-complete.

An analogous argument proves that Th $\mathcal{M}$ has no quantifier elimination whenever $\mathcal{M}$ has a right endpoint.                                    □

**Exercise 9.3.1.** Show that the countable dense linear ordering without endpoints is an elementarily prime model of DLO$_{--}$, an algebraically (but not elementarily) prime model of DLO, and a prime structure for (but not itself a member of) $\mathrm{Mod}(\mathrm{DLO}_{+-}) \cup \mathrm{Mod}(\mathrm{DLO}_{-+}) \cup \mathrm{Mod}(\mathrm{DLO}_{++})$.

**Exercise 9.3.2.** Exhibit formulas exemplifying that DLO, $\mathrm{DLO}_{+-}$, $\mathrm{DLO}_{-+}$, and $\mathrm{DLO}_{++}$ do not admit quantifier elimination. In case of DLO one can even find such a *sentence*. Why is this impossible in the other three cases?

The next exercise shows how far DLO is from having quantifier elimination.

**Exercise 9.3.3.** Let $\varphi_l$ be the formula $\forall y \, (x = y \vee x < y)$ and $\varphi_r$ the formula $\forall y \, (x = y \vee y < x)$. Set $\Delta = \mathbf{qf} \cup \{\varphi_l, \exists x \, \varphi_l, \varphi_r, \exists x \, \varphi_r\}$. Verify that DLO admits $\overline{\Delta}$-elimination.

## 9.4   Algebraically closed fields

We now turn to the most prominent example, which essentially motivated model-theoretic concepts and results. The quantifier elimination for algebraically closed fields, which we are to verify first, will prove to be a useful tool: it immediately yields a number of classical algebraic results, as we will see in the next section.

The theory we have to deal with is the theory ACF of **algebraically closed fields**, axiomatized by the set of axioms

$$\mathrm{TF} \cup \{\forall y_0 \dots y_{n-1} \exists x \, (x^n + y_{n-1} x^{n-1} + \dots + y_1 x + y_0 = 0) \; : \; 0 < n < \omega\}.$$

*As in §5.3, we work in the language* $L = \mathrm{L}(0, 1; +, -, \cdot)$ *whenever we deal with fields (or rings).*

The models of this theory are precisely the fields in which every polynomial of positive degree (in one indeterminate) and with coefficients from the given field have a root. Such fields are said to be **algebraically closed** . (Note that, in the axioms ACF, we required the existence of a root only for **monic** polynomials of positive degree, i. e. for polynomials of positive degree whose **leading coefficient** is 1; however, this is no restriction, as the coefficient ring is a field.)

**Remark.** ACF has only infinite models, for if $\mathcal{K}$ is a finite field with universe $\{k_i : i < n\}$, then the polynomial $1 + \prod_{i<n}(x - k_i)$ can have no root in $\mathcal{K}$.

Adding the axiom about a positive characteristic $p$ or infinitely many axioms about characteristic 0 (see §5.3), we obtain the extensions $\mathrm{ACF}_p = \mathrm{ACF} \cup \mathrm{TF}_p$ and $\mathrm{ACF}_0 = \mathrm{ACF} \cup \mathrm{TF}_0$.

**Example.** Gauss' so-called *fundamental theorem of algebra* says that the field of complex numbers $\mathbb{C}$ is a model of ACF (hence also of $ACF_0$). The fields $\mathbb{Q}$ and $\mathbb{R}$ (of the rationals and the reals, respectively) are not algebraically closed: e. g. $x^2 + 1$ has no root in them.

Let us recall some material concerning polynomials in one indeterminate **over** a field $\mathcal{K}$ (i. e. with coefficients from that field). Such a polynomial is a term of the form $k_n x^n + k_{n-1} x^{n-1} + \ldots + k_1 x + k_0$, where $k_i \in \mathcal{K}$ ($i \leq n$) (it is indeed a term in the formal sense of §2.2 if we regard it as being in the language $L(K)$). Under the usual operations these polynomials form the so-called **polynomial ring** $\mathcal{K}[x]$ over $\mathcal{K}$. If $f$ is the polynomial above and $k_n \neq 0$, then $n$ is the **degree** of $f$, denoted by $\deg f$. If $k_0 = \ldots = k_n = 0$, that is $f = 0$, we set $\deg f = -\infty$. Usually the polynomials of degree $\leq 0$ in $\mathcal{K}[x]$ are identified with the elements of the field $\mathcal{K}$, which is thus regarded as a subring of $\mathcal{K}[x]$. The constant polynomial $0 \in \mathcal{K}$ is the zero of the ring $\mathcal{K}[x]$, while the constant polynomial $1 \in \mathcal{K}$ is its one. If $a$ is a root of a polynomial of positive degree from $\mathcal{K}[x]$ in a certain extension field $\mathcal{E}$ of $\mathcal{K}$, then a **minimal polynomial** of $a$ over $\mathcal{K}$ is a monic polynomial $f \in \mathcal{K}[x]$ of minimal degree having $a$ as a root. (We will see shortly that this does not depend on $\mathcal{E}$.) Since the difference of two monic polynomials of the same degree is a polynomial of lesser degree, there is exactly one minimal polynomial of $a$. More can be said, once it is taken into account that $\mathcal{K}[x]$ is a **euclidean** ring, i. e., given nonzero polynomials $g, f \in \mathcal{K}[x]$, there are polynomials $q, r \in \mathcal{K}[x]$ such that $g = qf + r$ and $\deg r < \deg f$. From this one easily sees that a nonzero polynomial from $\mathcal{K}[x]$ with a root $a \in \mathcal{K}$ is divisible by $x - a$ in $\mathcal{K}[x]$. By induction on the degree one can further show that a nonzero polynomial $g \in \mathcal{K}[x]$ has at most $\deg g$ roots. (This implies in turn that over an *infinite* field $\mathcal{K}$ only the polynomials of degree $\leq 0$, i. e. only the elements from $\mathcal{K}$, can be **constant** (as a function from $\mathcal{K}$ to $\mathcal{K}$). In finite fields this is different: the so-called **Frobenius** map on $\mathbb{F}_p$, given by $x \mapsto x^p$, is the identical map, i. e., the polynomial $x^p - x$ is constantly 0.) It follows from $\mathcal{K}[x]$ being euclidean that the minimal polynomial $f$ of $a$ over $\mathcal{K}$ divides every polynomial $g \in \mathcal{K}[x]$ that has $a$ as a root. For if $g = qf + r$ as before, $a$ is a root also of $r$, hence $r = 0$, as $\deg r < \deg f$ (otherwise one could make $r$ monic, contradicting the minimality of $f$). This implies that *all* roots of the minimal polynomial $f$ of $a$ over $\mathcal{K}$ (in any extension field of $\mathcal{K}$) are roots also of every polynomial in $\mathcal{K}[x]$ that has $a$ as a root. Further, the minimal polynomial $f$ of $a$ over $\mathcal{K}$ is **irreducible** (in $\mathcal{K}[x]$), i. e., if $f$ is a product of polynomials $f_1, f_2 \in \mathcal{K}[x]$, then $f_1 \in \mathcal{K}$ or $f_2 \in \mathcal{K}$ (i. e. $\deg f_1 = 0$ or $\deg f_2 = 0$). This is so, since $f(a) = 0$ implies $f_1(a) = 0$ or $f_2(a) = 0$, as $\mathcal{E}$ has no zero-divisors; however, $\deg f = \deg f_1 + \deg f_2$, hence the minimality of $f$ yields $\deg f_1 = 0$ or $\deg f_2 = 0$. Together with the aforementioned feature of sets of roots of minimal polynomials, this shows that $f$ is a minimal polynomial of *all* its roots (in any extension of $\mathcal{K}$). This also implies that the set of roots of any other polynomial from $\mathcal{K}[x]$ in *any* field extension $\mathcal{K}'$ of $\mathcal{K}$ contains either *all* or *none* of the roots of $f$ in $\mathcal{K}'$.

Since in our signature substructures of fields are in general only sub *rings* (cf. §5.3), we need to know that there is a canonical way of passing to a field again, which is a generalization of the construction of the rationals from the integers. It works for every integral domain (i. e. for every commutative ring without zero-divisors, cf. Exercise 5.3.2), which is no problem, as every subring of a field is such an integral domain. More precisely, given any integral domain $\mathcal{A}$, there is a so-called **quotient field** (or **field of fractions**) of $\mathcal{A}$ which contains $\mathcal{A}$ as a subring and is embeddable in every field containing $\mathcal{A}$. In this sense the quotient field is the smallest field containing $\mathcal{A}$. Thus it is unique up to $\mathcal{A}$-isomorphism. It is denoted by $Q(\mathcal{A})$. (It has an even stronger feature: every isomorphism of rings, $\mathcal{A} \cong \mathcal{B}$, can be extended to an isomorphism of their quotient fields, $Q(\mathcal{A}) \cong Q(\mathcal{B})$. For more details see the quoted algebraic literature at the end of the book.

In §5.3 we convinced ourselves that, modulo the theory CR (of commutative rings), the $L$-terms $t(x, \bar{a})$ with parameters $\bar{a}$ from $\mathcal{K}$ are just the polynomials from $\mathcal{K}[x]$, and that correspondingly the atomic $L$-formulas $\varphi(x, \bar{a})$ with parameters $\bar{a}$ from $\mathcal{K}$ are the polynomial equations $f = 0$, where $f \in \mathcal{K}[x]$; see Lemma 5.3.3. Hence the sets definable in an extension field $\mathcal{E}$ of $\mathcal{K}$ by 1-place atomic formulas with parameters from $K$ are exactly the sets of roots of such polynomials. Thus these definable sets are either finite or—in the case of the zero polynomial—all of $\mathcal{E}$. Next we investigate what this looks like for arbitrary quantifier-free formulas instead of just atomic ones.

**Lemma 9.4.1.** *Let $\mathcal{A}$ be an integral domain and $\psi(x) \in L(\mathcal{A})$ a quantifier-free formula (in at most one free variable $x$ and with parameters from $\mathcal{A}$).*

*Then one of the next three cases takes place.*

(a)   *$\psi$ is satisfied in no field $\mathcal{K} \supseteq \mathcal{A}$.*

(b)   *For all fields $\mathcal{K} \supseteq \mathcal{A}$, the set $\mathcal{K} \setminus \psi(\mathcal{K}) = \{a \in K : \mathcal{K} \models \neg\psi(a)\}$ is finite.*

(c)   *There is a minimal polynomial $f \in Q(\mathcal{A})[x]$ (of an element satisfying $\psi$) such that, for all fields $\mathcal{K} \supseteq \mathcal{A}$ (which clearly contain also $Q(\mathcal{A})$), the definable set $\psi(\mathcal{K})$ contains all roots of $f$ in $\mathcal{K}$; i. e.*

$$\text{TF} \cup (\text{D}(\mathcal{A})) \models \forall x \, (f(x) = 0 \rightarrow \psi(x)).$$

*Further, $\psi$ is satisfied in every algebraically closed field $\mathcal{K}^* \supseteq \mathcal{A}$, provided $\psi$ is satisfied in some field $\mathcal{K} \supseteq \mathcal{A}$ at all.*

*Proof.* By the above remark about quotient fields we may assume that $\mathcal{A}$ itself is a field, i. e. $\mathcal{A} = Q(\mathcal{A})$.

Being a quantifier-free formula, $\psi$ is a boolean combination of atomic formulas. It is easy to see that it suffices to consider the case, where $\psi$ consists of one disjunct only. By Lemma 5.3.3(2), $\psi$ may be taken of the form

$$\bigwedge_{i<m} t_i(x) = 0 \wedge \bigwedge_{j<n} s_j(x) \neq 0 \qquad (t_i, s_j \in \mathcal{A}[x]).$$

A polynomial equation $a = 0$ with a constant polynomial $a \in \mathcal{A}$ is satisfied either simultaneously in all fields $\mathcal{K} \supseteq \mathcal{A}$ (namely if and only if $a = 0$) or else in no field $\mathcal{K} \supseteq \mathcal{A}$ at all. Since the same holds true also for boolean combinations of such equations, the lemma is proved for *sentences* $\psi$. Moreover, any conjunct of the form $a = 0$ or $a \neq 0$ is redundant or makes the formula $\psi$ inconsistent. Therefore we may assume that all occurring polynomials $t_i$ and $s_j$ have positive degree. We consider two cases.

*Case 1: $m = 0$.* This means that $\psi$ is logically equivalent to the formula $\neg \bigvee_{j<n} s_j(x) = 0$. Then, in any field $\mathcal{K} \supseteq \mathcal{A}$, the set $\mathcal{K} \setminus \psi(\mathcal{K})$ is the union of the (finitely many) sets of roots of the $s_j$, hence (b) holds.

*Case 2: $m > 0$.* Assume (a) and (b) do not hold. Then $\psi$ is satisfied by an element $c$ of some field extension $\mathcal{K} \supseteq \mathcal{A}$. As $m > 0$, the element $c$ is a root of a polynomial of positive degree and thus has a minimal polynomial $f$ over $\mathcal{A}$. The roots of $f$ are—as explained above—roots of all polynomials $t_i$ ($i < m$), but of no polynomial $s_j$ ($j < n$) (for $c$ is no root of any of the $s_j$). Consequently, all roots of $f$ satisfy $\psi$ (in all field extensions of $\mathcal{A}$), i. e., (c) takes place.

Now that this trichotomy is verified, suppose $\mathcal{A} \subseteq \mathcal{K}^* \models \mathrm{ACF}$ and $\psi$ is satisfied in some field $\mathcal{K} \supseteq \mathcal{A}$. Then we are in (b) or (c). In case (c), $\psi(\mathcal{K}^*)$ contains all roots of $f$ in $\mathcal{K}^*$. But this field is algebraically closed, hence $f$ does have a root therein. In case (b), $\mathcal{K}^* \setminus \psi(\mathcal{K}^*)$ is finite. But being an algebraically closed field, $\mathcal{K}^*$ is infinite (see Remark above), hence $\psi(\mathcal{K}^*)$ can't be empty either. □

This lemma is the essential technical ingredient of the proof of the main result of this section.

**Theorem 9.4.2.** ACF *admits quantifier elimination: every* $(0, 1; +, -, \cdot)$-*formula in the free variables $\bar{x}$ is ACF-equivalent to a boolean combination of polynomial equations $t = 0$, where $t \in \mathbb{Z}[\bar{x}]$.*

*The same is true for* $\mathrm{ACF}_q$ *in all characteristics $q$.*

Tarski (1948), unpublished, see

[Robinson, A. : **Introduction to Model Theory and to the Metamathematics of Algebra**, Studies in Logic and the Foundations of Mathematics 66, North-Holland, Amsterdam [2]1965, 284 pp.]

*Proof.* We verify (iii) of Theorem 9.2.2, i. e., we show for any $\mathcal{M}, \mathcal{N} \models$ ACF and any joint subring $\mathcal{A}$ of $\mathcal{M}$ and $\mathcal{N}$ (finitely generated or not) that $(\mathcal{M}, A) \equiv_{\exists^*} (\mathcal{N}, A)$.

To this end let $\varphi$ be a simply primitive $L(A)$-sentence true in $\mathcal{M}$. By reasons of symmetry we need only show that it is true also in $\mathcal{N}$. Write $\varphi$ as $\exists y \, \psi(y)$, where $\psi$ is a conjunction of $L(A)$-literals. Since $\mathcal{M} \models \varphi$, the quantifier-free formula $\psi$ is satisfied in the field $\mathcal{M} \supseteq \mathcal{A}$. By the preceding lemma it must be satisfied in all algebraically closed fields containing $\mathcal{A}$, in particular in $\mathcal{N}$. But then also $\varphi$ holds in $\mathcal{N}$.          □

Remarks (1) and (2) of §9.2 yield the following consequences of the theorem—for (1) just recall that the classes of (algebraically closed) fields of fixed characteristic have prime structures, the prime fields.

**Corollary 9.4.3.**
(1)   *The theories* $\mathrm{ACF}_0$ *and* $\mathrm{ACF}_p$ *$(p > 0)$ are complete.*
(2)   *Every embedding between algebraically closed fields is elementary.*   □

Thus the completions of ACF are exactly the theories $\mathrm{ACF}_q$, where $q$ is a prime or 0. In case $q = 0$, for instance, this implies that a $(0, 1; +, -, \cdot)$-sentence is true in any algebraically closed field of characteristic 0 if and only if it is true in $\mathbb{C}$.

We will see in §14.2 that the theories $\mathrm{ACF}_q$ are uncountably categorical and have elementarily prime models (Corollary 14.2.3 and Remarks (2) and (3) after 14.2.5).

**Exercise 9.4.1.** Assuming that every field is contained in an algebraically closed field (see before Lemma 9.5.1 below) prove that $\mathrm{ACF}_\forall$ is the deductive closure of $\mathrm{CR} \cup \{\forall xy \,(xy = 0 \rightarrow x = 0 \lor y = 0)\}$. (This is the theory of integral domains, cf. Exercise 5.3.2.)

**Exercise 9.4.2.** Show that $\mathrm{ACF}_0$ is not finitely axiomatizable.

**Exercise 9.4.3.** What about $\mathrm{ACF}_p$?

## 9.5   Field-theoretic applications

The quantifier elimination for ACF has far-reaching consequences. In particular it provides us with new and sometimes more transparent proofs of classical results, of which we consider a few in this section.

**Fact.** Every field is contained in an algebraically closed field.

On a more formal level, this means that for every $\mathcal{K} \models$ TF the set $\mathrm{ACF} \cup \mathrm{D}(\mathcal{K})$ is consistent. This is a well-known algebraic fact. Assuming it for the moment, we want to derive now some deeper algebraic results from the material of the previous section. We postpone a (self-contained) proof of the above fact until Exercise 12.1.4. (Moreover, in Exercise 12.1.5 a famous theorem of Steinitz about algebraic closure, which has the above fact as a consequence, is to be derived from our much more general—model-theoretic—considerations. This topic will be resumed in Chapter 11.)

**Lemma 9.5.1.** (Hilbert) *Suppose $\sigma$ is a finite system of polynomial equations and inequations with coefficients from the field $\mathcal{K}$.*

*If $\sigma$ has a solution in some field extending $\mathcal{K}$, then $\sigma$ has a solution in every algebraically closed field extending $\mathcal{K}$. (Equivalently, if $\mathrm{TF} \cup \mathrm{D}(\mathcal{K}) \cup \{\exists \bar{x}\, \sigma(\bar{x})\}$ is consistent, then $\mathrm{ACF} \cup \mathrm{D}(\mathcal{K}) \models \exists \bar{x}\, \sigma(\bar{x})$.)*

*Proof.* Suppose $\sigma$ has a solution in a certain field $\mathcal{K}' \supseteq \mathcal{K}$ and $\mathcal{K} \subseteq \mathcal{K}^* \models$ ACF. We have to show, $\sigma$ has a solution in $\mathcal{K}^*$.

By the above fact there is an algebraically closed field $\mathcal{K}'' \supseteq \mathcal{K}'$. Since $\sigma$ is quantifier-free with parameters in $\mathcal{K}$, to have a solution is an existential statement and thus passes up to $\mathcal{K}''$. Hence $\mathrm{ACF} \cup \mathrm{D}(\mathcal{K}) \cup \{\exists \bar{x}\, \sigma\}$ is consistent. But the substructure-completeness of ACF implies the completeness of $\mathrm{ACF} \cup \mathrm{D}(\mathcal{K})$. Consequently, $\mathrm{ACF} \cup \mathrm{D}(\mathcal{K}) \models \exists \bar{x}\, \sigma(\bar{x})$, as desired. $\square$

Now we can give a proof of a classical theorem that is fundamental in algebraic geometry. The previous lemma constitutes—in our treatment—the model-theoretic part of it, while the remaining part is purely algebraic (and known as *Kronecker's construction*).

**Theorem 9.5.2.** (Hilbert's Nullstellensatz)[2] *Suppose $\mathcal{K}$ is a field, $\mathcal{R} = \mathcal{K}[x_0, \ldots, x_{n-1}]$ is the ring of polynomials in the indeterminates $x_0, \ldots, x_{n-1}$ with coefficients in $\mathcal{K}$, and $f_0, \ldots, f_{m-1} \in \mathcal{R}$ .*

*If the ideal generated by the polynomials $f_0, \ldots, f_{m-1}$ in $\mathcal{R}$ is a proper ideal (i. e. different from $\mathcal{R}$), then the polynomials $f_0, \ldots, f_{m-1}$ have a joint root in every algebraically closed field extending $\mathcal{K}$, i. e.*

$$\mathrm{ACF} \cup \mathrm{D}(\mathcal{K}) \models \exists x_0 \ldots x_{n-1} \bigwedge_{i < n} f_i(x_0, \ldots, x_{n-1}) = 0.$$

---

[2] 'Nullstelle' [the stress is on 'Null'] is a German word for 'root', 'Satz' a German word for 'proposition'.

*Proof.* By the above lemma, it suffices to find an extension field of $\mathcal{K}$ in which $f_0, \ldots, f_{m-1}$ have a joint root.

Invoking Zorn's lemma, first choose a maximal ideal $M$ in $\mathcal{R}$ containing the $f_i$. Then $\mathcal{K}' =_{\text{def}} \mathcal{R}/M$ is a field. It is easily seen that $k \mapsto k + M$ defines a ring homomorphism $h : \mathcal{K} \to \mathcal{K}'$ with kernel $\mathcal{K} \cap M$. Note that $M$ contains no invertible elements (otherwise 1 would lie in $M$, contradicting $\mathcal{R} \neq M$). Hence $\mathcal{K} \cap M = \{0\}$, i. e. $h : \mathcal{K} \hookrightarrow \mathcal{K}'$. This embedding canonically induces an embedding $\mathcal{K}[\bar{x}] \hookrightarrow \mathcal{K}'[\bar{x}]$, which we denote by $h$ again.

Now we show that $(x_0 + M, \ldots, x_{n-1} + M)$ is a root of each of the polynomials $h(f_0), \ldots, h(f_{m-1})$ in $\mathcal{K}'$. Here we identify, for simplicity, polynomials from $\mathcal{K}[\bar{x}]$ with their $h$-images in $\mathcal{K}'[\bar{x}]$.

A typical element of $\mathcal{K}'$ is of the form $r + M$, where $r \in \mathcal{R}$. The arithmetical operations in $\mathcal{K}'$ are $(r + M) + (s + M) = (r + s) + M$ and $(r + M) \cdot (s + M) = (r \cdot s) + M$, where $r, s \in \mathcal{R}$. Hence we have $f(r_0 + M, \ldots, r_{n-1} + M) = f(r_0, \ldots, r_{n-1}) + M$ for every polynomial $f \in \mathcal{R}$. Thus $f_i(x_0 + M, \ldots, x_{n-1} + M) = f_i(x_0, \ldots, x_{n-1}) + M = f_i + M = 0 + M$ for every $i < m$ (as $f_i \in M$). But $0 + M$ is the zero of the field $\mathcal{K}'$. Consequently, $x_0 + M, \ldots, x_{n-1} + M$ is a joint root of the $f_i$ in $\mathcal{K}'$.

By isomorphic correction one can finally make $\mathcal{K}'$ contain $\mathcal{K}$. $\square$

**Remark.** The ideal generated by a (single) polynomial $f \in \mathcal{K}[x]$ is the entire ring $\mathcal{K}[x]$ if and only if there is a polynomial $g \in \mathcal{K}[x]$ such that $fg = 1$, i. e. if and only if $f$ has degree 0. Thus the Hilbert Nullstellensatz generalizes in two ways the defining property of algebraically closed fields (that every polynomial in $\mathcal{K}[x]$ of positive degree have a root in every algebraically closed field extension of $\mathcal{K}$): to finitely many polynomials and to (finitely) many indeterminates.

By the way, the Nullstellensatz does not require the full strength of quantifier elimination of the theory ACF, but only its so-called model-completeness, as is to be shown in Exercise 9.6.2 below. Our next application, however, would not follow from the model-completeness alone.

Suppose $\mathcal{K} \models$ ACF and $n < \omega$. By a **constructible set** of the affine space $\mathcal{K}^n$ we mean a set definable in $\mathcal{K}$ by a formula from **qf** $\cap \, L_n(\mathcal{K})$, i. e. a (finite) boolean combination of solution sets of polynomial equations in $n$ indeterminates with coefficients from $\mathcal{K}$.

**Theorem 9.5.3.** (Chevalley) *Suppose $\mathcal{K} \models$ ACF and $m < n < \omega$.*

*Every projection of a constructible set from $\mathcal{K}^n$ onto $\mathcal{K}^m$ is constructible (in $\mathcal{K}^m$).*

[D. Mumford, **Algebraic Geometry I (Complex Projective Varieties)**, Grundlehren der Mathematischen Wissenschaften 221, Springer, Berlin 1976]

*Proof.* If $X \subseteq \mathcal{K}^n$ is defined by $\varphi(x_0, \dots, x_{n-1})$, then its projection e. g. onto the last $n - 1$ components is defined by $\exists x_0 \varphi(x_0, \dots, x_{n-1})$. But the latter is, by quantifier elimination, ACF-equivalent to a quantifier-free formula, hence defines a constructible set in $\mathcal{K}^n$. □

**Remark.** The proof yields a strengthening of Chevalley's theorem: for every $\varphi(x_0, \dots, x_{n-1})$ there is a quantifier-free $\psi(x_1, \dots, x_{n-1})$ defining the projection of $\varphi(\mathcal{K}^n)$ onto $x_1, \dots, x_{n-1}$ not only in one, but simultaneously in all $\mathcal{K} \models \mathrm{ACF}$.

Let $\mathcal{K} \models \mathrm{ACF}$. A function $f : \mathcal{K}^n \longrightarrow \mathcal{K}^m$ is said to be **constructible** if so is its graph (as a subset of $\mathcal{K}^{n+m}$).

**Corollary 9.5.4.** *The image as well as the domain of injectivity of a constructible function is constructible.*

*Proof.* If $\varphi(\bar{x}, \bar{y})$ defines the graph of the given function, $\exists \bar{x}\, \varphi(\bar{x}, \bar{y})$ defines its image, while $\forall \bar{x}' \bar{y}\, (\varphi(\bar{x}, \bar{y}) \wedge \varphi(\bar{x}', \bar{y}) \rightarrow \bar{x} = \bar{x}')$ defines its domain of injectivity. □

**Remark.** A formula $\varphi(\bar{x}, \bar{y}, \bar{a})$ with parameters $\bar{a}$ from a field $\mathcal{K} \models \mathrm{ACF}$ defines in $\mathcal{K}$ a function $\bar{x} \mapsto \bar{y}$ if and only if it does so in any other field $\mathcal{K}' \models \mathrm{ACF}$ containing the field (or ring) generated by $\bar{a}$ in $\mathcal{K}$.

This is another consequence of the quantifier elimination for ACF, or rather its substructure-completeness, which implies

$$\mathcal{K} \models \forall \bar{x}\, \exists! \bar{y}\, \varphi(\bar{x}, \bar{y}, \bar{a}) \text{ iff } \mathcal{K}' \models \forall \bar{x}\, \exists! \bar{y}\, \varphi(\bar{x}, \bar{y}, \bar{a}).$$

We will see more field-theoretic applications in (the exercises of) Chapter 12 and in Chapter 14.

**Exercise 9.5.1.** (Simultaneous Solvability Criterion.) Given a finite system $\sigma(\bar{x}, \bar{y})$ of polynomial equations and inequations in the indeterminates $\bar{x}$ and $\bar{y}$ (with integer coefficients), prove that there is a boolean combination $\sigma^*(\bar{y})$ of polynomial equations in the indeterminates $\bar{y}$ (with integer coefficients) satisfying the following equivalence for all $\mathcal{K} \models \mathrm{ACF}$ and all $l(\bar{y})$-tuples $\bar{c}$ from $\mathcal{K}$: $\sigma(\bar{x}, \bar{c})$ has a solution in $\mathcal{K}$ if and only if $\mathcal{K} \models \sigma^*(\bar{c})$, i. e.,

$$\mathrm{ACF} \models \forall \bar{y}(\exists \bar{x}\sigma(\bar{x}, \bar{y}) \leftrightarrow \sigma^*(\bar{y})).$$

**Exercise 9.5.2.** Do the same as in the previous exercise, but allowing parameters from a certain field $\mathcal{K}_0$.

**Exercise 9.5.3.** Find a more down-to-earth proof of Lemma 9.5.1 using the first exercise above.

**Exercise 9.5.4.** Derive from Theorem 9.5.2 the following more usual formulation of Hilbert's Nullstellensatz.

If, in an algebraically closed field extension of $\mathcal{K}$, all joint roots of $f_0, \dots, f_{m-1} \in \mathcal{K}[\bar{x}]$ are also roots of $g \in \mathcal{K}[\bar{x}]$, then $g^k$, for some $k > 0$, is contained in the ideal generated by $f_0, \dots, f_{m-1}$ in $\mathcal{K}[\bar{x}]$.

Given a field $\mathcal{K}$, $\mathrm{GL}_2(\mathcal{K})$ denotes the so-called **general linear group** of $2 \times 2$ matrices over $\mathcal{K}$, i. e. the set of $2 \times 2$ matrices over $\mathcal{K}$ whose determinant is not 0, with the usual matrix multiplication. Two elements $a$ and $b$ of a group $G$ are said to be **conjugate** (in $G$) if there is $g \in G$ with $a = g^{-1}bg$.

**Exercise 9.5.5.** Show that two matrices from $\mathrm{GL}_2(\mathcal{K})$ with $\mathcal{K} \models \mathrm{ACF}$ are conjugate in $\mathrm{GL}_2(\mathcal{K})$ if and only if they are conjugate in $\mathrm{GL}_2(\mathcal{K}')$ for some $\mathcal{K}' \models \mathrm{ACF}$ extending $\mathcal{K}$.

## 9.6   Model-completeness

Consider the following weakening of substructure-completeness. A theory $T$ is said to be **model-complete** if it is $\mathrm{Mod}\,T$-complete (see §9.2).

[Robinson, A. : **Complete Theories**, Studies in Logic and the Foundations of Mathematics, North-Holland, Amsterdam, 1956]

We are going to show that model-completeness is equivalent to a weaker elimination property, namely to $\exists$-elimination, and, equivalently, also to $\forall$-elimination. There is another interesting equivalent, for which we introduce another concept.

Given $L$-structures $\mathcal{M} \subseteq \mathcal{N}$, we say that $\mathcal{M}$ is **existentially closed** in $\mathcal{N}$ if $\mathcal{N} \models \varphi(\bar{a})$ implies $\mathcal{M} \models \varphi(\bar{a})$ for every existential $L$-formula $\varphi(\bar{x})$ and every matching tuple $\bar{a}$ from $M$, i. e. if $\mathcal{M} \preccurlyeq_\exists \mathcal{N}$ (cf. notation in §8.3). $\mathcal{M}$ is said to be **existentially closed** in a given class of $L$-structures, $\mathbf{K}$, if $\mathcal{M} \in \mathbf{K}$ and $\mathcal{M}$ is existentially closed in every $\mathcal{N} \in \mathbf{K}$ with $\mathcal{M} \subseteq \mathcal{N}$. An **existentially closed model** of a theory $T$ is, by definition, a model of $T$ that is existentially closed for the class $\mathrm{Mod}\,T$.

**Remarks.**

(1)   As mentioned before Lemma 9.2.1, every $\exists$-formula is logically equivalent to a disjunction of primitive formulas. Therefore for the existential closedness of $\mathcal{M}$ in $\mathcal{N} \supseteq \mathcal{M}$ it suffices to check primitive formulas.

(2)   A primitive formula is of the form $\exists \bar{x}\, \sigma(\bar{x}, \bar{y})$, where $\sigma$ is a conjunction of literals. In languages without relation symbols the latter are just term

equations and inequations. Hence in this case the existential closedness of $\mathcal{M}$ in $\mathcal{N} \supseteq \mathcal{M}$ means that every finite system $\sigma$ of equations and inequations with parameters from $\mathcal{M}$ is solvable in $\mathcal{M}$ whenever it it is solvable in $\mathcal{N}$. Thus Lemma 9.5.1 implies that algebraically closed fields are existentially closed (in the class of all fields).[3] Since every polynomial of positive degree in one indeterminate with coefficients from a given field has a solution in an extension field (use e. g. Kronecker's construction from the proof of the Nullstellensatz), we infer that, conversely, every existentially closed field is algebraically closed. Thus for fields both notions coincide (where **existentially closed field** is to mean 'existentially closed in the class of fields').

Next we prove an extended analogue of Theorem 9.2.2. The implication (iv) $\Longrightarrow$ (i) is known as *Robinson's model-completeness test*.

**Theorem 9.6.1.** (A. Robinson) *For every theory $T$ the following are equivalent.*

(i)   *$T$ is model-complete.*
(ii)  *Every monomorphism between models of $T$ is elementary.*
(iii) *If $\mathcal{M}, \mathcal{N} \models T$ and $\mathcal{M} \subseteq \mathcal{N}$, then $\mathcal{M} \preccurlyeq \mathcal{N}$.*
(iv)  *If $\mathcal{M}, \mathcal{N} \models T$ and $\mathcal{M} \subseteq \mathcal{N}$, then $\mathcal{M} \preccurlyeq_\exists \mathcal{N}$.*
(v)   *All models of $T$ are existentially closed.*
(vi)  *$T$ has $\exists$-elimination.*
(vii) *$T$ has $\forall$-elimination.*

*Proof.* (i) $\Longrightarrow$ (ii). Clearly, $(\mathcal{M}, M)$ is a model of $T \cup \mathrm{D}(\mathcal{M})$. If $f : \mathcal{M} \hookrightarrow \mathcal{N}$ and $\mathcal{M}, \mathcal{N} \models T$, also $(\mathcal{N}, f[M])$ is a model of $T \cup \mathrm{D}(\mathcal{M})$. Its completeness (given by hypothesis) yields $(\mathcal{M}, M) \equiv (\mathcal{N}, f[M])$, whence $f$ is elementary.

(ii) $\Longrightarrow$ (iii) $\Longrightarrow$ (iv) $\Longrightarrow$ (v) $\Longrightarrow$ (iv) is trivial.

(iv) $\Longrightarrow$ (vi). Since the $\forall$-formulas are (up to logical equivalence) exactly the negations of $\exists$-formulas, $\mathcal{M} \preccurlyeq_\exists \mathcal{N}$ if and only if for all $\forall$-formulas $\psi(\bar{x})$ and all matching $\bar{a}$ from $M$ we have the implication: if $\mathcal{M} \models \psi(\bar{a})$, then $\mathcal{N} \models \psi(\bar{a})$. Corollary 6.2.6 then implies that all $\forall$-formulas are $T$-equivalent to $\exists$-formulas.

In order to show that *every* formula $\varphi$ is $T$-equivalent to an $\exists$-formula, we proceed by induction on the complexity of $\varphi$. The initial step is, as well as the quantifier step, is trivial. But so is also the conjunction step, as $\exists \bar{x}\, \varphi(\bar{x}) \wedge \exists \bar{y}\, \psi(\bar{y})$ is logically equivalent to $\exists \bar{x} \bar{y}\, (\varphi(\bar{x}) \wedge \psi(\bar{y}))$. We are left with the negation. So let $\varphi$ be of the form $\neg \psi$, where $\psi$ is already $T$-

---

[3]This can also be derived using *Rabinowitsch' trick*: TF $\models \forall x (x \neq 0 \leftrightarrow \exists y (xy = 1))$.

equivalent to an ∃-formula. Then $\varphi$ is $T$-equivalent to an ∀-formula, which in turn is, as mentioned above, equivalent to an ∃-formula.

(vi) ⟹ (vii) is immediate, as every formula can be written as a (double) negation.

(vii) ⟹ (iii) follows from Lemma 6.2.1 and Remark (6) in §6.1.

(iii) ⟹ (ii). Let $f : \mathcal{M} \hookrightarrow \mathcal{N}$ and $\mathcal{M}, \mathcal{N} \models T$. Then there is $\mathcal{M}'$ with $f : \mathcal{M} \cong \mathcal{M}' \subseteq \mathcal{N}$, and (iii) yields $\mathcal{M}' \prec \mathcal{N}$. Hence $f$ is elementary (Lemma 8.2.2(1)).

(ii) ⟹ (i). Any two models of $T \cup D(\mathcal{M})$ are of the form $(\mathcal{N}_0, f_0[M])$ and $(\mathcal{N}_1, f_1[M])$ for certain $f_i : \mathcal{M} \hookrightarrow \mathcal{N}_i$. By (ii) these latter maps are elementary, hence $(\mathcal{N}_0, f_0[M]) \equiv (\mathcal{M}, M) \equiv (\mathcal{N}_1, f_1[M])$.                    □

**Remarks.**

(3)  Also for model-complete theories the two notions of prime model coincide, and we just say **prime model**.

(4)  (Prime Model Test)

Every model-complete theory with a prime model is complete.

One of the exercises below shows that finite linear orderings with more than one element are examples of complete and model-complete theories that do not admit quantifier elimination.

**Example.** (A complete, model-complete theory with infinite models that does not have quantifier elimination.)

Let $L$ be the language whose only non-logical symbol is a binary relation symbol $R$. Consider the $L$-structure $\mathcal{M}$ on the universe $\omega$ defined by $R^{\mathcal{M}} = \{(1,2),(2,1)\} \cup \{(n,n) : 3 \le n < \omega\}$, i. e.

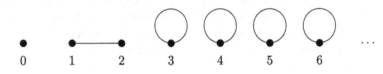

Let $T$ be the $L$-theory axiomatized by the sentences

(R0)    $\forall xy\,(R(x,y) \leftrightarrow R(y,x))$                    ($R$ is symmetrical),

(R1)    $\exists^{=1}x\,\forall y\,\neg R(x,y)$   (there is a unique isolated and irreflexive point),

(R2)    $\exists^{=3}x\,\neg R(x,x) \wedge \forall x\,(R(x,x) \rightarrow \forall y\,(R(x,y) \rightarrow x = y))$     (there are exactly three irreflexive points, and all reflexive points are isolated),

(R3)    $\exists xy\,(x \neq y \wedge R(x,y))$                    (there are nonisolated points).

It follows from (R1) and (R2) that there are at most two nonisolated points. Together with (R3) this yields exactly two nonisolated points. The models of $T$ thus differ from $\mathcal{M}$ at most in the number of *reflexive* points (related to themselves). Hence every model $\mathcal{N}$ of $T$ has power $\geq 3$ and, in case $\mathcal{N}$ is finite, exactly $|N| - 3$ reflexive points. Consequently, $T$ is totally categorical. Then $T^{\infty}$ is totally categorical, and hence complete. A point of a substructure $\mathcal{A}$ of a model $\mathcal{N}$ of $T$ is reflexive in $\mathcal{A}$ if and only if it is so in $\mathcal{N}$. Hence in case $\mathcal{A} \models T$ all three irreflexive points of $\mathcal{N}$ must be irreflexive also in $\mathcal{A}$. If, in addition, both models are infinite, Corollary 8.3.4 easily gives $\mathcal{A} \preccurlyeq \mathcal{N}$. Consequently, $T^{\infty}$ is model-complete.

But $T^{\infty}$ is not substructure-complete: we claim that $T \cup D(\mathcal{B})$ is not complete for the substructure $\mathcal{B}$ that consists of a single, irreflexive point. For, the substructures of $\mathcal{M}$ on the universe $\{0\}$ and on the universe $\{1\}$ are both isomorphic to $\mathcal{B}$, hence $(\mathcal{M}, 0)$ and $(\mathcal{M}, 1)$ are models of $T \cup D(\mathcal{B})$. However, they are not elementarily equivalent, as $\mathcal{M} \models \forall y \, \neg R(0, y) \wedge \exists y \, R(1, y)$.

**Exercise 9.6.1.** Why don't Lemma 8.2.2(2) and Corollary 8.4.5 imply that every complete theory is model complete?

**Exercise 9.6.2.** Derive Lemma 9.5.1 (and thus Hilbert's Nullstellensatz) from the model-completeness of ACF. (For ACF, though, model-completeness does not seem to be easier to prove than quantifier elimination.)

**Exercise 9.6.3.** Every incomplete theory with quantifier elimination (like ACF) is an example of a model-complete, but not complete theory. Find a model-complete theory in a language with only a binary relation symbol which is not complete.

**Exercise 9.6.4.** Consider, in the language $L$ of the above example, the structure $\mathcal{N}$ on $\omega$ such that $R^{\mathcal{N}} = \{(n, n + 1) \; : \; 3 \leq n < \omega\}$ and show that its complete theory is not model-complete.

**Exercise 9.6.5.** Which theories of dense linear orderings are model-complete?

**Exercise 9.6.6.** Prove that every complete theory with finite models is model-complete, and that no finite linear ordering with more than one element admits quantifier elimination.

**Exercise 9.6.7.** Prove that a theory $T$ admits quantifier elimination if and only if it is model-complete and has the following **amalgamation property**.

If $\mathcal{A}$ is a joint substructure of the models $\mathcal{M}, \mathcal{N} \models T$, then there is a model $\mathcal{M}' \supseteq \mathcal{M}$ of $T$ and an embedding $g : \mathcal{N} \hookrightarrow \mathcal{M}'$ such that $\mathrm{id}_M \mathrm{id}_A = g \, \mathrm{id}_A$.

**Exercise 9.6.8.** Show that $\mathrm{Th}(\mathbb{Z}; 0; +, -)$ is not model-complete.

# Chapter 10

# Chains

Chain constructions occur in many mathematical arguments. One such construction we met in Theorem 8.4.1 (Löwenheim-Skolem downward for $\preccurlyeq$).

## 10.1 Elementary chains

Let $\alpha \in \mathbf{On}$.

(1)  A **chain of $L$-structures** (of length $\alpha$) is a sequence of $L$-structures $(\mathcal{M}_i : i < \alpha)$ such that $\mathcal{M}_i \subseteq \mathcal{M}_j$ for all $i < j < \alpha$.

(2)  A chain like in (1) is said to be **continuous** if for all limit ordinals $\delta < \alpha$ we have $\bigcup_{i<\delta} \mathcal{M}_i = \mathcal{M}_\delta$.

(3)  The **union** of the chain in (1) is the canonical $L$-structure on the set $\bigcup_{i<\alpha} \mathcal{M}_i$; i. e., the non-logical symbols are interpreted on a tuple $\bar{a}$ from $\bigcup_{i<\alpha} \mathcal{M}_i$ as in the $\mathcal{M}_i$ with the smallest (or any other) index $i < \alpha$ containing $\bar{a}$. This structure is denoted by $\bigcup_{i<\alpha} \mathcal{M}_i$.

(4)  The chain from (1) is said to be **elementary** if it is continuous and for all $i < \alpha$ we have $\mathcal{M}_i \preccurlyeq \mathcal{M}_{i+1}$.

**Remark.** Since $\mathcal{M}_i \subseteq \mathcal{M}_j$ for all $i < j < \alpha$, we have $\mathcal{M}_j \subseteq \bigcup_{i<\alpha} \mathcal{M}_i$ for all $j < \alpha$.

Elementary chains behave particularly nicely.

**Theorem 10.1.1.** (On Elementary Chains)
*If $(\mathcal{M}_i : i < \alpha)$ is an elementary chain, then $\mathcal{M}_j \preccurlyeq \bigcup_{i<\alpha} \mathcal{M}_i$ for all $j < \alpha$.*

[Tarski, A. and Vaught, R. : Arithmetical extensions of relational systems, Compositio Math. **13** (1957) 81 – 102]

151

*Proof.* For the case $\alpha = \omega$; the general case is left as an exercise.

Let $\mathcal{M} = \bigcup_{n < \omega} \mathcal{M}_n$. By induction on the complexity of formulas we show for all $n < \omega$, all $\varphi \in L_n$, all $k < \omega$, and all $n$-tuples $\bar{a}$ from $M_k$ that $\mathcal{M}_k \models \varphi(\bar{a})$ if and only if $\mathcal{M} \models \varphi(\bar{a})$.

All induction steps besides the one with the existential quantifier are trivial. So let $\varphi \in L_n$ be of the form $\exists x \, \psi$, where $\psi \in L_{n+1}$ satisfies the induction hypothesis. Let $\bar{a}$ be any $n$-tuple from $M_k$.

If $\mathcal{M}_k \models \varphi(\bar{a})$, there is some $c \in M_k$ satisfying $\mathcal{M}_k \models \psi(c, \bar{a})$. Then the induction hypothesis yields $\mathcal{M} \models \psi(c, \bar{a})$, hence also $\mathcal{M} \models \varphi(\bar{a})$.

If, conversely, $\mathcal{M} \models \varphi(\bar{a})$, there is some $c \in M$ satisfying $\mathcal{M} \models \psi(c, \bar{a})$ and some $m < \omega$ such that $c$ and $\bar{a}$ are in $M_m$. Assume, without loss of generality, $k \leq m$. By induction hypothesis, $\mathcal{M}_m \models \psi(c, \bar{a})$, hence $\mathcal{M}_m \models \varphi(\bar{a})$. Then $\mathcal{M}_k \preccurlyeq \mathcal{M}_m$ implies $\mathcal{M}_k \models \varphi(\bar{a})$.                                      $\square$

**Corollary 10.1.2.** *If $(\mathcal{M}_i : i < \alpha)$ is an elementary chain and $\mathcal{M}_j \preccurlyeq \mathcal{N}$ for all $j < \alpha$, then $\bigcup_{i < \alpha} \mathcal{M}_i \preccurlyeq \mathcal{N}$.*

*Proof.* Left as an exercise.                                      $\square$

**Exercise 10.1.1.** Prove the preceding corollary.

**Exercise 10.1.2.** Find a proof for the general case of the above theorem.

**Exercise 10.1.3.** Show that the union of an *arbitrary* chain of groups (resp., fields) is again a group (resp., a field).

## 10.2   Inductive theories

The theorem on elementary chains implies that the union of an *elementary* chain of models of a certain theory is again a model of that theory. In this section we characterize the theories for which this is true even when 'elementary' is omitted—this amounts to finding another preservation theorem.

A theory $T$ is said to be **inductive** if $\mathrm{Mod}\, T$ is closed under taking unions of continuous chains (i. e., if $(\mathcal{M}_i : i < \alpha)$ is a continuous chain of models of $T$, then its union $\bigcup_{i < \alpha} \mathcal{M}_i$ is a model of $T$ too).

**Example.** $T_=$ and most of the theories from §§5.2–6 are inductive. Further, by the theorem on elementary chains, every model-complete theory is inductive, as every continuous chain of its models is elementary by Theorem 9.6.1.

**Lemma 10.2.1.** $\forall\exists$-*theories are inductive.*

*Proof.* Suppose $T$ is an $\forall\exists$-theory and $(\mathcal{M}_i : i < \alpha)$ is a continuous chain of models of $T$. Set $\mathcal{M} = \bigcup_{i<\alpha} \mathcal{M}_i$. We have to show $\mathcal{M} \models T$.

To this end let $\forall \bar{x}\, \exists \bar{y}\, \psi(\bar{x}, \bar{y}) \in T$ and $\psi \in \mathbf{qf}$. We have to verify $\mathcal{M} \models \exists \bar{y}\, \psi(\bar{a}, \bar{y})$ for all matching $\bar{a}$ from $M$. But every such $\bar{a}$ is already contained in some $M_i$, and $\mathcal{M}_i \models T$. Hence $\mathcal{M}_i \models \exists \bar{y}\, \psi(\bar{a}, \bar{y})$. Finally, $\mathcal{M} \models \exists \bar{y}\, \psi(\bar{a}, \bar{y})$ follows from $\mathcal{M}_i \subseteq \mathcal{M}$. $\square$

**Example.** Group theory, as a theory in the signature $(1; \cdot, ^{-1})$, is an $\forall$-theory (cf. §5.2). In the signature $(1; \cdot)$ we can use $\forall x\, \exists y\, xy = 1$ to express the existence of an inverse, which is an $\forall\exists$-sentence. We leave as an exercise to prove that in this signature the theory of groups is an $\forall\exists$-, but not an $\forall$-theory. The same is true for DR, TF, ACF, $\mathrm{TF}_q$, and $\mathrm{ACF}_q$, for all $q$.

The preservation theorem we are going to prove asserts that the converse is true too. As before, this is a harder task. In the proof of the (quite simple) Łoś-Tarski preservation theorem the first step for this was Lemma 6.2.2 saying that $T'_\forall \subseteq T$ if and only if every $\mathcal{M} \models T$ can be embedded in an $\mathcal{N} \models T'$. Analogously we have

**Lemma 10.2.2.** *The following are equivalent for all $L$-theories $T$ and $T'$.*
(i) $T'_{\forall\exists} \subseteq T$.
(ii) *For all $\mathcal{M} \models T$ and $\mathcal{N} \models T'$ there is $f : \mathcal{M} \xrightarrow{\forall} \mathcal{N}$. (Moreover, one can ensure $f = \mathrm{id}_M$, hence $\mathcal{N} \supseteq \mathcal{M}$.)*
(iii) *Every model of $T$ is existentially closed in some model of $T'$.*

Before we can prove this we need some preparation.

**Remarks.**
(1) If $M \subseteq N$, then $\mathrm{id}_M : \mathcal{M} \xrightarrow{\forall} \mathcal{N}$ is the same as saying that $\mathcal{M}$ is existentially closed in $\mathcal{N}$ (in the sense of §9.6).
(2) Since $f : \mathcal{M} \xrightarrow{\forall} \mathcal{N}$ is an embedding (as $\mathbf{qf} \subseteq \forall$), isomorphic correction yields an extension $\mathcal{N}' \cong \mathcal{N}$ of $\mathcal{M}$ in which $\mathcal{M}$ is existentially closed (whence the moreover clause in the above). Consequently, (ii) (with the clause in parentheses) is equivalent to (iii) in the lemma.

**Lemma 10.2.3.** *Let $\mathcal{M}$ and $\mathcal{N}$ be $L$-structures.*
*If $f : \mathcal{M} \xrightarrow{\forall} \mathcal{N}$, then $\mathcal{N} \Rightarrow_{\forall\exists} \mathcal{M}$.*

*Proof.* Assume $\mathcal{N} \models \varphi$, where $\varphi$ is the sentence $\forall \bar{x}\, \exists \bar{y}\, \psi(\bar{x}, \bar{y})$ and $\psi \in \mathbf{qf}$. To show: $\mathcal{M} \models \varphi$.

$\mathcal{N} \models \varphi$ says that $\mathcal{N} \models \exists \bar{y}\, \psi(\bar{b}, \bar{y})$ for all matching $\bar{b}$ from $N$. In particular, $\mathcal{N} \models \exists \bar{y}\, \psi(f[\bar{a}], \bar{y})$, i. e. $\mathcal{N} \not\models \forall \bar{y}\, \neg\psi(f[\bar{a}], \bar{y})$ for all matching $\bar{a}$ from $M$.

By hypothesis, we have $\mathcal{M} \not\models \forall \bar{y} \, \neg\psi(\bar{a}, \bar{y})$, hence $\mathcal{M} \models \exists \bar{y} \, \psi(\bar{a}, \bar{y})$, for all matching $\bar{a}$ from $M$, hence also $\mathcal{M} \models \varphi$. □

*Proof.* Of Lemma 10.2.2, $(i) \Longleftrightarrow (ii)$ (as the other equivalence was mentioned already in the remark above).

For $\Longleftarrow$, given $\mathcal{M} \models T$, we prove $\mathcal{M} \models T'_{\forall\exists}$. According to the hypothesis, choose $\mathcal{N} \models T'$ and $f : \mathcal{M} \xrightarrow{\vee} \mathcal{N}$. Then $\mathcal{M} \models \mathrm{Th}_{\forall\exists}(\mathcal{N})$ by Lemma 10.2.3, hence $\mathcal{M} \models T'_{\forall\exists}$.

For $\Longrightarrow$, assume $T'_{\forall\exists} \subseteq T$. It suffices to verify that, given $\mathcal{M} \models T$, the theory $T' \cup \mathrm{Th}_{\forall}(\mathcal{M}, M)$ is consistent (for if $\mathcal{N}^*$ is a model and $\mathcal{N}$ its $L$-reduct, then, on the one hand $\mathcal{N} \models T'$, and $f(a) = \underline{a}^{\mathcal{N}^*}$ defines the desired map from $\mathcal{M}$ to $\mathcal{N}$, on the other). As $\forall$ is closed under (finite) conjunction, we only need to check the consistency $T' \cup \{\forall \bar{x} \, \psi(\bar{x}, \bar{a})\}$, where $\bar{a}$ is from $M$, $\psi \in \mathbf{qf}$, and $\mathcal{M} \models \forall \bar{x} \, \psi(\bar{x}, \bar{a})$.

Assume this were not the case. Then $T' \models \exists \bar{x} \, \neg\psi(\bar{x}, \bar{a})$ and we may assume, without loss of generality, that the constants representing $\bar{a}$ are new, i. e. not in $L$ (or $T'$). Hence $T' \models \forall \bar{y} \, \exists \bar{x} \, \neg\psi(\bar{x}, \bar{y})$, whereby the $L$-sentence $\forall \bar{y} \, \exists \bar{x} \, \neg\psi(\bar{x}, \bar{y})$ lies in $T'_{\forall\exists}$. Then $\mathcal{M} \models T$ and $T'_{\forall\exists} \subseteq T$ yield $\mathcal{M} \models \forall \bar{y} \, \exists \bar{x} \, \neg\psi(\bar{x}, \bar{y})$, which contradicts $\mathcal{M} \models \forall \bar{x} \, \psi(\bar{x}, \bar{a})$. □

For the chain argument in the preservation theorem we need to know that we can always 'sandwich' a map of the kind above between an elementary map. We prove this next.

**Lemma 10.2.4.** *For all $f : \mathcal{M} \xrightarrow{\vee} \mathcal{N}$ there is $\mathcal{M}' \supseteq \mathcal{N}$ such that the composition map $\mathrm{id}_N f$ is elementary.*

*Proof.* By hypothesis, $(\mathcal{N}, f[M]) \models \mathrm{Th}_{\forall}(\mathcal{M}, M)$. Hence Lemma 6.2.2 yields an embedding $g$ from $(\mathcal{N}, f[M])$ in some $\mathcal{M}^* \models \mathrm{Th}(\mathcal{M}, M)$. Set $\mathcal{M}' = \mathcal{M}^* \upharpoonright L$. Consider the map $h : M \to M'$ given by $h(a) = \underline{a}^{\mathcal{M}^*}$ for $a \in M$. Then $\mathcal{M}^* = (\mathcal{M}', h[M])$, and the elementary diagram lemma (8.2.1) tells us that $h$ is an elementary map from $\mathcal{M}$ to $\mathcal{M}'$. Now $g : (\mathcal{N}, f[M]) \to (\mathcal{M}', h[M])$ implies $g(f(a)) = h(a)$ for all $a \in M$, whereby $gf = h$, and so $gf$ is elementary. Finally, by an isomorphic correction (Lemma 8.3.1) it can be achieved that $g = \mathrm{id}_N$. □

**Theorem 10.2.5.** (The Łoś-Suszko-Chang Preservation Theorem)
*A theory is inductive if and only if it is an $\forall\exists$-theory.*

[Łoś, J. and Suszko, R. : On the extending of models IV: Infinite sums of models, Fundamenta Math. **44** (1957) 52 – 60]

[Chang, C. C. : On unions of chains of models, Proceedings Am. Math. Soc. **10** (1959) 120 – 127]

*Proof.* The easy direction is treated in Lemma 10.2.1. For the converse, let $T$ be inductive. To prove $T \subseteq T_{\forall\exists}$, choose $\mathcal{M} \models T_{\forall\exists}$, and, using a chain argument, let us show that $\mathcal{M} \models T$.

Put $\mathcal{M}_0 = \mathcal{M}$. Since $T_{\forall\exists} \subseteq T_{\forall\exists}$, Lemma 10.2.2 yields $\mathcal{N}_0 \models T$ with $\mathcal{M}_0 \subseteq \mathcal{N}_0$ and $\mathrm{id}_{\mathcal{M}_0} : \mathcal{M}_0 \xrightarrow{\vee} \mathcal{N}_0$. Lemma 10.2.4, in turn, yields $\mathcal{M}_1 \supseteq \mathcal{N}_0$ with $\mathcal{M}_0 \preccurlyeq \mathcal{M}_1$. Successively we obtain a chain $\mathcal{M} = \mathcal{M}_0 \subseteq \mathcal{N}_0 \subseteq \mathcal{M}_1 \subseteq \mathcal{N}_1 \subseteq \ldots$, whose subchain $\mathcal{M}_0 \preccurlyeq \mathcal{M}_1 \preccurlyeq \mathcal{M}_2 \preccurlyeq \ldots$ is elementary. By the elementary chain theorem, $\mathcal{M}_\omega = \bigcup_{n<\omega} \mathcal{M}_n \succcurlyeq \mathcal{M}$. As $T$ is inductive, $\mathcal{M}_\omega = \bigcup_{n<\omega} \mathcal{N}_n \models T$, whereby $\mathcal{M} \models T$. $\qquad\square$

**Remark.** The proof shows that checking a theory for inductiveness it suffices to consider only countable chains (of order type $\omega$).

**Exercise 10.2.1.** Verify that the theory of groups in the signature $(1; \cdot)$ is an $\forall\exists$-theory, but not an $\forall$-theory. Do the same for DR, TF, ACF, $TF_q$, and $ACF_q$, for all $q$.

**Exercise 10.2.2.** Determine, which of the theories in §§5.2–6 are inductive.

**Exercise 10.2.3.** Show that the theory of divisible abelian groups introduced in §6.3 is inductive.

**Exercise 10.2.4.** Prove for an arbitrary theory $T$ that the models of $T_{\forall\exists}$ are exactly the substructures $\mathcal{M}$ of models $\mathcal{N}$ of $T$ for which $\mathrm{id}_M : \mathcal{M} \xrightarrow{\vee} \mathcal{N}$ (i. e., up to isomorphism, the structures that are existentially closed in models of $T$).

**Exercise 10.2.5.** Prove that every model $\mathcal{M}$ of an inductive theory $T$ is a substructure of a model $\mathcal{N}$ of $T$ such that for every $\exists$-sentence $\varphi$ from $L_0(M)$ we have: if $\mathcal{N} \subseteq \mathcal{N}' \models T$ and $\mathcal{N}' \models \varphi$, then $\mathcal{N} \models \varphi$. (Be aware that this does not yet mean that $\mathcal{N}$ is existentially closed, i. e. $\mathcal{N} \preccurlyeq_\exists \mathcal{N}'$ (which is the same as $(\mathcal{N}', \mathcal{N}) \Rrightarrow_\exists (\mathcal{N}, \mathcal{N})$), but only $(\mathcal{N}', \mathcal{M}) \Rrightarrow_\exists (\mathcal{N}, \mathcal{M})$.)

**Exercise 10.2.6.** Show that one can ensure that $|N| \leq |M| + |L|$ in the preceding exercise.

# 10.3*  Lyndon's preservation theorem

This section is devoted to a preservation theorem for homomorphic images. The proof involves a nested chain argument for which we prepare in the next two lemmas.

**Remark.** By analogy with Lemma 7.1.1, we have $f : \mathcal{M} \xrightarrow{\pm} \mathcal{N}$ if and only if $(\mathcal{M}, M) \Rrightarrow_+ (\mathcal{N}, f[M])$, if and only if $(\mathcal{N}, f[M]) \models \text{Th}_+(\mathcal{M}, M)$.

**Lemma 10.3.1.** *Let $\mathcal{M}$ and $\mathcal{N}$ be L-structures and $f : M \to N$.*
(1)  *If $f : \mathcal{M} \xrightarrow{\pm} \mathcal{N}$, then $f : \mathcal{M} \to \mathcal{N}$.*
(2)  *If $f : \mathcal{M} \to \mathcal{N}$ is a surjective homomorphism, then $f : \mathcal{M} \xrightarrow{\pm} \mathcal{N}$.*

*Proof.* (1) is clear, as **at** $\subseteq$ **+**. In order to prove (2) by induction on the complexity of positive formulas, note first that $f : \mathcal{M} \to \mathcal{N}$ iff $f : \mathcal{M} \xrightarrow{\text{at}} \mathcal{N}$ iff $f : \mathcal{M} \xrightarrow{\text{qf}\cap+} \mathcal{N}$. Hence we are left with the quantifier steps. That with the existential quantifier is clear again, as $f : \mathcal{M} \xrightarrow{\varphi} \mathcal{N}$ implies $f : \mathcal{M} \xrightarrow{\exists x\,\varphi} \mathcal{N}$. Thus it remains to show that $f : \mathcal{M} \xrightarrow{\varphi} \mathcal{N}$ implies $f : \mathcal{M} \xrightarrow{\forall x\,\varphi} \mathcal{N}$.

So let $\varphi \in L$ such that $\mathcal{M} \models \varphi(a, \bar{c}) \Rightarrow \mathcal{N} \models \varphi(f(a), f[\bar{c}])$ for all matching $a, \bar{c}$ from $M$. Assume further $\mathcal{M} \models \forall x\,\varphi(x, \bar{c})$. We have to show $\mathcal{N} \models \forall x\,\varphi(x, f[\bar{c}])$. But $\mathcal{M} \models \varphi(a, \bar{c})$ for all $a \in M$, hence $\mathcal{N} \models \varphi(f(a), f[\bar{c}])$ for all $a \in M$. The assertion now follows from the surjectivity of $f$.  $\square$

A theory $T$ is said to be **preserved** in (or **closed** under) **homomorphic images** if, along with every model $\mathcal{M}$ of $T$, every homomorphic image of $\mathcal{M}$ is a model of $T$.

From Lemma 10.3.1(2) we see that positive theories are preserved in homomorphic images. Again the converse is the hard direction, and we prove a technical lemma first. Recall from §2.4 that $-$ denotes the set of all negative (i. e. negated positive) formulas.

**Lemma 10.3.2.** *Let $\mathcal{M}$ and $\mathcal{N}$ be L-structures with $\mathcal{M} \Rrightarrow_+ \mathcal{N}$.*
(1)  *There are $\mathcal{N}' \succcurlyeq \mathcal{N}$ and $f : \mathcal{M} \xrightarrow{\pm} \mathcal{N}'$.*
(2)  *There are $\mathcal{M}' \succcurlyeq \mathcal{M}$ and $g : \mathcal{N} \xrightarrow{-} \mathcal{M}'$.*

*Proof.* We prove only (1) and leave the proof of (2) as an exercise. Choose, in a disjoint fashion, new constants for $L(M)$ and $L(N)$. All we have to do is show that $\text{Th}(\mathcal{N}, N) \cup \text{Th}_+(\mathcal{M}, M)$ is consistent. For if $\mathcal{N}'$ is the $L$-reduct of a model, we have $\mathcal{N} \xrightarrow{\equiv} \mathcal{N}'$; then setting $f(a) = \underline{a}^{\mathcal{N}^*}$ for all $a \in M$ yields $f : \mathcal{M} \xrightarrow{\pm} \mathcal{N}'$ and (after an appropriate isomorphic correction) also (1).

Now, to show the consistency of the above set, it is enough to show that of $\text{Th}(\mathcal{N}, N) \cup \{\varphi(\bar{a})\}$ for every $\varphi(\bar{a}) \in \text{Th}_+(\mathcal{M}, M)$ (as $\text{Th}_+(\mathcal{M}, M)$ is $\bigwedge$-closed). Assume, not. Then we get $\text{Th}(\mathcal{N}, N) \models \neg\varphi(\bar{a})$. By the initial disjoint choice, the constants $\underline{a}_i$ (where $a_i \in \bar{a}$) do not occur in $L(N)$. Hence $\text{Th}(\mathcal{N}, N) \models \forall \bar{x}\,\neg\varphi(\bar{x})$, and so $\mathcal{N} \models \neg\exists \bar{x}\,\varphi(\bar{x})$. But $\varphi \in \mathbf{+}$, hence also $(\exists \bar{x}\, \varphi) \in +$. Then $\mathcal{M} \Rrightarrow_+ \mathcal{N}$ gives $\mathcal{M} \models \neg \exists \bar{x}\, \varphi(\bar{x})$, which contradicts $\mathcal{M} \models \varphi(\bar{a})$. □

**Remark.** $g : \mathcal{N} \xrightarrow{-} \mathcal{M}'$ does not need to be a homomorphism from $\mathcal{N}$ to $\mathcal{M}'$, but solely a map preserving $-$.

**Theorem 10.3.3.** (Lyndon's Preservation Theorem)
*A theory is preserved in homomorphic images if and only if it is positive.*

[Lyndon, R. C. : Properties preserved under homomorphism, Pacific J. Math. **9** (1959) 143 – 154]

*Proof.* $\Longleftarrow$ is Lemma 10.3.1(2).
$\Longrightarrow$. We have to show $T \subseteq T_+$. By Lemma 6.2.8 this amounts to showing $\mathcal{N} \models T$, whenever $\mathcal{M} \models T$ and $\mathcal{N} \models \mathrm{Th}_+\mathcal{M}$. So suppose $\mathcal{M} \models T$ and $\mathcal{M} \Rrightarrow_+ \mathcal{N}$. In order to verify $\mathcal{N} \models T$ we form chains of models according to the following prescriptions. (Note that the asterisks indicate certain expansions by constants, but never change the universe of the structure under consideration, which is the reason that we omit the asterisks at the symbols for the underlying universes altogether.)
Start with
$$\mathcal{M}_0^{**} =_{\mathrm{def}} \mathcal{M} \quad \text{and} \quad \mathcal{N}_0^{***} =_{\mathrm{def}} \mathcal{N}.$$
In view of $\mathcal{M} \Rrightarrow_+ \mathcal{N}$ Lemma 10.3.2(1) yields
$$\mathcal{N}_1^* \succcurlyeq \mathcal{N}_0^{***} \quad \text{and} \quad f_0 : \mathcal{M}_0^{**} \xrightarrow{+} \mathcal{N}_1^*.$$
Next we expand $\mathcal{M}_0^{**}$ and $\mathcal{N}_1^*$ by new constants from $M_0$ (the universe of $\mathcal{M}$) and $f_0[M_0]$, respectively, and obtain the $L(M_0)$-structures
$$\mathcal{M}_0^{***} = (\mathcal{M}_0^{**}, M_0) \quad \text{and} \quad \mathcal{N}_1^{**} = (\mathcal{N}_1^*, f_0[M_0]).$$
Obviously, $\mathcal{M}_0^{***} \Rrightarrow_+ \mathcal{N}_1^{**}$, hence Lemma 10.3.2(2) yields
$$\mathcal{M}_1^* \succcurlyeq \mathcal{M}_0^{***} \quad \text{and} \quad g_1 : \mathcal{N}_1^{**} \xrightarrow{-} \mathcal{M}_1^*.$$
Expand again, this time to $L(M_0 \cup N_1)$-structures
$$\mathcal{M}_1^{**} = (\mathcal{M}_1^*, g_1[N_1]) \quad \text{and} \quad \mathcal{N}_1^{***} = (\mathcal{N}_1^{**}, N_1).$$
Again we have $\mathcal{M}_1^{**} \Rrightarrow_+ \mathcal{N}_1^{***}$ and find
$$\mathcal{N}_2^* \succcurlyeq \mathcal{N}_1^{***} \quad \text{and} \quad f_1 : \mathcal{M}_1^{**} \xrightarrow{+} \mathcal{N}_2^*,$$
and so on.
Altogether we get the following picture.

$$\mathcal{M} = \mathcal{M}_0^{**}$$

$$\mathcal{M}_0^{***} = (\mathcal{M}_0^{**}, M_0) \preccurlyeq \mathcal{M}_1^*$$

$$\mathcal{M}_1^{**} = (\mathcal{M}_1^*, g_1[N_1]) \qquad \cdots$$

$$\mathcal{M} = \mathcal{M}_0^{**} \; \overset{+}{\Longrightarrow} \; \mathcal{N} = \mathcal{N}_0^{***} \preccurlyeq \mathcal{N}_1^*$$

$$\overset{f_0 \; +}{\nearrow}$$

$$\overset{g_1 \; -}{\nwarrow}$$

$$\overset{f_1 \; +}{\nearrow}$$

$$\mathcal{M}_0^{***} \; \overset{+}{\Longrightarrow} \; \mathcal{N}_1^{**} = (\mathcal{N}_1^*, f_0[M_0])$$

$$\mathcal{M}_1^{**} \; \overset{+}{\Longrightarrow} \; \mathcal{N}_1^{***} = (\mathcal{N}_1^{**}, N_1) \preccurlyeq \mathcal{N}_2^* \quad \cdots$$

in $L$

in $L(M_0)$

in $L(M_0 \cup N_1)$

$$\cdots \; \mathcal{M}_k^{**}$$

$$\mathcal{M}_k^{***} = (\mathcal{M}_k^{**}, M_k) \preccurlyeq \mathcal{M}_{k+1}^*$$

$$\mathcal{M}_{k+1}^{**} = (\mathcal{M}_{k+1}^*, g_{k+1}[N_{k+1}]) \quad \cdots$$

$$\cdots \; \mathcal{N}_k^{***} \preccurlyeq \mathcal{N}_{k+1}^*$$

$$\overset{f_k \; +}{\nearrow}$$

$$\overset{g_{k+1} \; -}{\nwarrow}$$

$$\overset{f_{k+1} \; +}{\nearrow}$$

$$\mathcal{M}_k^{**} \; \overset{+}{\Longrightarrow} \; \mathcal{N}_{k+1}^{**} = (\mathcal{N}_{k+1}^*, f_k[M_k])$$

$$\mathcal{M}_{k+1}^{**} \; \overset{+}{\Longrightarrow} \; \mathcal{N}_{k+1}^{***} = (\mathcal{N}_{k+1}^{**}, N_{k+1}) \preccurlyeq \mathcal{N}_{k+2}^* \quad \cdots$$

in $L(M_{k-1} \cup N_k)$

in $L(M_k \cup N_k)$

in $L(M_k \cup N_{k+1})$

For all $k < \omega$ we have $\mathcal{M}_k^* \restriction L = \mathcal{M}_k^{**} \restriction L = \mathcal{M}_k^{***} \restriction L$. Denote this structure by $\mathcal{M}_k$ and define $\mathcal{N}_k$ analogously. Then we obtain two elementary chains, $\mathcal{M} = \mathcal{M}_0 \preccurlyeq \mathcal{M}_1 \preccurlyeq \mathcal{M}_2 \preccurlyeq \ldots \preccurlyeq \mathcal{M}_k \preccurlyeq \ldots$ and $\mathcal{N} = \mathcal{N}_0 \preccurlyeq \mathcal{N}_1 \preccurlyeq \mathcal{N}_2 \preccurlyeq \ldots \preccurlyeq \mathcal{N}_k \preccurlyeq \ldots$, and the maps $f_k : \mathcal{M}_k \overset{+}{\to} \mathcal{N}_{k+1}$ and $g_{k+1} : \mathcal{N}_{k+1} \overset{-}{\to} \mathcal{M}_{k+1}$ for all $k < \omega$. (Be aware that $g_k$ need not be a homomorphism.) Setting $\mathcal{M}_\omega = \bigcup_{k<\omega} \mathcal{M}_k$ and $\mathcal{N}_\omega = \bigcup_{k<\omega} \mathcal{N}_k$ the elementary chain theorem implies $\mathcal{M} \preccurlyeq \mathcal{M}_\omega$ and $\mathcal{N} \preccurlyeq \mathcal{N}_\omega$.

Consider the $L(M_{k-1} \cup N_k)$-structures

$$\mathcal{M}_k^{**} = (\mathcal{M}_k^*, M_0 \cup g_1[N_1] \cup M_1 \cup \ldots \cup M_{k-1} \cup g_k[N_k]) \quad \text{and}$$

$$\mathcal{N}_k^{***} = (\mathcal{N}_k^*, f_0[M_0] \cup N_1 \cup \ldots \cup f_{k-1}[M_{k-1}] \cup N_k).$$

The new constants correspond to each other: e. g., for $a \in M_1$, $b \in N_2$, and $k \geq 2$, we have $\underline{a}^{\mathcal{M}_k^{**}} = a$, $\underline{a}^{\mathcal{N}_k^{***}} = f_1(a)$, $\underline{b}^{\mathcal{M}_k^{**}} = g_2(b)$, and $\underline{b}^{\mathcal{N}_k^{***}} = b$. In view of $f_k : \mathcal{M}_k^{**} \to \mathcal{N}_{k+1}^*$ we also have $f_k(\underline{c}^{\mathcal{M}_k^{**}}) = \underline{c}^{\mathcal{N}_{k+1}^*}$, and, by $\mathcal{N}_k^{***} \preccurlyeq \mathcal{N}_{k+1}^*$, also $\underline{c}^{\mathcal{N}_{k+1}^*} = \underline{c}^{\mathcal{N}_k^{***}}$. Hence $f_k(\underline{c}^{\mathcal{M}_k^{**}}) = \underline{c}^{\mathcal{N}_k^{***}}$ for all $c \in M_{k+1}$.

In the case $c = a$ this yields $f_1(a) = \underline{a}^{\mathcal{N}_k^{***}} = f_k(\underline{a}^{\mathcal{M}_k^{**}}) = f_k(a)$, whence $f_1 \subseteq f_k$. More generally, one can show that $f_1 \subseteq f_2 \subseteq \ldots \subseteq f_k \subseteq \ldots$. As these are homomorphisms, setting $f_\omega = \bigcup_{k<\omega} f_k$ we obtain a homomorphism from $\mathcal{M}_\omega$ to $\mathcal{N}_\omega$.

In the case $c = b$ we analogously get $f_k(g_2(b)) = f_k(\underline{b}^{\mathcal{M}_k^{**}}) = \underline{b}^{\mathcal{N}_k^{***}} = b$, i. e. $f_k g_2 = \mathrm{id}_{N_2}$, and more generally $f_k g_k = \mathrm{id}_{N_k}$, whereby $f_\omega$ is surjective. But $\mathcal{M} \models T$, hence also $\mathcal{M}_\omega \models T$, and $T$ is preserved in homomorphic images. Thus $\mathcal{N}_\omega \models T$, and consequently $\mathcal{N} \models T$, as desired. □

A preservation theorem for strong homomorphisms (i. e. factor structures, see the fine print at the end of §1.3) can be found in

[Keisler, H. J. : Some applications of infinitely long formulas, J. Symbolic Logic **30** (1965) 339 – 349].

The literature of the 60s and 70s is full of preservation theorems, and it is hard to even get a complete picture. Quite a few can be found in the books of Chang & Keisler and Hodges.

**Exercise 10.3.1.** Prove Lemma 10.3.2(2).

**Exercise 10.3.2.** In the formulation of Lyndon's theorem it is essential that theories are consistent. Why?

**Exercise 10.3.3.** Show that the one-element $(0, 1; +, -, \cdot)$-structure (in which every term is evaluated by the same element) is the only homomorphic image of a ring (or a field) which is not a ring (resp., a field).

# Part IV

# Theories and types

A theory is a collection of *sentences* that prescribes certain properties to its models. Analogously, certain properties can be prescribed to *elements* of a model by demanding that they satisfy particular collections of *formulas* (or equivalently that they be contained in certain definable sets of the model). This leads to the concept of type and thus to a finer description of the models of a theory, types being the characteristics of elements very much the same way as theories are the characteristics of models (Chapter 11). (In a way the complete type of an element can be regarded as 'the complete theory of that element.') We may distinguish between two extremes of structures, those that realize a minimal amount of types and those that realize a maximal amount (Chapter 12). All this was first applied to countable models of countable theories. We present these applications in Chapter 13.

*Again $\sigma$ will be an arbitrary signature and $L = L(\sigma)$ the corresponding language, and $\mathcal{M}$ and $\mathcal{N}$ will be L-structures, unless we say otherwise.*

# Chapter 11

# Types

To explain what types are about we first consider an example, which is of interest in its own right.

## 11.1 The Prüfer group

As in §5.2, let $L_{\mathbb{Z}}$ denote the language of signature $(0; +, -)$.

Given a prime number $p$, consider the multiplicative group of all $p^n$th complex roots of unity where $n$ runs over the natural numbers. Written additively (i. e. as an $L_{\mathbb{Z}}$-structure) this group is the **Prüfer group** (for $p$); we denote it by $\mathbb{Z}_{p^\infty}$.

**Remarks.** Let us summarize a few obvious properties of $\mathbb{Z}_{p^\infty}$.

(1) $\mathbb{Z}_{p^\infty}$ is the union of the chain $\mathbb{Z}_p \subseteq \mathbb{Z}_{p^2} \subseteq \mathbb{Z}_{p^3} \subseteq \ldots \subseteq \mathbb{Z}_{p^n} \subseteq \ldots$ of cyclic subgroups of order $p^n$ ($n < \omega$) (where the latter are identified with the corresponding groups of roots of unity). (Hence $\mathbb{Z}_{p^\infty}$ is locally cyclic and locally finite, cf. 6.3.)

(2) $\mathbb{Z}_{p^\infty}$ is a periodic group (cf. §6.3). More concretely, for every $a \in \mathbb{Z}_{p^\infty}$ there is $n < \omega$ such that $p^n a = 0$.

(3) $\mathbb{Z}_{p^\infty}$ is not of **bounded exponent**, i. e., there is no $n < \omega$ annihilating all $a \in \mathbb{Z}_{p^\infty}$ (no uniform bound on the orders of its elements): for all $n < \omega$ we have $\mathbb{Z}_{p^\infty} \models \exists x\, (p^n x \neq 0)$.

Thus, in the language $L_{\mathbb{Z}}(c)$ (where $c$ is a new constant), the set of sentences $\text{Th}(\mathbb{Z}_{p^\infty}) \cup \{p^n c \neq 0 : n < \omega\}$ is consistent. Even $\text{Th}(\mathbb{Z}_{p^\infty}, \mathbb{Z}_{p^\infty}) \cup \{p^n c \neq 0 : n < \omega\}$ is consistent (in the expansion $L_{\mathbb{Z}}(\mathbb{Z}_{p^\infty} \cup \{c\})$). Hence there an element $a$ in an elementary extension $\mathcal{G} \succcurlyeq \mathbb{Z}_{p^\infty}$ such that $p^n a \neq 0$ for all $n < \omega$ (Lemmas 8.2.1 and 8.3.1). The set of formulas $\{p^n x \neq 0 : n < \omega\}$

163

we shall call a *type* of $\mathbb{Z}_{p^\infty}$ below (see §11.2). It is, more precisely, part of the *complete type* of the element $a$, which will be defined to be the set of all formulas (in the corresponding language) that are satisfied by $a$. Thus it really *is* the (first-order) 'type' of that element. Let us state this in greater generality.

**Remark.** Let $\mathcal{G}$ be a group, not of bounded exponent, i. e., $n\mathcal{G} \neq 0$ for all $n < \omega$. Then the 'type' $\{nx \neq 0 : 0 < n < \omega\}$ is satisfied (we shall say *realized*) in an elementary extension $\mathcal{G}'$ of $\mathcal{G}$, i. e., there is $g \in G'$ such that $ng \neq 0$ for all $n < \omega$.

We use the opportunity to completely describe the model class of (the theory of) $\mathbb{Z}_{p^\infty}$, i. e. all groups elementarily equivalent to $\mathbb{Z}_{p^\infty}$. (This doesn't have much to do with types yet, but is a nice exercise in classifying models of a theory, a central objective of model theory.)

First of all, $\mathbb{Z}_{p^\infty}$ is a divisible abelian group, since $\mathbb{Z}_{p^\infty} \models \forall x \, \exists y \, (x = ny)$ for all $0 < n < \omega$. Therefore all other models of $\mathrm{Th}(\mathbb{Z}_{p^\infty})$ must be divisible abelian groups too. Hence, by a well-known classification theorem for divisible abelian groups,[1] all the models have the form $\bigoplus_{q\in\mathbb{P}} \mathbb{Z}_{q^\infty}^{(\alpha_q)} \oplus \mathbb{Q}^{(\beta)}$, where we use $\mathcal{A}^{(\gamma)}$ to denote the direct sum $\bigoplus_{\alpha<\gamma} \mathcal{A}_\alpha$ of a family $\{\mathcal{A}_\alpha : \alpha < \gamma\}$, indexed by $\gamma \in \mathbf{On}$, of groups $\mathcal{A}_\alpha$ isomorphic to $\mathcal{A}$. (In common mathematical jargon, '$\mathcal{A}^{(\gamma)}$ is the direct sum of $\gamma$ copies of $\mathcal{A}$'; here one usually takes only cardinals $\gamma$, for the simple reason that $\mathcal{A}^{(\gamma)}$ is isomorphic to $\mathcal{A}^{(|\gamma|)}$.)

Further, the formula $\exists^{=p}x \, px = 0$ holds in $\mathbb{Z}_{p^\infty}$ and is thus contained in $\mathrm{Th}(\mathbb{Z}_{p^\infty})$. Let $\psi_p(x)$ denote the formula $px = 0$. Then $|\psi_p(\mathcal{G})| = p$ for every $\mathcal{G} \models \mathrm{Th}(\mathbb{Z}_{p^\infty})$. The formula $\psi_p$ can be pulled into the direct sum:

$$\psi_p\left(\bigoplus_{i<\gamma} \mathcal{A}_i\right) = \bigoplus_{i<\gamma} \psi_p(\mathcal{A}_i)$$

(cf. first exercise below). Note that $\psi_p(\mathbb{Z}_{p^\infty}) \cong \mathbb{Z}_p$ and $\psi_p(\mathbb{Q}) = \{0\}$, while $\psi_p(\mathbb{Z}_{q^\infty}) = \{0\}$ in case $p \neq q$. Hence we obtain

$$\psi_p\left(\bigoplus_{q\in\mathbb{P}} \mathbb{Z}_{q^\infty}^{(\alpha_q)} \oplus \mathbb{Q}^{(\beta)}\right) = \bigoplus_{q\in\mathbb{P}} \psi_p(\mathbb{Z}_{q^\infty})^{(\alpha_q)} \oplus \psi_p(\mathbb{Q})^{(\beta)} \cong \psi_p(\mathbb{Z}_{p^\infty})^{(\alpha_p)}.$$

Therefore we must have $\alpha_p = 1$ for all models $\mathcal{G} = \bigoplus_{q\in\mathbb{P}} \mathbb{Z}_{q^\infty}^{(\alpha_q)} \oplus \mathbb{Q}^{(\beta)}$ of $\mathrm{Th}(\mathbb{Z}_{p^\infty})$ (remember $|\psi_p(\mathcal{G})| = p$). If, on the other hand, $q \neq p$, then

---

[1]See the literature on abelian groups cited in Appendix F.

$\mathbb{Z}_{p^\infty} \models \forall x\, (qx = 0 \rightarrow x = 0)$. But, as mentioned before, $\psi_q(\mathbb{Z}_{q^\infty}) \neq \{0\}$, and we get $\alpha_q = 0$ for all models of $\mathrm{Th}(\mathbb{Z}_{p^\infty})$.

Thus the possible models of $\mathrm{Th}(\mathbb{Z}_{p^\infty})$ must have the form

$$\mathcal{G}_\beta =_{\mathrm{def}} \mathbb{Z}_{p^\infty} \oplus \mathbb{Q}^{(\beta)},$$

where $\beta \in \mathbf{On}$. We are going to show that all of these are indeed models of $\mathrm{Th}(\mathbb{Z}_{p^\infty})$. First of all, we need only consider *cardinals* $\beta$ here, for $\mathcal{G}_\beta \cong \mathcal{G}_\alpha$ if and only if $|\alpha| = |\beta|$. Given $\kappa \in \mathbf{Cn}$, the group $\mathcal{G}_\kappa$ has power $|\mathcal{G}_\kappa| = |\mathbb{Z}_{p^\infty}| \cdot |\mathbb{Q}^{(\kappa)}| = \max\{|\mathbb{Z}_{p^\infty}|, |\mathbb{Q}^{(\kappa)}|\} = \max\{\aleph_0, \kappa\}$ (cf. Exercise 7.6.2). Hence $|\mathcal{G}_\kappa| = \aleph_0$ if $\kappa \leq \aleph_0$, and $|\mathcal{G}_\kappa| = \kappa$ otherwise. In an uncountable cardinality $\kappa$ the theory $\mathrm{Th}(\mathbb{Z}_{p^\infty})$ can therefore have only $\mathcal{G}_\kappa$ as a model (up to isomorphism, as always). Since by Löwenheim-Skolem there must be a model of every power $\kappa > \aleph_0$, we conclude that $\mathcal{G}_\kappa \models \mathrm{Th}(\mathbb{Z}_{p^\infty})$ for all $\kappa > \aleph_0$.

Finally we show that $\mathcal{G}_\lambda \models \mathrm{Th}(\mathbb{Z}_{p^\infty})$ also for all $\lambda \leq \aleph_0$. We have $\psi_p(\mathcal{G}_\lambda) \cong \psi_p(\mathbb{Z}_{p^\infty})$ and $\psi_q(\mathcal{G}_\lambda) = \{0\}$, whenever $q \neq p$. Hence all models of $\mathrm{Th}\,\mathcal{G}_\lambda$ have also the form $\mathcal{G}_\kappa$. Then in any uncountable cardinality $\kappa$, this latter theory has only the model $\mathcal{G}_\kappa$. But this is also a model of $\mathrm{Th}(\mathbb{Z}_{p^\infty})$, whereby $\mathcal{G}_\lambda \equiv \mathcal{G}_\kappa \equiv \mathbb{Z}_{p^\infty}$. Consequently, $\mathcal{G}_\lambda \models \mathrm{Th}(\mathbb{Z}_{p^\infty})$.

We have proved

**Proposition 11.1.1.** *Up to isomorphism, the models of* $\mathrm{Th}(\mathbb{Z}_{p^\infty})$ *are exactly the groups* $\mathcal{G}_\kappa = \mathbb{Z}_{p^\infty} \oplus \mathbb{Q}^{(\kappa)}$, *where* $\kappa \in \mathbf{Cn}$. *These have cardinality* $|\mathcal{G}_\kappa| = \kappa + \aleph_0$. *Consequently,* $\mathrm{Th}(\mathbb{Z}_{p^\infty})$ *is uncountably categorical, but not countably categorical, as all the* $\mathcal{G}_\lambda$ *with* $\lambda \leq \aleph_0$ *are pairwise nonisomorphic countable models.* $\square$

**Exercise 11.1.1.** Recall the definition of positive primitive formula from p.45.
(1) Show that $\psi_p$ above is positive primitive.
(2) Prove that, in every abelian group $\mathcal{A}$, the set $\psi(\mathcal{A})$ defined by any positive primitive $L_{\mathbb{Z}}$-formula $\psi$ is a subgroup.
(3) Prove that $\psi(\bigoplus_{i<\gamma} \mathcal{A}_i) = \bigoplus_{i<\gamma} \psi(\mathcal{A}_i)$ for *every* positive primitive $L_{\mathbb{Z}}$-formula $\psi$ and any family of abelian groups $\mathcal{A}_i$ $(i < \gamma)$.
(4) Formulate and prove a similar result for direct *products* of abelian groups.

**Exercise 11.1.2.** Find all groups elementarily equivalent to $\mathbb{Z}_{p^\infty}^{(\lambda)}$, where $\lambda > 1$ is arbitrary but fixed.

**Exercise 11.1.3.** Prove that the complete theory of a Prüfer group has quantifier elimination.

The following is a special case of a much more general phenomenon, see Corollary 14.5.4 below (and the remarks following about the Baldwin-Lachlan theorem)

**Exercise 11.1.4.** Show that the canonical embeddings between the models $\mathcal{G}_\kappa$ of the theory $\mathrm{Th}(\mathbb{Z}_{p^\infty})$ are elementary and that the models of this theory thus form (up to isomorphism) an elementary chain $\mathbb{Z}_{p^\infty} = \mathcal{G}_0 \preccurlyeq \mathcal{G}_1 \preccurlyeq \ldots \preccurlyeq \mathcal{G}_n \preccurlyeq \ldots \preccurlyeq \mathcal{G}_{\aleph_0} \preccurlyeq \mathcal{G}_{\aleph_1} \preccurlyeq \ldots \preccurlyeq \mathcal{G}_\kappa \preccurlyeq \ldots$, where $\mathbb{Z}_{p^\infty}$ is a(n elementarily) prime model of $\mathrm{Th}(\mathbb{Z}_{p^\infty})$.

## 11.2 Types and their realization

Let $n < \omega$. An $n$-**type** of $\mathcal{M}$ is a set $\Phi \subseteq L_{\bar{x}}(M)$ which is simultaneously satisfied by an $n$-tuple $\bar{c}$ in some elementary extension $\mathcal{N} \succcurlyeq \mathcal{M}$, i. e., $\bar{c} \in N^n$ and $\mathcal{N} \models \Phi(\bar{c})$ (i. e. $\mathcal{N} \models \varphi(\bar{c})$ for every $\varphi \in \Phi$, cf. notation from §3.1[2]). Here $\bar{x}$ is an arbitrary $n$-tuple of variables. The tuple $\bar{c}$ is called a **realization** of the type $\Phi$ in $\mathcal{N}$. Sometimes we write $\bar{c} \models_{\mathcal{N}} \Phi$ instead of $\mathcal{N} \models \Phi(\bar{c})$, or just $\bar{c} \models \Phi$, if no confusion arises. $\Phi$ is also said to be **realized** (by $\bar{c}$) in $\mathcal{N}$.

If all parameters of the type $\Phi$ are in a subset $A \subseteq M$, i. e. if $\Phi \subseteq L_{\bar{x}}(A)$, then $\Phi$ is said to be (an $n$-type of $\mathcal{M}$) **over** $A$. If we need not specify the arity $n$, we just say **type**.

The usage of the term 'over' comes from field theory: it always indicates where the occurring parameters are located. We apply this to types as well as to formulas. So from now on we often prefer to say, a type or a formula is **over** $A$, as opposed to indicating its being a subset or an element of $L(A)$.

Clearly, every type of $\mathcal{M}$ is over $M$. Further, every type over $\mathcal{M}$ is also a type of every $\mathcal{M}' \succcurlyeq \mathcal{M}$ (formally one needs Corollary 8.4.5 to prove this), and every type of $\mathcal{M}$ over $A \subseteq M$ is also a type over every $\mathcal{M}' \preccurlyeq \mathcal{M}$ such that $A \subseteq M'$.

Note that a type of $\mathcal{M}$ is nothing more than a set of $L(M)$-formulas in finitely many variables[3] that is consistent with the elementary diagram of $\mathcal{M}$. Then it is clear from the finiteness theorem that a set $\Phi$ as above is a type of $\mathcal{M}$ if and only if every finite subset of $\Phi$ is a type of $\mathcal{M}$. The latter can be stated more simply by saying that the conjunction of any finite collection of formulas from $\Phi$ is satisfied in (some elementary extension of, hence in) $\mathcal{M}$. Thus a type of $\mathcal{M}$ is just a set of $L(M)$-formulas (in finitely many variables) that is **finitely satisfied** in $\mathcal{M}$.

We can say a little more if $\Phi$ is over a (proper) subset $A \subseteq M$ to begin with. Namely, then being consistent with $\mathrm{Th}(\mathcal{M}, M)$ is equivalent to being consistent with $\mathrm{Th}(\mathcal{M}, A)$ as an easy argument involving the lemma on new constants shows. (An alternative argument would consist in an application of Corollary 8.4.5 to the situation $(\mathcal{N}, B) \equiv (\mathcal{M}, A)$, where $\Phi$ is realized in

---

[2]One can also think of $\Phi(\bar{c})$ as $\{\varphi_{\bar{x}}(\bar{c}) : \varphi(\bar{x}) \in \Phi\}$.

[3]There is no problem—and in many cases it is very convenient—to consider types with infinitely many free variables. We refrain from doing so in this text.

$\mathcal{N}$, in order to obtain a realization of $\Phi$ in an elementary extension of $\mathcal{M}$.) We summarize this discussion in

**Proposition 11.2.1.** *Suppose $A \subseteq M$, $\bar{x}$ is an $n$-tuple of variables, $\Phi \subseteq L_{\bar{x}}(A)$, and $\bar{c}$ is an $n$-tuple of new constants (i. e., disjoint from $L_{\bar{x}}(A)$).*
*Then the following are equivalent.*
(i)   *$\Phi$ is a type of $\mathcal{M}$ (over $A$).*
(ii)   *$\mathcal{M} \models \exists \bar{x} \bigwedge_{i < k} \varphi_i(\bar{x})$ for all finite subsets $\{\varphi_0, \dots, \varphi_{k-1}\}$ of $\Phi$.*
(iii)   *$\mathrm{Th}(\mathcal{M}, M) \cup \Phi(\bar{c})$ is consistent.*
(iv)   *$\mathrm{Th}(\mathcal{M}, A) \cup \Phi(\bar{c})$ is consistent.* $\qquad\square$

(This neatly illustrates that there is no essential difference between variables and new constants.[4])

**Examples.**
(1)   $\{p^n x \neq 0 : n < \omega\}$ is a 1-type of $\mathbb{Z}_{p^\infty}$ over $\emptyset$. It is realized in $\mathcal{G}_\kappa$ exactly by the elements not lying in $\mathbb{Z}_{p^\infty}$ (cf. the analysis in the preceding section).
(2)   Given an infinite structure $\mathcal{M}$, the set $\{x \neq a : a \in M\}$ is a 1-type of $\mathcal{M}$ that is realized in an $\mathcal{N} \succ \mathcal{M}$ exactly by the elements that are not in $M$.
(3)   $\{x > n : n \in \omega\}$ is a 1-type of the standard model of Peano arithmetic.
(4)   Every Dedekind cut in $\eta$ (the countable model of DLO$_{--}$, see 8.3.5(2)) is a 1-type of $\eta$.

A **type** of a theory $T$ is, by definition, a type of some model of $T$; similarly for $n$-types. The first of the assertions following is immediate from the preceding proposition.

**Corollary 11.2.2.**
(1)   *Suppose $T$ is an arbitrary $L$-theory, $\bar{x}$ an $n$-tuple of variables, $\Phi \subseteq L_{\bar{x}}$, and $\bar{c}$ an $n$-tuple of new constants.*
  *$\Phi$ is a type of $T$ (over $\emptyset$) if and only if $T \cup \Phi(\bar{c})$ is consistent.*
(2)   *If $T$ is complete, $\Phi$ is a type of $\mathcal{M} \models T$ if and only if $\Phi$ is a type of any other $\mathcal{N} \models T$.*
(3)   *Every type of $\mathcal{M} \models T$ over $A \subseteq M$ can be realized in a model of $T$ of power $\leq |A| + |L|$. (In particular, every type over $\emptyset$ is realized in a model of power $\leq |L|$.)*
(4)   *Every type of $\mathcal{M}$ can be realized in an elementary extension of $\mathcal{M}$ of power $\leq |M| + |L|$.*

---

[4]See footnote on p.172.

*Proof.* (2) follows from $\mathrm{Th}(\mathcal{M}, \emptyset) = \mathrm{Th}(\mathcal{N}, \emptyset) = T$.

(3) follows from $|\mathrm{Th}(\mathcal{M}, A) \cup \Phi(\bar{c})| \leq |L_0(A)| + |L_0(A \cup \bar{c})| = |L| + |A| + |A \cup \bar{c}| = |L| + |A|$ (and Löwenheim-Skolem downward).

(4) is a special case of (3). $\qquad\qquad\qquad\qquad\qquad\qquad\qquad\qquad\qquad$ □

**Remark.** We see from this also that being a type of $\mathcal{M}$ over $A \subseteq M$ is the same as being a type of the theory $\mathrm{Th}(\mathcal{M}, A)$, which is, in turn, the same as being a type of any other $\mathcal{N}$ containing $A$, provided $(\mathcal{N}, A) \equiv (\mathcal{M}, A)$.

Next we want to extend the last assertion of the above corollary to arbitrary *sets* of types. In the proof we make use of the following concept.

Given a type $\Phi$ of $\mathcal{M}$ over $A \subseteq M$ and any map $f$ from $A$ to some set of parameters (typically some subset of a structure), we set $f(\Phi) = \{\varphi(\bar{x}, f[\bar{c}]) : \varphi(\bar{x}, \bar{c}) \in \Phi\}$. The set $f(\Phi)$ is called the $f$-**conjugate** of $\Phi$. As we see next, the conjugate of a type with respect to an elementary map is again a type—we therefore use also the term **conjugate type**.

**Lemma 11.2.3.** *Suppose* $\mathcal{M} \preccurlyeq \mathcal{N}$ *and* $f : \mathcal{N} \overset{\equiv}{\hookrightarrow} \mathcal{N}'$.

*If* $\Phi$ *is a type of* $\mathcal{M}$ *(over* $A \subseteq M$*) that is realized in* $\mathcal{N}$ *by some tuple* $\bar{a}$, *then the conjugate* $f(\Phi)$ *is a type of* $f(\mathcal{M}) \preccurlyeq \mathcal{N}'$ *(over* $f[A]$*) realized by* $f[\bar{a}]$.

*Proof.* The first assertion is immediate from the second, which in turn follows from the elementariness of $f$. $\qquad\qquad\qquad\qquad\qquad\qquad\qquad\qquad$ □

**Corollary 11.2.4.** *Suppose* $A \subseteq M, N$ *and* $f : \mathcal{M} \cong_A \mathcal{N}$.

*If* $\Phi$ *is a type of* $\mathcal{M}$ *over* $A$, *then* $\Phi$ *is also a type of* $\mathcal{N}$ *over* $A$. *Further,* $\bar{a} \models \Phi$ *if and only if* $f[\bar{a}] \models \Phi$, *for all matching* $\bar{a}$ *from* $M$.

*Proof.* One only has to take into account that isomorphisms are elementary and that $f(\Phi) = \Phi$ follows from $f \restriction A = \mathrm{id}_A$. $\qquad\qquad\qquad\qquad$ □

We turn now to the aforementioned extension of assertion (4) above.

**Proposition 11.2.5.** *If* $S$ *is an arbitrary set of types of* $\mathcal{M}$, *then there is an elementary extension* $\mathcal{M}_S \succcurlyeq \mathcal{M}$ *of a power* $\leq |\mathcal{M}| + |L| + |S|$ *realizing all types from* $S$.

*Proof.* To keep notation simple let $S$ consist of 1-types only.

The case $|S| = 1$ is assertion (4) of Corollary 11.2.2. It provides us with $\mathcal{M}_\Phi \succcurlyeq \mathcal{M}$ of power $|\mathcal{M}| + |L|$ and realizations $a_\Phi$ of $\Phi$ in $\mathcal{M}_\Phi$ for all $\Phi \in S$.

Using Corollary 8.4.5 choose $\mathcal{N}^* \models \mathrm{Th}(\mathcal{M}, M)$ with $(\mathcal{M}, M) \preccurlyeq \mathcal{N}^*$ and maps $f_\Phi : (\mathcal{M}_\Phi, M) \overset{\equiv}{\hookrightarrow} \mathcal{N}^*$ for all $\Phi \in S$. Then $f_\Phi \restriction M = \mathrm{id}_M$ (and this is the only reason for applying 8.4.5 to $\text{Th}(\mathcal{M}, M)$ rather than to $\text{Th}\,\mathcal{M}$). Reducing to the language $L$ we obtain $f_\Phi : \mathcal{M}_\Phi \overset{\equiv}{\hookrightarrow} \mathcal{N}$, where $\mathcal{N} = \mathcal{N}^* \restriction L$. Since $\Phi$ is over $M$ and $f_\Phi \restriction M = \text{id}_M$, we see that $f_\Phi(\Phi) = \Phi$ and thus $f_\Phi(a_\Phi) \models \Phi$ in $\mathcal{N}$ by Lemma 11.2.3.

Now all $\Phi \in S$ are realized in $\mathcal{N}$. Using Löwenheim-Skolem 8.4.1 (downward for $\preccurlyeq$) we choose $\mathcal{M}_S \preccurlyeq \mathcal{N}$ of power $|M \cup \{a_\Phi \; : \; \Phi \in S\}| + |L|$ containing $M \cup \{a_\Phi \; : \; \Phi \in S\}$. Then $\mathcal{M}_S$ realizes all types from $S$. Further, $\mathcal{M} \subseteq \mathcal{M}_S \preccurlyeq \mathcal{N}$ and $\mathcal{M} \preccurlyeq \mathcal{N}$ imply $\mathcal{M} \preccurlyeq \mathcal{M}_S$ (Lemma 8.2.2(2)). Finally we estimate $|\mathcal{M}_S| \leq |\mathcal{M}| + |L| + |S|$. $\qquad\square$

**Exercise 11.2.1.** Replace the application of Corollary 8.4.5 in the proof of Proposition 11.2.5 by a chain argument.

**Exercise 11.2.2.** Show that $\{x \neq 0\} \cup \{\exists y \; x = ny \; : \; 0 < n < \omega\}$ is a 1-type (over $\emptyset$) of the abelian group $(\mathbb{Z}; 0; +, -)$ of integers.

**Exercise 11.2.3.** Find, for every set of primes, $X$, a 1-type (over $\emptyset$) of the standard model of Peano arithmetic (cf. §3.5) whose realizations are divisible by a prime if and only if it is in $X$.

**Exercise 11.2.4.** Show that conjugation of types constitutes an equivalence relation on the class of all types.

## 11.3 Complete types and Stone spaces

Suppose $A \subseteq M$, $n < \omega$, and $\bar{x}$ is an $n$-tuple of variables.

An $n$-type $\Phi \subseteq L_{\bar{x}}(A)$ of $\mathcal{M}$ over $A$ is said to be **complete** over $A$ if for all $\varphi \in L_{\bar{x}}(A)$ (either) $\varphi \in \Phi$ or $\neg\varphi \in \Phi$. Given $\bar{a}$ in $M$, the set $\{\varphi \in L_{\bar{x}}(A) \; : \; \mathcal{M} \models \varphi(\bar{a})\}$ is called **the type** of $\bar{a}$ over $A$ in $\mathcal{M}$ and denoted by $\text{tp}^\mathcal{M}(\bar{a}/A)$. If the context $\mathcal{M}$ is understood, we omit the superscript $^\mathcal{M}$. We may also omit $A$ if it is empty: for $\text{tp}^\mathcal{M}(\bar{a}/\emptyset)$ we simply write $\text{tp}^\mathcal{M}(\bar{a})$ or $\text{tp}(\bar{a})$. Further we use the shorthand **(complete) type over** $\mathcal{M}$ to mean '(complete) type of $\mathcal{M}$ over $M$'.

The type $\text{tp}^\mathcal{M}(\bar{a}/A)$ is easily seen to be complete. It is therefore sometimes called the *complete* type of $\bar{a}$ over $A$ in $\mathcal{M}$; however, the usage of the definite article is precise enough. Note that the choice of the variables $\bar{x}$ is somewhat ambiguous in this definition. To be more precise, one would have to specify them. However, they really don't matter in most of the cases, all that matters usually is that they are in 1–1 correspondence with the entries of $\bar{a}$. And, to be really precise, one tacitly assumes such a correspondence as being fixed, usually the one given by the indices of the entries in $\bar{a}$ and $\bar{x}$. (This is analogous to the arbitrariness of the choice of free variables in formulas—or just any mathematical equations—and never really causes problems, as our focus is on semantics.)

Given an arbitrary $n$-type $\Psi$ of $\mathcal{M}$ over $A \subseteq M$ and a set $B \subseteq M$ containing $A$, a **complete extension** (or a **completion**) of $\Psi$ over $B$ is, by definition, a complete $n$-type of $\mathcal{M}$ over $B$ containing $\Psi$.

There is an easy way to find such completions. Just realize the type $\Psi$ by a tuple $\bar{a}$ in some elementary extension $\mathcal{N}$ of $\mathcal{M}$ and take $\mathrm{tp}^{\mathcal{N}}(\bar{a}/B)$. (Note, this latter is also a type of $\mathcal{M}$.) In fact, every complete type is of this form—just consider a realization. Hence, given two *distinct* complete $n$-types $\Phi$ and $\Psi$ over the same set $A$ in $\mathcal{M}$ (in the same variables), there must be an $L_n(A)$-formula in $\Phi$, whose negation lies in $\Psi$. (Here the variables do matter, but one can always rename them.)

The intuition behind the concept of type is this. In the same way as a theory describes its model class, a type describes the class of tuples realizing it. And a complete type, i. e. the type of some tuple $\bar{a}$ over some set $A$ in a structure $\mathcal{M}$, 'says' everything one can say using parameters from $A$ about this tuple in $\mathcal{M}$ in a first-order way.

If $\bar{y}$ is an $m$-subtuple of $\bar{x}$ (where $m \leq n$), then $L_{\bar{y}} \subseteq L_{\bar{x}}$, and every complete $n$-type over $A$ (a subset of $L_{\bar{x}}(A)$) contains a complete $m$-type over $A$ (a subset of $L_{\bar{y}}(A)$). In other 'words', if $\bar{b} \subseteq \bar{a}$ then $\mathrm{tp}^{\mathcal{M}}(\bar{b}/A) \subseteq \mathrm{tp}^{\mathcal{M}}(\bar{a}/A)$. (One could, alternatively, introduce $n$-types over $A$ as subsets of $L_n(A)$ and would then have the inclusion 'up to introduction of dummy variables', cf. §3.2.) The special case $m = 0$ is worth noticing; we make it the first of the following list of statements summarizing the preceding discourse.

**Remarks.** Let $n < \omega$.

(1) If $\Phi$ is a complete type of $\mathcal{M}$ over $A \subseteq M$, then $\mathrm{Th}(\mathcal{M}, A) \subseteq \Phi$. In particular, if $\Phi$ is complete over $\emptyset$, then $\mathrm{Th}\,\mathcal{M} \subseteq \Phi$. Thus $\mathrm{tp}^{\mathcal{M}}(\bar{a}) = \mathrm{tp}^{\mathcal{N}}(\bar{b})$ implies $\mathcal{M} \equiv \mathcal{N}$.

(2) $f : \mathcal{M} \to \mathcal{N}$ is elementary if and only if *every* tuple $\bar{a}$ from $M$ has the same type over $\emptyset$ as its image $f[\bar{a}]$. Moreover, given any $A \subseteq M$, the map $f : \mathcal{M} \to \mathcal{N}$ is elementary if and only if, for every $\bar{a}$ from $M$, we have $f(\mathrm{tp}^{\mathcal{M}}(\bar{a}/A)) = \mathrm{tp}^{\mathcal{N}}(f[\bar{a}]/f[A])$; in other words, for every tuple $\bar{a}$ from $M$, its image $f[\bar{a}]$ realizes the $f$-conjugate of $\mathrm{tp}^{\mathcal{M}}(\bar{a}/A)$. In particular, $\mathrm{tp}^{\mathcal{M}}(\bar{a}/A) = \mathrm{tp}^{\mathcal{N}}(\bar{a}/A)$, whenever $A \subseteq M$, $\mathcal{M} \preccurlyeq \mathcal{N}$, and $\bar{a}$ is an $n$-tuple from $M$.

(3) Suppose $A \subseteq M \cap N$, $\bar{b}$ is an $n$-tuple from $M$, and $\bar{c}$ is an $n$-tuple from $N$.

Then $\bar{c} \models \mathrm{tp}^{\mathcal{M}}(\bar{b}/A)$ if and only if $\mathrm{tp}^{\mathcal{M}}(\bar{b}/A) \subseteq \mathrm{tp}^{\mathcal{N}}(\bar{c}/A)$, if and only if $\mathrm{tp}^{\mathcal{M}}(\bar{b}/A) = \mathrm{tp}^{\mathcal{N}}(\bar{c}/A)$, if and only if $(\mathcal{M}, A, \bar{b}) \equiv (\mathcal{N}, A, \bar{c})$. (Notice the special case $A = \emptyset$.)

(4)  If the type $\Psi$ of $\mathcal{M}$ over $A$ is realized by the tuple $\bar{a}$ in $\mathcal{N} \succcurlyeq \mathcal{M}$, then $\mathrm{tp}^{\mathcal{N}}(\bar{a}/A)$ is a complete extension of $\Psi$ over $A$.

(5)  Every type of $\mathcal{M}$ over $A \subseteq M$ can be extended to a complete type over any set $B \subseteq M$ containing $A$. In particular, every type can be completed.

(6)  Every conjugate of a complete type (over $A$) is complete (over the image of $A$).

We have mentioned already that the complete types are exactly the types of the form $\mathrm{tp}^{\mathcal{M}}(\bar{a}/A)$. We state this—together with some other ways of looking at complete types—as a separate proposition.

**Proposition 11.3.1.** *Suppose $A \subseteq M$, $\bar{x}$ is an $n$-tuple of variables, and $\Phi \in L_{\bar{x}}(A)$.*

*Then the following are equivalent.*

(i)   *$\Phi$ is a complete $n$-type of $\mathcal{M}$ over $A$.*

(ii)  *There is $\mathcal{N} \succcurlyeq \mathcal{M}$ and an $n$-tuple $\bar{a}$ from $N$ such that $\Phi = \mathrm{tp}^{\mathcal{N}}(\bar{a}/A)$.*

(iii) *If $\bar{c}$ is an $n$-tuple of new constants, then $\Phi(\bar{c})$ is a complete $L(A \cup \bar{c})$-theory containing $\mathrm{Th}(\mathcal{M}, A)$.*

(iv)  *$\Phi$ is maximal (with respect to inclusion) among the $n$-types of $\mathcal{M}$ over $A$.*

*Proof.* This is the analogue for types of the corresponding result for complete theories, Lemma 3.5.1. In fact, it partly is a special case of the latter if we replace the variables $\bar{x}$ by constants $\bar{c}$ as mentioned. The details are left as an exercise (and the reader is strongly advised to take this seriously). □

From now on we usually don't specify the variables of a type, but just assume that they are chosen 'reasonably'. This applies in particular to the next definition (where we specify only the *number* of free variables, but not the variables themselves).

Let $A \subseteq M$. We use $\mathrm{S}_n^{\mathcal{M}}(A)$ to denote the set of all complete $n$-types of $\mathcal{M}$ over $A$ (in the same $n$ variables), i. e. $\mathrm{S}_n^{\mathcal{M}}(A) = \{\mathrm{tp}^{\mathcal{N}}(\bar{a}/A) : \mathcal{N} \succcurlyeq \mathcal{M}, \bar{a} \in N^n\}$. Given $\varphi \in L_n(A)$, we further use $\langle\varphi\rangle$ to denote the subset $\{\Phi \in \mathrm{S}_n^{\mathcal{M}}(A) : \varphi \in \Phi\}$. As $\mathrm{S}_n^{\mathcal{M}}(\emptyset) = \mathrm{S}_n^{\mathcal{N}}(\emptyset)$ for all models $\mathcal{M}$ and $\mathcal{N}$ of a complete theory $T$, we simply write $\mathrm{S}_n(T)$ for these. Their elements are called **complete types** of the (complete) theory $T$ (over the empty set).

As in §5.7, the sets $\langle\varphi\rangle$ form a basis of clopen sets and make $\mathrm{S}_n^{\mathcal{M}}(A)$ a topological space that is (not empty and) hausdorff, for, given two different $\Phi$ and $\Psi$ in $\mathrm{S}_n^{\mathcal{M}}(A)$, there is $\varphi \in L_n(A)$ such that $\varphi \in \Phi$ and $\neg\varphi \in \Psi$, i. e.

such that $\Phi \in \langle \varphi \rangle$ and $\Psi \in \langle \neg \varphi \rangle$. We further have $\langle \neg \varphi \rangle = S_n^{\mathcal{M}}(A) \smallsetminus \langle \varphi \rangle$. In due course we will see that $S_n^{\mathcal{M}}(A)$ is also compact, in fact, it will turn out to be the Stone space of the $n$th Lindenbaum-Tarski algebra of the theory $\mathrm{Th}(\mathcal{M}, A)$.

Let $\bar{c}$ be an $n$-tuple of new constants. Then $\Phi(\bar{x}) \mapsto \Phi(\bar{c})$ defines a map between the $n$-types over $\emptyset$ and the consistent subsets of $L_0(\bar{c})$, where, by Corollary 11.2.2, the $n$-types of any complete $L$-theory $T$ get mapped to the subsets of $L_0(\bar{c})$ that are consistent with $T$. Further, the **deductively closed** $n$-types of $T$ (i. e. the types that contain $\top$ and, along with every $\varphi$, also contain every $\psi \geq_T \varphi$, cf. notation from §5.6) get mapped to the $L(\bar{c})$-theories extending $T$. Finally, the complete $n$-types over $\emptyset$ of $T$ are mapped to the complete $L(\bar{c})$-extensions of $T$, i. e. to the elements of $S_{L(\bar{c})}$ which contain $T$ (cf. §5.7; notice, the set of all these latter elements is nothing more than the closed set $\bigcap_{\varphi \in T} \langle \varphi \rangle$).

It is not hard to see that this map is a homeomorphism. For this reason we often identify $\Phi(\bar{x})$ and $\Phi(\bar{c})$.[5] This means that, given an $L$-theory $T$, we identify

| | | |
|---|---|---|
| $n$-types over $\emptyset$ of $T$ | and | subsets of $L_0(\bar{c})$ consistent with $T$, |
| deductively closed $n$-types over $\emptyset$ of $T$ | and | $L(\bar{c})$-theories containing $T$, |
| complete $n$-types over $\emptyset$ of $T$ | and | complete $L(\bar{c})$-extensions of $T$. |

If $T$ is complete, we identify

$$S_n(T) \quad \text{and} \quad \{T' \in S_{L(\bar{c})} : T \subseteq T'\}.$$

(Here $S_0(T)$ corresponds to the singleton $\{T\}$.)

$n$-types over $\emptyset$ can also be regarded as filters in the $n$th Lindenbaum-Tarski algebra. For this recall the case $n = 0$ first: from §5.6 we know that theories are deductively closed and consistent subsets, i. e. filters in in $\mathcal{B}_L$, the (0th) Lindenbaum-Tarski algebra of $L$. A set of $L$-sentences $T$ thus is a theory if it satisfies

(i)  if $\varphi \in T$ and $\varphi \leq \psi$, then $\psi \in T$,

(ii)  if $\varphi, \psi \in T$ then $\varphi \wedge \psi \in T$, and

(iii)  $\bot \notin T$,

for all $\varphi, \psi \in L_0$. And the complete theories are precisely the ultrafilters in $\mathcal{B}_L$.

Generalizing this to arbitrary $n$, we see that $\Phi \mapsto \{\varphi / \sim_T : \varphi \in \Phi\}$ determines a map that brings

---

[5]See footnote on p.167.

| | | |
|---|---|---|
| $n$-types of $T$ over $\emptyset$ | to | subsets of $\mathcal{B}_n(T)$ not containing $\bot/\sim_T$, |
| deductively closed $n$-types of $T$ over $\emptyset$ | to | filters of $\mathcal{B}_n(T)$, |
| complete $n$-types of $T$ over $\emptyset$ | to | ultrafilters of $\mathcal{B}_n(T)$. |

The Stone representation theorem (5.6.1) says that the ultrafilters of $\mathcal{B}_n(T)$ form a compact hausdorff space, the Stone space of $\mathcal{B}_n(T)$. In case, $T$ is complete, this space is homeomorphic to $S_n(T)$, and we identify the two. (Here $\mathcal{B}_0(\emptyset)$ is nothing more than the Lindenbaum-Tarski algebra $\mathcal{B}_L$, whose (ultra) filters correspond to the (complete) $L$-theories and whose Stone space is $S_L$, cf. Exercise 5.6.6.)

For a structure $\mathcal{M}$ and a subset $A \subseteq M$ the set $S_n^{\mathcal{M}}(A)$ obviously is just the Stone space $S_n(\text{Th}(\mathcal{M}, A))$ of $\mathcal{B}_n(\text{Th}(\mathcal{M}, A))$, which shows that it is a compact hausdorff space.

After this lengthy discourse about the formal nature of types let us finally turn to some of their more vivid features.

**Remarks.** Suppose $T$ is a complete theory, $\mathcal{M} \models T$, and $A \subseteq M$.

(7) A deductively closed $n$-type $\Phi$ of $\mathcal{M}$ over $A$ is complete if and only if $\varphi \vee \psi \in \Phi$ implies $\varphi \in \Phi$ or $\psi \in \Phi$, for all $\varphi, \psi \in L_n(A)$.

(8) If every formula from $L_n(A)$ is $\text{Th}(\mathcal{M}, A)$-equivalent to a boolean combination of formulas from a class $\Delta$, then $\Delta \cap \text{tp}^{\mathcal{M}}(\bar{a}/A) = \Delta \cap \text{tp}^{\mathcal{M}}(\bar{b}/A)$ implies $\text{tp}^{\mathcal{M}}(\bar{a}/A) = \text{tp}^{\mathcal{M}}(\bar{b}/A)$ (for any $\bar{a}$ and $\bar{b}$ from $M^n$).

(9) Suppose $\Delta$ is a class of formulas that is closed under substitution by (possibly new) constants.
If $T$ admits $\Delta$-elimination, then $\Delta \cap \text{tp}^{\mathcal{M}}(\bar{a}/A) = \Delta \cap \text{tp}^{\mathcal{M}}(\bar{b}/A)$ implies $\text{tp}^{\mathcal{M}}(\bar{a}/A) = \text{tp}^{\mathcal{M}}(\bar{b}/A)$ (for all tuples $\bar{a}$ and $\bar{b}$ from $M$ of equal length).

(10) If $T$ has quantifier elimination, any two complete $n$-types over $A$ are equal, whenever they contain the same atomic formulas.

We want to convince ourselves that the converses of the latter three remarks are true too. In fact, up to the new terminology, we have shown this already in Chapter 9. To see this, first another bit of new notation.

Let $\Delta$ be any class of formulas. A $\Delta$-**type** is a type containing only formulas from $\Delta$. A $\Delta$-$n$-**type** is a $\Delta$-type which is also an $n$-type. The $\Delta$-**part** of the type $\text{tp}^{\mathcal{M}}(\bar{a}/A)$ is, by definition, the $\Delta$-type $\Delta \cap \text{tp}^{\mathcal{M}}(\bar{a}/A)$, which we denote by $\text{tp}_\Delta^{\mathcal{M}}(\bar{a}/A)$.

As announced in §9.1, we are now able to give nicer formulations of Proposition 9.1.2 and Corollary 9.1.3. These also constitute the converses mentioned above.

**Proposition 9.1.2′.** *Suppose* $\Sigma \subseteq L_0$, $\varphi \in L_n$, *and* $\Delta$ *is a class of formulas with boolean closure* $\widetilde{\Delta}$.

*The formula* $\varphi$ *is* $\Sigma$-*equivalent to* $\top, \bot$, *or a formula from* $\widetilde{\Delta} \cap L_n$ *if and only if, for all* $\mathcal{M}, \mathcal{N} \models \Sigma$ *and all* $\bar{a} \in M^n$ *and* $\bar{b} \in N^n$, *the following statement is true.*

*If* $\mathrm{tp}_{\Delta}^{\mathcal{M}}(\bar{a}) = \mathrm{tp}_{\Delta}^{\mathcal{N}}(\bar{b})$, *then* $\mathcal{M} \models \varphi(\bar{a})$ *iff* $\mathcal{N} \models \varphi(\bar{b})$.

**Corollary 9.1.3′.** *Let* $T$ *be an* $L$-*theory and* $\Delta$ *a class of formulas containing* $\top$ *and* $\bot$ *with boolean closure* $\widetilde{\Delta}$.

$T$ *has* $\widetilde{\Delta}$-*elimination if* $\mathrm{tp}_{\Delta}^{\mathcal{M}}(\bar{a}) = \mathrm{tp}_{\Delta}^{\mathcal{N}}(\bar{b})$ *implies* $\mathrm{tp}^{\mathcal{M}}(\bar{a}) = \mathrm{tp}^{\mathcal{N}}(\bar{b})$, *for all* $\mathcal{M}, \mathcal{N} \models T$, *and all* $\bar{a} \in M^n$ *and* $\bar{b} \in N^n$.

**Exercise 11.3.1.** Prove all remarks and Proposition 11.3.1.

**Exercise 11.3.2.** Describe all types in $S_1^{\eta}(\eta)$ (as before $\eta$ is the countable model of DLO$_{--}$).

**Exercise 11.3.3.** Let $\mathcal{M}$ be a structure and $A \subseteq B \subseteq M$.

Show that $\mathrm{tp}(\bar{a}/B) \mapsto \mathrm{tp}(\bar{a}/A)$, where $\bar{a}$ runs over the $n$-tuples in elementary extensions of $\mathcal{M}$, defines a continuous map from $S_n^{\mathcal{M}}(B)$ onto $S_n^{\mathcal{M}}(A)$, called **restriction** onto $A$ and denoted by $\rho_{B,A}$.

The series of exercises following deals with continuous sections of this restriction map in a situation as above, but with $A$ a model, where a **section** of a map $f : X \to Y$ is, as usual, a map $g : Y \to X$ such that $fg = \mathrm{id}_Y$. So suppose, for the remaining exercises, $\mathcal{N} \preccurlyeq \mathcal{M}$ and $N \subseteq B \subseteq M$. Note that *any* section of $\rho_{B,N}$ brings an algebraic type $\mathrm{tp}(\bar{a}/N)$ (where $\bar{a}$ is in $\mathcal{N}$) to $\mathrm{tp}(\bar{a}/B)$: the former contains the formula $\bar{x} = \bar{a}$, hence, being an extension, its image also has to contain it.

In the next exercise we deal with material due to Daniel Lascar and Bruno Poizat (see their books). Given a type $\Phi \in S_n^{\mathcal{M}}(N)$, a **coheir** of $\Phi$ over $B$ is a type $\Psi \in S_n^{\mathcal{M}}(B)$ extending $\Phi$ which is **finitely satisfied** in $\mathcal{N}$, i. e., every finite subset of $\Psi$ is satisfied in $\mathcal{N}$ (more precisely—as $\Psi$ contains parameters not in $\mathcal{N}$— satisfied in $\mathcal{M}$ by some tuple from $N$). Note that this is equivalent to saying that every *single* formula in $\Psi$ is satisfied in $\mathcal{N}$, for $\Psi$ is complete and thus closed under taking finite conjunctions.

**Exercise 11.3.4.** Prove that the set of types from $S_n^{\mathcal{M}}(B)$ that are coheirs of their restrictions to $N$ is the closure in $S_n^{\mathcal{M}}(B)$ of the set $\{\mathrm{tp}(\bar{a}/B) : \bar{a} \in N^n\}$.

The remainder of this section contains material from unpublished joint work of Ivo Herzog and the author.

**Exercise 11.3.5.** Show that there can be at most one continuous section of $\rho_{B,N}$.

**Exercise 11.3.6.** Prove that if $\sigma : S_n^{\mathcal{M}}(N) \to S_n^{\mathcal{M}}(B)$ is a continuous section of $\rho_{B,N}$, then, for each $\Phi \in S_n^{\mathcal{M}}(N)$, the type $\sigma(\Phi)$ is a coheir of $\Phi$.

**Exercise 11.3.7.** Show that the image of $S_n^{\mathcal{M}}(N)$ under a continuous section of $\rho_{B,N}$ contains the set of types from $S_n^{\mathcal{M}}(B)$ that are coheirs of their restrictions to $N$. Derive that, if there is such a continuous section, every type in $S_n^{\mathcal{M}}(N)$ has (at most) one coheir.

This yields a new definition of stability (see §15.4 below): a theory is stable if and only if , for any two models $\mathcal{N} \preccurlyeq \mathcal{M}$, the restriction map $\rho_{M,N}$ has a (n automatically unique) continuous section. One direction follows from the last exercise and the known fact that uniqueness of coheirs implies stability, see Théorème 12.29 in Poizat's *Cours*. The converse is mentioned explicitly before Corollaire 16.07 of the same text.

# 11.4   Isolated and algebraic types

Let $\bar{x}$ be an $n$-tuple of variables. An $n$-type $\Phi \subseteq L_{\bar{x}}(M)$ of $\mathcal{M}$ is said to be **isolated** over $A$ (respectively, **principal**) if there is $\varphi \in L_{\bar{x}}(A)$ (respectively, $\varphi \in \Phi$) satisfiable in $\mathcal{M}$ such that $\mathcal{M} \models \forall \bar{x}\,(\varphi \to \psi)$ for all $\psi \in \Phi$. In this case we say $\varphi$ **isolates** $\Phi$. A complete type over $A$ is said to be **isolated** if it is isolated over $A$.

Extend the notation $\varphi \leq_T \psi$ from §5.6 to (possibly infinite) sets on the right hand side in the obvious way: $\varphi \leq_T \Psi$ is to mean that $\varphi \leq_T \psi$ for every $\psi \in \Psi$. (We omit the braces around $\psi$ in $\varphi \leq_T \{\psi\}$ so that this is compatible with our old notation from §5.6.) Further, instead of $\leq_{\mathrm{Th}(\mathcal{M},M)}$ we simply write $\leq_{\mathcal{M}}$. This notation is handy for isolation, for $\varphi \leq_{\mathcal{M}} \Phi$ just means that $\varphi$ isolates $\Phi$ (where $\varphi$ and $\Phi$ are as above). We may omit the subscript $T$ or $\mathcal{M}$ if it is clear from the context.

**Remarks.** Let $A \subseteq M$ and $n < \omega$.

(1)   Let $\bar{x}$ be an $n$-tuple of variables.
      A formula $\varphi \in L_{\bar{x}}$ satisfiable in $\mathcal{M}$ isolates the $n$-type $\Phi \subseteq L_{\bar{x}}(M)$ if and only if $\bar{a} \models_{\mathcal{N}} \Phi$ for all $\mathcal{N} \succcurlyeq \mathcal{M}$ and all $\bar{a} \in \varphi(\mathcal{N})$.

(2)   If we consider complete types as ultrafilters, the principal complete types are precisely the principal ultrafilters (whence the name).

(3)   Every principal type over $A$ is isolated over $A$. For complete types over $A$ also the converse holds. Moreover, a complete $n$-type $\Phi$ of $\mathcal{M}$ over $A$ is isolated iff it is principal iff it is an isolated point (whence the name) in the Stone space $S_n^{\mathcal{M}}(A)$ (for $\varphi \leq_{\mathcal{M}} \Phi$ iff $\langle \varphi \rangle = \{\Phi\}$) iff it contains a formula whose equivalence class in the Lindenbaum-Tarski algebra $\mathcal{B}_n(\mathrm{Th}(\mathcal{M},A))$ is an atom (cf. §5.6)).

(4)   Every isolated type of $\mathcal{M}$ is realized in $\mathcal{M}$ (for $\varphi \leq_{\mathcal{M}} \Phi$ and $\mathcal{M} \models \exists \bar{x}\,\varphi$ yield a realization of $\Phi$ in $\mathcal{M}$).

(5)   Given a type $\Phi$ over $\emptyset$ of $\mathcal{M}$ and $\varphi \in L$, we have $\varphi \leq_{\mathcal{M}} \Phi$ iff $\varphi \leq_{\mathrm{Th}\,\mathcal{M}} \Phi$.
Thus isolation of types of $\mathcal{M}$ over the empty set does not depend on
$\mathcal{M}$, but only on $\mathrm{Th}\,\mathcal{M}$; similarly for types over $A \subseteq M$ and $\mathrm{Th}(\mathcal{M}, A)$.

A basic property of isolation is its transitivity, which we need in the next
chapter.

**Lemma 11.4.1.** (Transitivity of Isolation)
*Suppose $\bar{a}$ and $\bar{b}$ are tuples from some structure (and $\bar{a}{}^{\frown}\bar{b}$ their concatenation).*
*Then $\mathrm{tp}(\bar{a}{}^{\frown}\bar{b})$ is isolated if and only if $\mathrm{tp}(\bar{a}/\bar{b})$ and $\mathrm{tp}(\bar{b})$ are isolated.*

*Proof.* Let $\bar{a}$ be an $m$- and $\bar{b}$ an $n$-tuple. Correspondingly, let $\bar{x}$ be an $m$-
and $\bar{y}$ an $n$-tuple of variables such that $\mathrm{tp}(\bar{a}{}^{\frown}\bar{b}) \subseteq L_{\bar{x}{}^{\frown}\bar{y}}$. Regard $\mathrm{tp}(\bar{a}/\bar{b})$ as
a subset of $L_{\bar{x}}(\bar{b})$ and $\mathrm{tp}(\bar{b})$ as a subset of $L_{\bar{y}}$.

$\Longrightarrow$. Let $\psi(\bar{x}, \bar{y}) \leq \mathrm{tp}(\bar{a}{}^{\frown}\bar{b})$ (where the context is omitted as it stays
unchanged). We are going to show that $\psi(\bar{x}, \bar{b}) \leq \mathrm{tp}(\bar{a}/\bar{b})$ and $\exists \bar{x}\,\psi(\bar{x}, \bar{y}) \leq$
$\mathrm{tp}(\bar{b})$.

Since, by hypothesis, $\mathrm{tp}(\bar{a}{}^{\frown}\bar{b})$ is isolated by $\psi$, $\mathcal{M} \models \psi(\bar{a}', \bar{b}')$ implies
$\mathrm{tp}(\bar{a}'{}^{\frown}\bar{b}') = \mathrm{tp}(\bar{a}{}^{\frown}\bar{b})$. This immediately yields $\psi(\bar{x}, \bar{b}) \leq \mathrm{tp}(\bar{a}/\bar{b})$: if $\mathcal{M} \models$
$\psi(\bar{a}', \bar{b})$, then $\mathcal{M} \models \theta(\bar{a}', \bar{b})$ for all $\theta(\bar{x}, \bar{y}) \in \mathrm{tp}(\bar{a}{}^{\frown}\bar{b})$, hence also for all
$\theta(\bar{x}, \bar{b}) \in \mathrm{tp}(\bar{a}/\bar{b})$. This proves the first of the two assertions to prove. For
the other notice that every $\varphi(\bar{y}) \in \mathrm{tp}(\bar{b})$ is contained in $\mathrm{tp}(\bar{a}{}^{\frown}\bar{b})$. Then we
have $\mathcal{M} \models \forall \bar{x} \bar{y}\,(\psi(\bar{x}, \bar{y}) \rightarrow \varphi(\bar{y}))$, hence also $\mathcal{M} \models \forall \bar{y}\,(\exists \bar{x}\,\psi(\bar{x}, \bar{y}) \rightarrow \varphi(\bar{y}))$,
i. e. $\exists \bar{x}\,\psi(\bar{x}, \bar{y}) \leq \mathrm{tp}(\bar{b})$.

$\Longleftarrow$. Suppose $\psi(\bar{x}, \bar{b}) \leq \mathrm{tp}(\bar{a}/\bar{b})$ and $\varphi(\bar{y}) \leq \mathrm{tp}(\bar{b})$. We claim $\psi(\bar{x}, \bar{y}) \wedge$
$\varphi(\bar{y}) \leq \mathrm{tp}(\bar{a}{}^{\frown}\bar{b})$, i. e., given $\mathcal{M} \models \psi(\bar{a}', \bar{b}') \wedge \varphi(\bar{b}')$, we have to show $\mathrm{tp}(\bar{a}'{}^{\frown}\bar{b}') =$
$\mathrm{tp}(\bar{a}{}^{\frown}\bar{b})$, i. e. $(\mathcal{M}, \bar{a}', \bar{b}') \equiv (\mathcal{M}, \bar{a}, \bar{b})$.

Now $\varphi \leq \mathrm{tp}(\bar{b})$ and $\mathcal{M} \models \varphi(\bar{b}')$ imply $\mathrm{tp}(\bar{b}') = \mathrm{tp}(\bar{b})$, hence

(*)   $(\mathcal{M}, \bar{b}') \equiv (\mathcal{M}, \bar{b})$.

Consider the set $\Phi = \{\theta(\bar{x}, \bar{b}) : \theta(\bar{x}, \bar{b}') \in \mathrm{tp}(\bar{a}'/\bar{b}')\}$. Since $\mathrm{tp}(\bar{a}'/\bar{b}')$ is
closed under finite conjunction, so is $\Phi$. Thus, by Proposition 11.2.1, $\Phi$ is
a type of $\mathcal{M}$, whenever $\mathcal{M} \models \exists \bar{x}\,\theta(\bar{x}, \bar{b})$ for all $\theta(\bar{x}, \bar{b}) \in \Phi$. But the latter
follows from $\mathcal{M} \models \exists \bar{x}\,\theta(\bar{x}, \bar{b}')$ and (*).

Thus $\Phi$ is a type of $\mathcal{M}$ realized by some $\bar{c}$ in some $\mathcal{N} \succcurlyeq \mathcal{M}$. The
completeness of $\mathrm{tp}(\bar{a}'/\bar{b}')$ entails that of $\Phi$, whence $\Phi = \mathrm{tp}^{\mathcal{N}}(\bar{c}/\bar{b})$.

For all $\theta(\bar{x}, \bar{b}') \in \mathrm{tp}(\bar{a}'/\bar{b}')$ we have $\mathcal{N} \models \theta(\bar{c}, \bar{b})$. Hence $\mathcal{M} \models \theta(\bar{a}', \bar{b}')$
implies $\mathcal{N} \models \theta(\bar{c}, \bar{b})$, and we get the converse on taking negations. Thus
$(\mathcal{M}, \bar{a}', \bar{b}') \equiv (\mathcal{N}, \bar{c}, \bar{b})$.

Since $\equiv$ is transitive, we are left with showing that $(\mathcal{N}, \bar{c}, \bar{b}) \equiv (\mathcal{M}, \bar{a}, \bar{b})$,
or equivalently that $\mathrm{tp}^{\mathcal{N}}(\bar{c}/\bar{b}) = \mathrm{tp}^{\mathcal{M}}(\bar{a}/\bar{b})$. As $\psi(\bar{x}, \bar{b}') \in \mathrm{tp}^{\mathcal{M}}(\bar{a}'/\bar{b}')$, we

have $\psi(\bar{x}, \bar{b}) \in \Phi$, hence $\psi(\bar{x}, \bar{b}) \in \mathrm{tp}^{\mathcal{N}}(\bar{c}/\bar{b})$. But we also have $\psi(\bar{x}, \bar{b}) \leq \mathrm{tp}(\bar{a}/\bar{b})$, hence $\mathrm{tp}^{\mathcal{M}}(\bar{a}/\bar{b}) \subseteq \mathrm{tp}^{\mathcal{N}}(\bar{c}/\bar{b})$. Inasmuch as the type $\mathrm{tp}^{\mathcal{M}}(\bar{a}/\bar{b})$ is complete, we in fact have equality here, as desired. $\square$

Field theory inspired the introduction of a particular kind of isolated type, which we investigate in the remainder of this section.

A type is said to be **algebraic** if it has only finitely many realizations. A formula $\varphi \in L_n(M)$ is **algebraic** in $\mathcal{M}$ if it is either unsatisfiable in $\mathcal{M}$ or else $\{\varphi\}$ is an algebraic type of $\mathcal{M}$.

Note that $\varphi$ is algebraic in $\mathcal{M}$ if and only if $\varphi(\mathcal{M})$ is finite. The direction from left to right is clear. For the converse just notice that if $\varphi(\mathcal{M})$ is finite, of power $n$ say, then $\mathcal{M}$ satisfies $\exists^{=n} \bar{x}\, \varphi$, hence the same is true for any elementary extension of $\mathcal{M}$. (Therefore the two definable sets are the same, i. e., no new solution to $\varphi$ can come in when passing to an elementary extension, see Lemma 11.4.3 below.) From this and the next lemma it is not hard to derive that algebraic types are isolated (exercise!).

There is a tight connection between the algebraicity of types and of that formulas.

**Lemma 11.4.2.** *A type of a structure $\mathcal{M}$ is algebraic if and only if its deductive closure contains a formula that is algebraic in $\mathcal{M}$. In particular, a complete type is algebraic if it contains an algebraic formula.*

*Proof.* For the nontrivial direction consider a deductively closed type $\Phi$ not containing an algebraic formula. Then no finite conjunction of formulas from $\Phi$ is algebraic. Hence the set $\Phi(\bar{c}_i) \cup \{\bar{c}_i \neq \bar{c}_j : i < j < \omega\}$ (where the $\bar{c}_i$ are new constants) is consistent (with $\mathrm{Th}\,\mathcal{M}$). However, any model of the latter contains infinitely many realizations of $\Psi$. $\square$

Let us now look at the motivating

**Example.** Every nontrivial polynomial equation $t(x) = 0$ (i. e., $t$ has positive degree) has finitely many solutions and is thus, as a formula, algebraic in *any* field. The same applies to polynomial equations with parameters from a field.

Given $\mathcal{K} \models \mathrm{ACF}$, by quantifier elimination, every 1-place formula (possibly with parameters from $\mathcal{K}$) defines in $\mathcal{K}$ a boolean combination of solution sets of polynomials from $\mathcal{K}[x]$. Now boolean combinations of finite subsets of an infinite set are finite if and only if every disjunct contains at least one unnegated conjunct. Consequently, a set definable in $\mathcal{K}$ by a 1-place formula is finite if and only if it is contained in a union of finitely many solution sets of polynomials of positive degree from $\mathcal{K}[x]$. On taking the product of these polynomials we infer that a definable subset of $\mathcal{K}$ is finite if and only if it consists of roots of a *single* polynomial of positive degree from $\mathcal{K}[x]$. Correspondingly, a 1-place formula is algebraic in $\mathcal{K}$ if and only if it implies (in $\mathcal{K}$) a single nontrivial polynomial equation.

Corollary 9.1.3' tells us that any type of $\mathcal{K} \models \text{ACF}$ is determined by the polynomial equations it contains. But by the preceding lemma a complete type is algebraic if and only if it contains an algebraic formula. Hence we can conclude from the above discussion that a complete 1-type of $\mathcal{K}$ (over any subset) is algebraic if and only if it contains a nontrivial polynomial equation.

The situation for 2-types in ACF is not at all that simple: using 2-place formulas one can define curves, and after all, these are the object of investigation of algebraic geometry. It should be noted though that the so-called *geometric stability theory* has relevant things to say also in this case.

We conclude this section with a list of basic properties of algebraic types (and formulas).

**Lemma 11.4.3.** *Let $\mathcal{M}$ be an arbitrary structure.*

(1)  *A type of $\mathcal{M}$ is algebraic if and only if each of its realizations completely lies in $\mathcal{M}$. In particular, a formula $\varphi(\bar{x})$ over $M$ is algebraic (in $\mathcal{M}$) if and only if $\varphi(\mathcal{N}) \subseteq M^{l(\bar{x})}$ for all $\mathcal{N} \succcurlyeq \mathcal{M}$ if and only if $\varphi(\mathcal{N}) = \varphi(\mathcal{M})$ for all $\mathcal{N} \succcurlyeq \mathcal{M}$.*

(2)  *Let $\mathcal{N} \succcurlyeq \mathcal{M}$ and $\bar{c} = (c_0, \ldots, c_{n-1}) \in N^n$. The complete type $\text{tp}^{\mathcal{N}}(\bar{c}/M)$ is algebraic if and only if $\bar{c}$ is completely contained in $M$ if and only if every $\text{tp}^{\mathcal{N}}(c_i/M)$ is algebraic ($i < n$) if and only if $\text{tp}^{\mathcal{M}}(\bar{c}/M)$ is isolated (equivalently, principal).*

(3)  *Let $A \subseteq B \subseteq M$. Every nonalgebraic $n$-type over $A$ of $\mathcal{M}$ can be extended to a nonalgebraic type in $\text{S}_n(B)$.*

(4)  *$\mathcal{M}$ is infinite if and only if, for every $n$, there are nonalgebraic $n$-types of $\mathcal{M}$ over the empty set (and hence over any subset of $M$).*

*Proof.* Ad (1). The argument for formulas was given before the previous lemma. The assertion for arbitrary types then follows from that lemma.

Ad (2). The first equivalence is immediate from (1). Further, $\bar{c}$ lies completely in $M$ if and only if each of the $c_i$ is in $M$, hence the second equivalence follows from the first. Finally, $\bar{c}$ lies completely in $M$ if and only if the formula $\bar{x} = \bar{c}$ is contained in the type $\text{tp}^{\mathcal{N}}(\bar{c}/M)$. Now, for the third equivalence it remains to notice that formulas of this type (where $\bar{c}$

is an $n$-tuple from $M$)—or rather their equivalence classes—are exactly the atoms of the Lindenbaum-Tarski algebra $\mathcal{B}_n(\mathrm{Th}(\mathcal{M}, M))$.

Ad (3). If $\Phi$ is a nonalgebraic $n$-type of $\mathcal{M}$ over $A$, every finite subset of the set $\Phi \cup \{\neg\psi : \psi \in L_n(B)$ is algebraic$\}$ is satisfiable in $\mathcal{M}$. So this set is a type of $\mathcal{M}$ over $B$ and can thus be extended to a type from $S_n(B)$. But no extension of the above set is algebraic.

Ad (4). $\mathcal{M}$ is infinite if and only if the formula $\bar{x} = \bar{x}$ is not algebraic in $\mathcal{M}$. Now apply (3) to the type consisting of this formula alone. $\qquad \square$

We showed that an isolated complete type of $\mathcal{M}$ over the entire $M$ must be algebraic. This is far from being true for arbitrary types (see the exercises).

In §15.4 we investigate in detail the types of $\mathrm{Th}(\mathbb{Z}; 0; +, -)$. It is no problem to study this with the present knowledge; however, §§15.1-3 would be a necessary prerequisite.

**Exercise 11.4.1.** Show that every algebraic type is isolated.

**Exercise 11.4.2.** Prove that $S_n(A)$ is finite if and only if every type therein is isolated.

**Exercise 11.4.3.** Find an example where $S_1^{\mathcal{M}}(\emptyset)$ is a singleton. Is this possible for $S_2^{\mathcal{M}}(\emptyset)$ if $\mathcal{M}$ has at least 2 elements?

**Exercise 11.4.4.** Use the previous exercise to find an isolated type that is not algebraic.

**Exercise 11.4.5.** Let $A \subseteq \mathcal{K} \models \mathrm{ACF}$. Show that $S_1^{\mathcal{K}}(A)$ has exactly one nonalgebraic (1-) type $\Phi$. Prove that this type is not isolated either. What does $\Phi$ express? Is the same true for 2-types?

**Exercise 11.4.6.** Consider the same questions for the theories $T_=$, $T_\mathcal{K}$ (of $\mathcal{K}$-vector spaces), and $\mathrm{Th}(\mathbb{Z}_{p^\infty})$.

**Exercise 11.4.7.** Prove an analogue of Lemma 11.4.1 on the transitivity of algebraicity.

## 11.5 Algebraic closure

This section is used only in Chapter 14 (which the reader could read directly after this section). But it belongs here and is interesting in its own right. It is about another useful generalization of an important field-theoretic concept.

Let $A \subseteq M$. The **algebraic closure** of $A$ in $\mathcal{M}$, in symbols $\mathrm{acl}_{\mathcal{M}} A$ or, if the context $\mathcal{M}$ is understood, simply $\mathrm{acl}\, A$, is, by definition, the union of all finite sets that are definable with formulas over $A$ in $\mathcal{M}$, i. e. the union of all finite sets of the form $\varphi(\mathcal{M})$, where $\varphi \in L_1(A)$. The elements from and the subsets of $\text{acl}_{\mathcal{M}} A$ are said to be **algebraic** over $A$ (in $\mathcal{M}$). A subset $A$ of $M$ is said to be **algebraically closed** if $\text{acl}_{\mathcal{M}} A = A$.

**Example.** By what was said in the last section about algebraicity in ACF, the elements of $\text{acl}_{\mathbb{C}} \mathbb{Q}$ are exactly the so-called **algebraic numbers** from $\mathbb{C}$, i. e. the complex numbers which are roots of polynomials (in one indeterminate) of positive degree over $\mathbb{Q}$.

Since every rational number is the unique root of some linear equation over $\mathbb{Z}$ and all of $\mathbb{Z}$ is generated by the (interpretation of the non-logical) constant 1 in $\mathbb{C}$, every polynomial equation with coefficients from $\mathbb{Q}$ can be expressed in the language of ACF without any parameters. Therefore, $\text{acl}_{\mathbb{C}} \emptyset = \text{acl}_{\mathbb{C}} \mathbb{Q}$.

More generally, given $K \models \text{ACF}$ and $\mathcal{A} \subseteq \mathcal{K}$ (a subring), the algebraic closure $\text{acl}_{\mathcal{K}} A$ is, by the above discussion, exactly the set of roots in $\mathcal{K}$ of polynomials of positive degree from $\mathcal{A}[x]$. This shows that our model-theoretic notion of algebraicity for ACF coincides with the usual, field-theoretic one. (In particular, a field is algebraically closed in the field-theoretic sense of §9.4 if and only if it is a field that is an algebraically closed subset of all of its field extensions.) Analogously to the above, the subfield generated by $\mathcal{A}$ in $\mathcal{K}$ (which is isomorphic to the quotient field $Q(\mathcal{A})$) is contained in $\text{acl}_{\mathcal{K}} A$, and we have $\text{acl}_{\mathcal{K}} A = \text{acl}_{\mathcal{K}} Q(\mathcal{A})$.

In field theory the nonalgebraic numbers are called **transcendental**. More generally, '**transcendental** over a subfield $\mathcal{K}$' means 'nonalgebraic over $\mathcal{K}$,' and hence 'transcendental' just means 'transcendental over the prime field.'

Let us summarize this discussion.

**Proposition 11.5.1.** *In algebraically closed fields, the model-theoretic and the field-theoretic concepts of algebraic closure coincide. Hence the field elements transcendental over a subfield $\mathcal{K}$ are precisely the ones not in the algebraic closure* $\text{acl}\, \mathcal{K}$. *In particular,* $\text{acl}_{\mathbb{C}}\, \emptyset$ *is the field of algebraic numbers.* □

In Exercise 12.1.4 it was to be proved that *every* field is contained in an algebraically closed extension field (i. e., the ∀-part of ACF is the theory of integral domains from Exercise 5.3.2). This entails the existence of an algebraic closure of *every* field. Together with its uniqueness (which is to follow from the discussion in Exercise 11.5.4), this constitutes one of Steinitz' celebrated theorems, see Exercise 12.1.5.

**Remarks.**

(1) Lemma 11.4.2 shows that $\mathrm{acl}_{\mathcal{M}}A$ is the set of all realizations of algebraic 1-types of $\mathcal{M}$ over $A$. Lemma 11.4.3 in turn shows that $\mathrm{acl}_{\mathcal{M}}A$ is also the union (i. e. the set of all entries) of all realizations of algebraic $n$-types of $\mathcal{M}$ over $A$.

(2) Every algebraically closed set is closed under functions (from the signature) and thus forms (the universe of) a substructure: just consider formulas of the sort $x = c$ and $x = f(\bar{a})$, where $c \in \mathbf{C}$, $f \in \mathbf{F}$, and $\bar{a}$ is from the set under consideration.

(3) Therefore $\mathrm{acl}_{\mathcal{M}}A$ contains the substructure generated by $A \subseteq M$ in $\mathcal{M}$.

Part (2) below generalizes (the first two equivalences of) part (2) of Lemma 11.4.3 above to arbitrary complete types.

**Lemma 11.5.2.** *Let $\mathcal{M} \preccurlyeq \mathcal{N}$ and $A \subseteq M$.*

(1) $\mathrm{acl}_{\mathcal{N}}A = \mathrm{acl}_{\mathcal{M}}A$ $(\subseteq M)$; *in particular, $\mathrm{acl}_{\mathcal{N}}M = M$.*

(2) *Let $\bar{c} = (c_0, \dots, c_{n-1}) \in N^n$.*
$\mathrm{tp}^{\mathcal{N}}(\bar{c}/A)$ *is algebraic if and only if $\{c_0, \dots, c_{n-1}\} \subseteq \mathrm{acl}_{\mathcal{M}}A$.*

(3) *If $\mathcal{M} \equiv \mathcal{M}'$ and $A \subseteq M \cap M'$, then $\mathrm{acl}_{\mathcal{M}}A = \mathrm{acl}_{\mathcal{M}'}A$.*

*Proof.* Ad (1). By Lemma 11.4.3(1), every algebraic element over $A$ in $\mathcal{N}$ is contained in $M$, hence also algebraic over $A$ in $\mathcal{M}$.

Ad (2). Each $c_i \in \mathrm{acl}_{\mathcal{M}}A$ realizes an algebraic formula $\varphi_i \in L_1(A)$ in $\mathcal{M}$. Then $\bigwedge_{i<n} \varphi_i(x_i)$ is also algebraic and clearly contained in the type $\mathrm{tp}^{\mathcal{N}}(\bar{c}/A)$. Hence the latter is algebraic too.

If, conversely, this latter type is algebraic, it contains an algebraic formula $\varphi(\bar{x})$. Then the formula

$$\exists x_0 \dots x_{i-1} x_{i+1} \dots x_{n-1}\, \varphi(x_0, \dots, x_{i-1}, x, x_{i+1}, \dots, x_{n-1})$$

is algebraic all the more so. But $c_i$ satisfies it in $\mathcal{N}$, whence $c_i \in \mathrm{acl}_{\mathcal{N}}A = \mathrm{acl}_{\mathcal{M}}A$.

For (3) first choose a joined elementary extension $\mathcal{N}$ of $\mathcal{M}$ and $\mathcal{M}'$ (Corollary 8.4.5) and then apply (1) twice. $\square$

The next collection of properties expresses that $\mathrm{acl}_{\mathcal{M}}$ is what is known as a (**finitary**) **closure operator** on $\mathfrak{P}(M)$.

**Proposition 11.5.3.** *Let $A, B, C \in \mathfrak{P}(M)$.*

(1) (Finite Character) $\mathrm{acl}_{\mathcal{M}}A = \bigcup\{\mathrm{acl}_{\mathcal{M}}X \,:\, X \Subset A\}$.

(2) (Reflexivity) $A \subseteq \mathrm{acl}_{\mathcal{M}}A$.

(3)   (Transitivity) *If $A \subseteq \mathrm{acl}_{\mathcal{M}} B$ and $B \subseteq \mathrm{acl}_{\mathcal{M}} C$, then $A \subseteq \mathrm{acl}_{\mathcal{M}} C$.*

(4)   (Monotonicity) *If $A \subseteq B$, then $\mathrm{acl}_{\mathcal{M}} A \subseteq \mathrm{acl}_{\mathcal{M}} B$.*

(5)   (Idempotence) $\mathrm{acl}_{\mathcal{M}} \mathrm{acl}_{\mathcal{M}} A = \mathrm{acl}_{\mathcal{M}} A$, *i. e., the (algebraic) closure is (algebraically) closed.*

*Proof.* Ad (1). Every algebraic element over $A$ satisfies an algebraic formula from $L_1(A)$ and is thus algebraic already over the finitely many parameters occurring in that formula.

Ad (2). $x = a$ is algebraic in $\mathcal{M}$ for every $a \in A$.

Ad (3). Let $a$ satisfy the algebraic formula $\varphi(x, \bar{b})$, where $\bar{b}$ is from $B$. Set $k = |\varphi(\mathcal{M}, \bar{b})|$.

By hypothesis, every entry of $\bar{b}$ is algebraic over $C$. Hence, by Lemma 11.5.2(2), $\mathrm{tp}^{\mathcal{M}}(\bar{b}/C)$ is algebraic. Choose an algebraic formula $\psi(\bar{x}, \bar{c})$ in this type that isolates it (i. e. an atom in $\mathcal{B}_n(\mathrm{Th}(\mathcal{M}, C)))$, cf. Exercise 11.4.1. Denote by $\theta(x, \bar{c})$ the formula $\exists \bar{y}\,(\varphi(x, \bar{y}) \wedge \psi(\bar{y}, \bar{c}))$. As $a$ satisfies it, it suffices to verify that this formula is algebraic.

By its choice, the formula $\psi(\bar{y}, \bar{c})$ is algebraic and all elements satisfying it have the same type over $C$ as $\bar{b}$. But $\exists^{=k} x\, \varphi(x, \bar{b})$ is true in $\mathcal{M}$, whereby $\exists^{=k} x\, \varphi(x, \bar{y})$ is in the type of $\bar{b}$, hence also in the (equal) type of $\bar{b}'$ for every $\bar{b}' \in \psi(\mathcal{M}, \bar{c})$. Thus $\exists^{=k} x\, \varphi(x, \bar{b}')$ is true too, whereby $|\varphi(\mathcal{M}, \bar{b}')| = k$ for all such $\bar{b}'$, and consequently $|\theta(\mathcal{M}, \bar{c})| \leq k \cdot |\psi(\mathcal{M}, \bar{c})| < \aleph_0$, as desired.

Ad (4) Set $X = \mathrm{acl}\, A$ (in $\mathcal{M}$), $Y = A$, and $Z = B$. Trivially, $X \subseteq \mathrm{acl}\, Y$, and $A \subseteq B \subseteq \mathrm{acl}\, B$ implies $Y \subseteq \mathrm{acl}\, Z$ (note, we used (2) here). Now (3) yields the assertion.

Ad (5). The inclusion from right to left follows from (2). For the converse put $X = \mathrm{acl}(\mathrm{acl}\, A)$, $Y = \mathrm{acl}\, A$, and $Z = A$, and apply (3) to obtain $X \subseteq \mathrm{acl}\, Z$.   $\square$

We conclude this section with some important cardinality estimates.

**Proposition 11.5.4.** *Let $A \subseteq M$.*

(1)   $|A| \leq |\mathrm{acl}_{\mathcal{M}} A| \leq |A| + |L|$.

(2)   *There are no more than $|A| + |L|$ algebraic types in $\bigcup_{n < \omega} S_n^{\mathcal{M}}(A)$.*

*Proof.* (1) is clear, since there are no more formulas altogether (Corollary 7.6.8). For the same reason (2) follows from Lemma 11.4.2.   $\square$

In the case $A = \mathbb{Q}$ (or $\emptyset$) in $\mathbb{C} \models \mathrm{ACF}$, assertion (1) above says that there are only countably many algebraic numbers. This is Cantor's original argument for the existence of transcendental numbers (for $\mathbb{C}$ is uncountable).

Let $A \subseteq M$ in the following exercises.

**Exercise 11.5.1.** Prove $f[\mathrm{acl}_\mathcal{M} A] = \mathrm{acl}_\mathcal{M} A$ for all $f \in \mathrm{Aut}_A \mathcal{M}$, and, more generally, $f[\mathrm{acl}_\mathcal{M} A] = \mathrm{acl}_\mathcal{M} f[A]$ for all $f \in \mathrm{Aut} \mathcal{M}$.

**Exercise 11.5.2.** Find a formula defining the set $\mathrm{acl}_\mathcal{M} A$ in $\mathcal{M}$ with parameters from $A$, provided $\mathrm{acl}_\mathcal{M} A$ is finite. Show that if, on the other hand, $\mathrm{acl}_\mathcal{M} A$ is infinite, there is an elementary extension $\mathcal{N}$ of $\mathcal{M}$ in which $\mathrm{acl}_\mathcal{M} A$ ($= \mathrm{acl}_\mathcal{N} A$) is not (parametrically) definable.

**Exercise 11.5.3.** Prove that in vector spaces algebraic closure is the usual span, i. e. that the algebraic closure of a subset of a vector space is the subspace spanned by that subset.

In the next exercises we deal with **partial elementary maps**, i. e. maps $f$ from a subset of an $L$-structure $\mathcal{M}$ to an $L$-structure $\mathcal{N}$ such that $\mathcal{M} \models \varphi(\bar{a})$ iff $\mathcal{N} \models \varphi(f[\bar{a}])$, for all $\varphi$ from $L$ and matching tuples $\bar{a}$ from $\mathrm{dom}\, f$. (The elementary maps considered usually are just the *total* elementary maps, i. e. the partial elementary maps defined everywhere.)

**Exercise 11.5.4.** Show that every partial elementary map $f$ from $\mathcal{M}$ to $\mathcal{N}$ can be extended to a partial elementary map (from $\mathcal{M}$ to $\mathcal{N}$) with domain $\mathrm{acl}_\mathcal{M}(\mathrm{dom}\, f)$ and image $\mathrm{acl}_\mathcal{N}(f[\mathrm{dom}\, f])$.

**Exercise 11.5.5.** Consider a partial elementary map $f$ from $\mathcal{M}$ to $\mathcal{N}$. Show that $|\mathrm{acl}_\mathcal{M}(\mathrm{dom}\, f)| = |\mathrm{acl}_\mathcal{N}(f[\mathrm{dom}\, f])|$.

The next exercises deal with other, abstract, closure operators derived from Galois correspondences as introduced in the exercises to §3.4.

**Exercise 11.5.6.** Show that a Galois correspondence between two classes $X$ and $Y$ gives rise to (not necessarily finitary) closure operators (in the sense of Proposition 11.5.3) such that 'closed' in the sense of the Galois correspondence (cf. p. 35) is the same as 'closed' in the sense of the closure operator.

**Exercise 11.5.7.** Consider the Galois correspondence from Exercise 3.4.4. Which of the two arising closure operators are finitary?

# Chapter 12

# Thick and thin models

In the previous part we investigated structures with respect to the sentences they satisfy and with respect to their definable sets. With the notion of type we have a finer way of doing this. In this chapter we look at two extreme kinds of structures, those that realize as many types as possible—the so-called *saturated* structures—and those that realize as few types as possible—the so-called *atomic* structures. The former will turn out to be 'thick' inasmuch as many other structures can be elementarily embedded into them (cf. the *universality* from Theorem 12.1.1), while the latter are 'thin' in the sense that they can be elementarily embedded into many other structures (cf. Theorem 12.2.1).

## 12.1 Saturated structures

One can certainly not expect every model to realize all of its types. Only for finite models is this (trivially) the case. In an infinite structure $\mathcal{M}$, however, the 1-type $\{x \neq a : a \in M\}$ is not realized. It is therefore necessary to restrict to reasonable sets of parameters.

Let $\kappa$ be an infinite cardinal. A structure $\mathcal{M}$ is said to be $\kappa$-**saturated** if it realizes all of its 1-types over subsets $A \subseteq M$ of power $|A| < \kappa$. The structure $\mathcal{M}$ is said to be **saturated** if it is $|\mathcal{M}|$-saturated.

**Remarks.**
(1) A countably infinite structure $\mathcal{M}$ is saturated if and only if $\mathcal{M}$ realizes all of its 1-types having only finitely many parameters.
(2) Finite structures are trivially saturated, even $\kappa$-saturated for all $\kappa \in$ **Cn**, for they do not have proper elementary extensions. Every infinite $\kappa$-saturated structure $\mathcal{M}$ ($\kappa \in$ **Cn**) has cardinality $\geq \kappa$, as for every

185

set $A \subseteq M$ with $|A| < \kappa$, the type $\{x \neq a : a \in A\}$ must be realized in $M$. Moreover, considering the type $\{\varphi\} \cup \{\bar{x} \neq \bar{a} : \bar{a} \in A^n\}$ for any nonalgebraic formula $\varphi \in L_n(M)$, the same argument shows that every infinite definable set of a $\kappa$-saturated structure $M$ has power $\geq \kappa$. Similarly, considering the type $\{\varphi\} \cup \{\neg\psi(\bar{x}) : \psi \in L_n(A) \text{ is algebraic}\}$ instead, shows that a nonalgebraic formula $\varphi \in L_n(M)$ has a solution in $M$ that is not algebraic over $A$, i. e., $\varphi(M) \not\subseteq (\operatorname{acl} A)^n$ (provided $|A| < \kappa$).

(3)   If $M$ is $\kappa$-saturated (with $\kappa \geq \aleph_0$) and $A \subseteq M$ with $|A| < \kappa$, then also $(M, A)$ is $\kappa$-saturated, as $B \subseteq M$ implies $S_1^{(M,A)}(B) = S_1^M(A \cup B)$.

(4)   If $\kappa > |L|$, it suffices to consider parameter sets that are elementary substructures: then the $L$-structure $M$ is $\kappa$-saturated, whenever, for all $N \preccurlyeq M$ with $|N| < \kappa$, all types from $S_1^M(N)$ are realized in $M$. (For, every set $A \subseteq M$ with $|A| < \kappa$ is, by Löwenheim-Skolem, contained in an $N \preccurlyeq M$ of power $< \kappa$, hence every type over $A$ can be extended to a type over $N$.)

**Example.** Countable models of DLO$_{--}$ are saturated (cf. Exercise 11.3.2).

To make things a little simpler we investigate only countable saturated structures, although most of our results are true also for uncountable ones (cf. Exercise 12.1.2). The first assertion of our first result says that saturated models of complete theories are 'thick'.

**Theorem 12.1.1.** *Let $M$ be a countably infinite saturated structure.*

(1)   (Universality) *Every countable $N \equiv M$ can be elementarily embedded in $M$.*

(2)   (Uniqueness) *If $N \equiv M$ is countable and saturated too, then $N \cong M$.*

(3)   (Homogeneity) *Suppose $A$ is a finite subset of $M$ and $\bar{a}$ and $\bar{b}$ are tuples of the same length from $M$.*
*Then $\operatorname{tp}^M(\bar{a}/A) = \operatorname{tp}^M(\bar{b}/A)$ if and only if there is $\sigma \in \operatorname{Aut}_A M$ such that $\sigma[\bar{a}] = \bar{b}$.*

[Morley, M. and Vaught, R. L. : Homogeneous universal models, Math. Scandinavica **11** (1962) 37 – 57]

*Proof.* Ad (1). Let $a_0, a_1, \ldots$ be an enumeration of $N$ (i. e. $N = \{a_i : i < \omega\}$). Inductively on $i$ we choose $b_i \in M$ such that $(N, a_0, \ldots, a_n) \equiv (M, b_0, \ldots, b_n)$ for all $n < \omega$. Then, setting $f(a_i) = b_i$, we obtain the desired map $f : N \overset{\equiv}{\hookrightarrow} M$.

The initial step of the induction is just $\mathcal{M} \equiv \mathcal{N}$. Assume (as induction hypothesis) that $\bar{b} = b_0, \ldots, b_{n-1}$ has already been chosen so that $(\mathcal{M}, \bar{b}) \equiv (\mathcal{N}, \bar{a})$, where $\bar{a} = (a_0, \ldots, a_{n-1})$.

Set $\Phi_n = \{\theta(x, \bar{b}) : \theta(x, \bar{a}) \in \text{tp}^{\mathcal{N}}(a_n/\bar{a})\}$. This set is closed under finite conjunctions, as so is $\text{tp}^{\mathcal{N}}(a_n/\bar{a})$. Further, since $\mathcal{N} \models \exists x \, \theta(x, \bar{a})$ for all $\theta(x, \bar{a}) \in \text{tp}^{\mathcal{N}}(a_n/\bar{a})$, the induction hypothesis gives $\mathcal{M} \models \exists x \, \theta(x, \bar{b})$. Thus, by Proposition 11.2.1, $\Phi_n$ is a 1-type of $\mathcal{M}$ (over the finite set $\bar{b}$).

As $\mathcal{M}$ is saturated, $\Phi_n$ is realized in $\mathcal{M}$ by some $b_n$. Consequently, $(\mathcal{M}, b_0, \ldots, b_n) \equiv (\mathcal{N}, a_0, \ldots, a_n)$.

Ad (2). Suppose also $\mathcal{N}$ is saturated and $b_0, b_1, \ldots$ is an enumeration of $M$ (i. e. $M = \{b_i : i < \omega\}$). Using back-and-forth (cf. §7.3) we successively construct an isomorphism as follows.

In every step with even index $n$—assuming $(\mathcal{N}, \bar{a}) \equiv (\mathcal{M}, \bar{b})$, where $\bar{a} = (a_{i_0}, \ldots, a_{i_{n-1}})$ and $\bar{b} = (b_{i_0}, \ldots, b_{i_{n-1}})$—let $a_{i_n}$ be the element with the smallest index in $N \smallsetminus \{a_{i_0}, \ldots, a_{i_{n-1}}\}$ and set $\Phi_n = \{\theta(x, \bar{b}) : \mathcal{N} \models \theta(a_{i_n}, \bar{a})\}$. We then choose $b_{i_n} \models \Phi_n$ in $\mathcal{M}$ as above.

In case $n$ is odd, we choose, the other way around, $b_{i_n}$ of smallest index in $M \smallsetminus \{b_{i_0}, \ldots, b_{i_{n-1}}\}$, consider $\Psi_n = \{\theta(x, \bar{a}) : \mathcal{M} \models \theta(b_{i_n}, \bar{b})\}$, and realize this type by some $a_{i_n}$ in $\mathcal{N}$.

Letting $a_{i_j} \mapsto b_{i_j}$ we obtain an elementary map $f : \mathcal{N} \overset{\equiv}{\hookrightarrow} \mathcal{M}$ which is surjective due to back-and-forth. But surjective elementary maps are isomorphisms.

Ad (3). Write $\bar{a} = (a_0, \ldots, a_{n-1})$ and $\bar{b} = (b_0, \ldots, b_{n-1})$. If $\sigma \in \text{Aut}_A \mathcal{M}$ and $\sigma[\bar{a}] = \bar{b}$, then $\text{tp}^{\mathcal{M}}(\bar{a}/A) = \text{tp}^{\mathcal{M}}(\bar{b}/A)$ by Corollary 11.2.4. The proof of the converse is similar to that of (2) in the special case $\mathcal{M} = \mathcal{N}$, starting, however, with $\sigma(a) = a$ if $a$ is an element of $A$, and with $\sigma(a) = b_i$ if $a = a_i$ is an entry from $\bar{a}$. (This is reasonable, as $\text{tp}^{\mathcal{M}}(\bar{a}/A) = \text{tp}^{\mathcal{M}}(\bar{b}/A)$ means nothing more than $(\mathcal{M}, A, \bar{a}) \equiv (\mathcal{M}, A, \bar{b})$.) Then we choose enumerations $\{a_i : n \leq i < \omega\}$ of $M \smallsetminus (A \cup \bar{a})$ and $\{b_i : n \leq i < \omega\}$ of $M \smallsetminus (A \cup \bar{b})$ and continue as in (2). This again yields an isomorphism—this time an automorphism with the desired property. $\qquad \Box$

An alternative and shorter proof of (3) is this. $(\mathcal{M}, A, \bar{a})$ and $(\mathcal{M}, A, \bar{b})$ are, by the above Remark (3), also (countable and) saturated. Further, $(\mathcal{M}, A, \bar{a}) \equiv (\mathcal{M}, A, \bar{b})$ says the same as $\text{tp}^{\mathcal{M}}(\bar{a}/A) = \text{tp}^{\mathcal{M}}(\bar{b}/A)$. Hence uniqueness from (2) yields an isomorphism $f : (\mathcal{M}, A, \bar{a}) \cong (\mathcal{M}, A, \bar{b})$, i. e. $f : \mathcal{M} \cong_A \mathcal{M}$ such that $f[\bar{a}] = \bar{b}$, as desired.

**Remarks.**

(5) Using transfinite induction, similar properties can be verified in higher cardinalities, where $\aleph_0$ has to be replaced by $\kappa \in \mathbf{Cn}$, while 'finite' has to be replaced by $< \kappa$ (exercise!).

(6) If a complete theory $T$ without finite models has a countable saturated model $\mathcal{M}$, then, for all countable $\mathcal{N} \models T$ and all finite subsets $A$ of $N$, the equality of types, $\mathrm{tp}^{\mathcal{N}}(\bar{a}/A) = \mathrm{tp}^{\mathcal{N}}(\bar{b}/A)$ (where $\bar{a}$ and $\bar{b}$ are in $N$), is equivalent to the existence of $\mathcal{N}' \succcurlyeq \mathcal{N}$ and $f \in \mathrm{Aut}_A \mathcal{N}'$ such that $f[\bar{a}] = \bar{b}$.

*Proof.* Ad 6. Universality yields $\mathcal{N} \xhookrightarrow{\equiv} \mathcal{M}$. Then homogeneity (plus isomorphic correction) implies the assertion. □

Now that we have derived the three most important properties of saturated structures, it is natural to ask about their existence. Our main technical lemma in this direction is the following. It is not the most general result possible, but suffices for our purposes. Its proof is based on a double chain argument, which is typical for this kind of construction.

**Lemma 12.1.2.** *Suppose $T$ is an $L$-theory and $\kappa \in \mathbf{Cn}$ is a bound on the cardinalities $|\mathrm{S}_1^{\mathcal{M}}(A)|$, where $\mathcal{M}$ runs over all models of $T$ and $A$ runs over all finite subsets of $M$.*

*Then $T$ has an $\aleph_0$-saturated model of power $\leq \kappa$.*

*Proof.* If $\kappa$ is finite, all models of $T$ must be finite and even of power at most $\kappa$. Since these are trivially $\aleph_0$-saturated, we may assume that $\kappa$ is infinite.

Given $\mathcal{N}$, let $\mathrm{S}_{\mathcal{N}}$ denote the set of all 1-types of $\mathcal{N}$ over finite subsets of $N$. According to Proposition 11.2.5, choose $\mathcal{N}' \succcurlyeq \mathcal{N}$ of power $\leq |N| + |L| + |\mathrm{S}_{\mathcal{N}}|$ realizing all types from $\mathrm{S}_{\mathcal{N}}$.

We claim, $\mathcal{N}'$ has power at most $|N| + |L| + \kappa$. There are $|N|^{<\omega} = |N|$ finite subsets of $N$ (Proposition 7.6.7), and by hypothesis over each of these at most $\kappa$ relevant types. Thus $|\mathrm{S}_{\mathcal{N}}| \leq |N| \cdot \kappa = |N| + \kappa$, and hence $|N'| \leq |N| + |L| + \kappa$.

We start our first chain construction with an arbitrary model $\mathcal{M}_0$ of $T$ of power $|L|$. Then, for any model $\mathcal{M}_n$ already chosen, we let $\mathcal{M}_{n+1}$ be a model $(\mathcal{M}_n)' \succcurlyeq \mathcal{M}_n$ of power at most $|M_n| + |L| + \kappa$ realizing all the types from $\mathrm{S}_{\mathcal{M}_n}$ ( we've just proved the existence of such a model). In this way we obtain an elementary chain $\mathcal{M}_0 \preccurlyeq \mathcal{M}_1 \preccurlyeq \ldots \preccurlyeq \mathcal{M}_n \preccurlyeq \ldots$, whose union we denote by $\mathcal{M}_\omega$. The elementary chain theorem shows that $\mathcal{M}_\omega$ is a model of $T$.

$\mathcal{M}_\omega$ is $\aleph_0$-saturated, for if $A$ is a finite subset of $M_\omega$, then $A \subseteq M_n$ for some $n < \omega$, hence all $\Phi \in \mathrm{S}_1^{\mathcal{M}_\omega}(A)$ are realized already in $\mathcal{M}_{n+1}$, hence also in the elementary extension $\mathcal{M}_\omega$ (apply the elementary chain theorem again).

Nevertheless, $M_\omega$ may not be the desired model yet—it might be too big. Although $|M_0| = |L|$ and the above estimate yield $|M_1| = |M_0'| \leq |M_0|+|L|+\kappa = |L|+\kappa$, and inductively also $|M_{n+1}| \leq |M_n|+|L|+\kappa = |L|+\kappa$, hence, by Lemma 7.6.6, also $|\mathcal{M}_\omega| = \aleph_0 \cdot (|L| + \kappa) = |L| + \kappa$, this is good enough only in case $\kappa \geq |L|$. We are done, in particular, if $L$ is countable. However, in the general case we have to find a smaller such model. This model will be an elementary substructure of $M_\omega$. But since Löwenheim-Skolem says nothing about the existence of elementary substructures of a power smaller than the language, we have to do something more subtle. What we will do is construct an ascending chain of 'small' sub*sets* of $\mathcal{M}_\omega$ realizing appropriate types (utilizing the $\aleph_0$-saturation of $\mathcal{M}_\omega$) and show that their union is an elementary substructure of $\mathcal{M}_\omega$—invoking the Tarski-Vaught test in the form of Corollary 8.3.3. (This exhibits the relevance of this form of the test, for we cannot use the elementary chain theorem here, as we do not even know if the chosen sets are structures, let alone elementary ones.) This can be accomplished as follows.

Let $A_0$ be a set of realizations of all $\Phi \in S_1^{\mathcal{M}_\omega}(\emptyset)$. As $|S_1^{\mathcal{M}_\omega}(\emptyset)| = |S_1^{\mathcal{M}}(\emptyset)| \leq \kappa$, we can ensure that $|A_0| \leq \kappa$. If $A_n \supseteq A_{n-1} \supseteq \ldots \supseteq A_0$ have already been constructed, we choose $A_{n+1} \supseteq A_n$ as to contain a realization of every $\Phi \in S_1^{\mathcal{M}_\omega}(C)$, where $C$ runs over the finite subsets of $A_n$. As for $A_0$, we can inductively ensure that $A_{n+1} \leq \kappa$. Let $N$ be the union of this chain.

We claim, $N$ can be made an elementary substructure of $M_\omega$. For this we apply Tarski-Vaught (in the form of Corollary 8.3.3). So suppose $\varphi \in L$ and $\bar{a}$ is from $N$ such that $\mathcal{M}_\omega \models \exists x\, \varphi(x, \bar{a})$. Then there is $c \in M_\omega$ with $\mathcal{M}_\omega \models \varphi(c, \bar{a})$. Since $\bar{a}$ is finite, there is $A_k$ (for some $k < \omega$) containing $\bar{a}$. By construction, the type $\text{tp}^{\mathcal{M}_\omega}(c/\bar{a})$ is realized by some $b \in A_{k+1} \subseteq N$. Then $\mathcal{M}_\omega \models \varphi(b, \bar{a})$, and the restriction of $\mathcal{M}_\omega$ to $N$ is an elementary substructure $\mathcal{N} \preccurlyeq \mathcal{M}_\omega$, as desired.

That $\mathcal{N}$ is $\aleph_0$-saturated now follows as for $\mathcal{M}_\omega$ above. The cardinality estimate is analogous too—the only difference being that right at the beginning we get a better bound for $|A_1|$, since there are only $|A_0|^{<\omega} \leq \kappa$ finite subsets of $A_0$ and hence only $\kappa \cdot \kappa = \kappa$ types to consider, whence $|A_1| \leq \kappa$. Inductively we then obtain $|A_{n+1}| \leq \kappa$, hence altogether $|N| \leq \aleph_0 \cdot \kappa = \kappa$, as desired. $\qquad\square$

In Chapter 13 we will make use of a description to be derived next of the complete theories (without finite models) that have a countable saturated model. These will turn to be the theories that are **small** in the sense that $S_n(T)$ is countable for every $n < \omega$ (and thus also $\bigcup_{n<\omega} S_n(T)$ is countable). Note that the complete theory of a finite structure is trivially small.

**Theorem 12.1.3.** *A complete theory has a countable saturated model if and only if it is small.*

[Vaught, R. : Denumerable models of complete theories, **Infinitistic Methods Pergamon**, London and Panstwowe Wydawnictwo Naukowe, Warszawa 1961, 303 – 321]

*Proof.* $\Longrightarrow$ is clear, for all $\Phi \in S_n(T)$ are realized in every saturated model (and in a countable structure there are only countably many different $n$-tuples).

$\Longleftarrow$. We first convince ourselves that in a model $\mathcal{M}$ of a small theory even $S_1^{\mathcal{M}}(\bar{a})$ is countable (for all tuples $\bar{a}$ in $M$): since $\mathrm{tp}^{\mathcal{M}}(b/\bar{a}) \neq \mathrm{tp}^{\mathcal{M}}(c/\bar{a})$ implies $\mathrm{tp}^{\mathcal{M}}(b\hat{\ }\bar{a}) \neq \mathrm{tp}^{\mathcal{M}}(c\hat{\ }\bar{a})$, we see that $|S_1^{\mathcal{M}}(\bar{a})| \leq |S_{n+1}(T)| \leq \aleph_0$.

Now we may apply the preceding lemma with $\kappa = \aleph_0$ to obtain an $\aleph_0$-saturated model of $T$ of power at most (hence equal to) $\aleph_0$, i. e. a countable saturated model, as desired.                                                              $\square$

It is very interesting to note that this existence result needs no assumptions on the power of the language: a small theory, countable or not, has always a countable saturated model. (This was established by the second chain argument using Tarski-Vaught, which is stronger than Löwenheim-Skolem downward.) The point is that not the *formal* cardinality of the language, i. e. the cardinality of its syntactic alphabet, counts, but rather its *actual* semantic expressibility—how many different elements one can distinguish using it, i. e. how many types there are. (This discussion leads directly to *stability theory*, where similar phenomena are observed; see Exercise 15.4.4.)

**Remarks.**
(7)  The bound $|S_n(A)| \leq 2^{|L|+|A|}$ always exists, since every $n$-type over $A$ is a subset of $L_n(A)$ and there are at most $2^{|L_n(A)|} = 2^{|L|+|A|}$ such subsets (Lemma 7.6.4(1) and Corollary 7.6.8(2)).

(8)  This together with the above lemma yields an $\aleph_0$-saturated model of power $\leq 2^{\aleph_0}$ for every countable theory. (This can be strengthened, see Exercise 12.1.8.)

(9)  No completion $T$ of Peano arithmetic PA is small. For if, as before, $\mathbb{P}$ denotes the set of prime numbers, for every subset $X \subseteq \mathbb{P}$ the set

$\Phi_X = \{p|x \; : \; p \in X\} \cup \{p \restriction x \; : \; p \in \mathbb{P} \smallsetminus X\}$ is a type of the standard model $\omega$ of PA. This type is over $\emptyset$, for $p$ can be written as $1+1+\ldots+1$ ($p$ times). As for different sets these types are different, we get as many types in $S_1(T)$ as there are subsets of $\mathbb{P}$, hence continuum many.

The reader may be interested to know that Chapter 15 (on $\mathbb{Z}$) is more or less independent of the rest of the material to follow.

**Exercise 12.1.1.** Show that every $\mathcal{K}$-vector space of dimension $> |\mathcal{K}|$ is saturated.

**Exercise 12.1.2.** Give a proof of the generalization of Theorem 12.1.1 to arbitrary cardinalities as mentioned in Remark (5).

**Exercise 12.1.3.** Prove that every model $M$ of an inductive $L$-theory $T$ can be embedded in an existentially closed model of $T$ of power $\leq |\mathcal{M}| + |L|$.

**Exercise 12.1.4.** Show that every field $\mathcal{K}$ is embeddable in an algebraically closed field (of power $|\mathcal{K}| + \aleph_0$).

**Exercise 12.1.5.** Prove Steinitz' theorem asserting the existence and uniqueness of an algebraic closure of *every* field.

**Exercise 12.1.6.** Let $U$ be a nonprincipal ultrafilter on $\omega$ and let $L$ be countable.
Show that every ultraproduct $\prod_{i<\omega} \mathcal{M}_i / U$ of $L$-structures (with respect to $U$) is $\aleph_1$-saturated (i. e. every 1-type over a countable subset is realized).

**Exercise 12.1.7.** Suppose $\mathcal{M}$ is $\kappa$-saturated, $A \subseteq M$ with $|A| < \kappa$, and $n > 1$ is a natural number.
Prove that all $n$-types of $\mathcal{M}$ over $A$ are realized in $\mathcal{M}$.

**Exercise 12.1.8.** Modify the proof of Lemma 12.1.2 as to yield an $\aleph_1$-saturated model of power $\leq 2^{\aleph_0}$ for every countable theory.

**Exercise 12.1.9.** Verify that a theory $T$ is small if and only if the Stone spaces $S_1(A)$ are countable for all finite subsets $A$ of models of $T$.

**Exercise 12.1.10.** Suppose $T$ is a countable complete theory without finite models such that over every finite subset of a model of $T$ there are at most countably many non-algebraic 1-types.
Prove that $T$ has a countable saturated model.

## 12.2 Atomic structures

After having realized as many types as possible, let us now try and realize as few as possible. Since isolated types are always realized (see §11.4), we cannot avoid those however.

A structure $\mathcal{M}$ is said to be **atomic** if it realizes only isolated types (over the empty set), i. e., every tuple $\bar{a}$ over $M$ has an isolated type $\mathrm{tp}^{\mathcal{M}}(\bar{a})$.

**Remarks.**

(1)   Since, as mentioned, isolated types are always realized, atomic struc-
      tures realize as few types as possible. Note that this says nothing about
      how often each type is realized, i. e. how big the model is in terms of
      cardinality.

(2)   The transitivity of isolation (Lemma 11.4.1) implies that $\mathcal{M}$ is atomic
      if and only if every (1-) type of an element of $\mathcal{M}$ over a finite subset of
      $M$ is isolated.

The next theorem contains assertions about countable atomic structures
that are dual to those of Theorem 12.1.1. The first of these says that—
dually—countable atomic structures are 'thin'.

**Theorem 12.2.1.** *Let $T$ be a complete theory without finite models.*

(1)   (Embeddability) *Every countable atomic model of $T$ is an elementarily
      prime model of $T$.*

(2)   (Uniqueness) *All countable atomic models of $T$ are isomorphic.*

(3)   (Homogeneity) *Suppose $\mathcal{N}$ is a countable atomic model of $T$, $A$ is a
      finite subset of $N$, and $\bar{a}$ and $\bar{b}$ are tuples of the same length from $N$.
      Then $\mathrm{tp}^{\mathcal{N}}(\bar{a}/A) = \mathrm{tp}^{\mathcal{N}}(\bar{b}/A)$ if and only if there is $\sigma \in \mathrm{Aut}_A \mathcal{N}$ such
      that $\sigma[\bar{a}] = \bar{b}$.*

This was proved by L. Svenonius (unpublished) as well as in

[Vaught, R. : Denumerable models of complete theories, **Infinitistic Methods
Pergamon**, London and Panstwowe Wydawnictwo Naukowe, Warszawa 1961, 303
– 321].

*Proof.* Also the proof is dual to that of Theorem 12.1.1—with the following
difference. If $\mathcal{N} \models T$ with $N = \{a_i : i < \omega\}$ is to be elementarily embedded
in $\mathcal{M} \models T$ (in 12.1.1, $\mathcal{N}$ arbitrary and $\mathcal{M}$ countable and saturated—here, $\mathcal{N}$
countable and atomic and $\mathcal{M}$ arbitrary) and is the type $\Phi_n = \{\theta(x, f[\bar{a}]) :
\theta(x, \bar{a}) \in \mathrm{tp}^{\mathcal{N}}(a_n/\bar{a})\}$ to be realized in $\mathcal{M}$ assuming $(\mathcal{N}, \bar{a}) \equiv (\mathcal{M}, f[\bar{a}])$, then
in 12.1.1 this was possible due to the saturation of $\mathcal{M}$. While here we are
to show that $\Phi_n$ is isolated and *therefore* realized in $\mathcal{M}$.

This can be seen as follows. As $\mathcal{N}$ is atomic, $\mathrm{tp}^{\mathcal{N}}(a_n/\bar{a})$ is isolated (apply
transitivity of isolation). But if $\varphi(x, \bar{a}) \leq \mathrm{tp}^{\mathcal{N}}(a_n/\bar{a})$, then also $\varphi(x, f[\bar{a}]) \leq
\Phi_n$, for $\mathcal{N} \models \forall x \, (\varphi(x, \bar{a}) \to \theta(x, \bar{a}))$ implies $\mathcal{M} \models \forall x \, (\varphi(x, f[\bar{a}]) \to \theta(x, f[\bar{a}]))$
for all $\theta(x, \bar{a}) \in \mathrm{tp}^{\mathcal{N}}(a_n/\bar{a})$ (remember: $(\mathcal{N}, \bar{a}) \equiv (\mathcal{M}, f[\bar{a}])$). So $\Phi_n$ is indeed
isolated.                                                                      □

In contrast to Theorem 12.1.1 there is no canonical generalization of this
to uncountable powers. The concept of atomic model is not adequate for

this, because the transitivity of isolation says nothing about the isolation of elements over (already chosen) infinite sets—see the example following—and therefore the embedding argument goes no longer through. However, there is another adequate notion, as will be shown in the exercises.

**Example.** Every model $\mathcal{M}$ of $T_{\equiv}^{\infty}$ is by quantifier elimination (Exercise 9.2.6) atomic. If, however, $A$ is an infinite subset of $\mathcal{M}$ and $c \in M \smallsetminus A$, then the type of $c$ over $A$, which essentially says that $c \neq a$ for all $a \in A$, cannot be isolated anymore.

The adequate replacement for uncountable atomic structures is the following. A structure $\mathcal{M}$ is said to be **constructible** if it has an enumeration $\{a_i : i < |M|\}$ of its universe $M$ such that the type $\operatorname{tp}(a_i/\{a_j : j < i\})$ is isolated for all $i < |M|$.

**Exercise 12.2.1.** Show that any constructible model of a theory is an elementarily prime model (and hence is 'thin').

**Exercise 12.2.2.** Prove that countable structures are constructible if and only if they are atomic. (This explains why we could do without explicitly mentioning constructibility.)

**Exercise 12.2.3.** Show that every **purely algebraic** structure, i. e. a structure $\mathcal{M}$ such that $\operatorname{acl}_{\mathcal{M}}\emptyset = M$, is constructible and atomic.

**Exercise 12.2.4.** Show that all models of DLO are atomic.
(This yields another instance of the lack of good properties of uncountable atomic structures: uncountable dense linear orderings can be extremely complicated, and there are very many of them!)

**Exercise 12.2.5.** Prove that the standard model of Peano arithmetic is atomic (and constructible).

**Exercise 12.2.6.** Prove that the Prüfer group $\mathbb{Z}_{p^\infty}$ is atomic (and constructible).

**Exercise 12.2.7.** Show that elementary substructures of atomic structures as well as unions of elementary chains of atomic structures are again atomic.

A structure $\mathcal{M}$ is said to be a **minimal model** (of its theory) if it has no proper elementary substructures, i. e. $\mathcal{N} \preccurlyeq \mathcal{M}$ implies $\mathcal{N} = \mathcal{M}$.

**Exercise 12.2.8.** Prove that a theory has an atomic model of power $\aleph_1$, provided it has a countable atomic model that is not a minimal model.

# Chapter 13

# Countable complete theories

In this chapter the material of the text is to culminate in Vaught's theorem saying that a countable complete theory cannot have, up to isomorphism, exactly two countable models (Theorem 13.4.1). This is a curious anomaly,[1] for all other finite numbers do occur, as will be seen in §13.4. In itself the result is a good deal less consequential than its proof, which really uses almost everything we have done so far (and requires yet some more preparation). Thus we reach a certain highlight in this chapter in view of which it could well serve as the final topic of a one semester course. (The remaining two chapters, especially the last one, have a rather separate character.)

*Throughout this chapter, unless stated otherwise, $T$ is a theory in a* **countable** *language L, which, from §13.2 on, is assumed to be in addition also* **complete without finite models.**

## 13.1 Omitting types

A structure is said to **omit** a type, simply if it does not realize it.

Following a *bon mot* cited by Gerald Sacks,[2] *any fool can realize a type, but it takes a model theorist to omit one.* Let's prove ourselves worthy of the name, then.

The omitting types result we are after works as well for incomplete theories (and types over the empty set), once we extend our terminology concerning types to this case in an appropriate way. We do this only for types without parameters. An $n$-**type** (over the empty set) of a (not necessarily complete) theory $T$ is, by definition, simply a subset of $L_n$ consistent with

---

[1]showing once again that 2 is an odd number

[2]Cf. Sacks' *Saturated Model Theory*, §18, p.96.

$T$, i. e. a set of formulas from $L_n$ that can be realized in some model of $T$. (This means that it is an $n$-type, in the old terminology of p.171, of some completion of $T$.) Such a type $\Phi$ of $T$ is said to be **isolated** if there is a formula $\varphi \in L_n$ consistent with $T$ and **isolating** it with respect to all of $T$, i. e. such that $T \cup \{\exists \bar{x}\varphi\}$ is consistent and $\varphi \leq_T \Phi$.

If $T$ is complete we get back the original definition, as then $T \models \exists \bar{x}\varphi$. Note, however, that to be an isolated type of an incomplete theory $T$ does not require that it be isolated in *every* completion of $T$ (that it is consistent with), but only in every completion of $T$ *containing* $\exists \bar{x}\varphi$. Therefore it may happen that an isolated type of an incomplete theory is omitted in some model (but only in a model satisfying $\neg\exists \bar{x}\varphi$, of course). In a complete theory, therefore, isolated types cannot be omitted. The following utterly useful result says that, in the countable case, all other types (over the empty set) can indeed be omitted, whether the theory is complete or not.

**Theorem 13.1.1.** (Ehrenfeucht's Omitting Types Theorem)
*Every nonisolated type of (a countable theory) $T$ over $\emptyset$ can be omitted in some countably infinite model of $T$.*

Andrzej Ehrenfeucht did not publish this result, but see

[Vaught, R. : Denumerable models of complete theories, **Infinitistic Methods Pergamon**, London and Panstwowe Wydawnictwo Naukowe, Warszawa 1961, 303 – 321],

where the bulk of this chapter's material was published for the first time; see also

[Grzegorczyk, A., Mostowski, A., and Ryll-Nardzewski, C. : Definability of sets in models of axiomatic theories, Bulletin Acad. Polon. Sci. Sér. Sci. Math. Astronom. Phys. **9** (1961) 163 – 167].

*Proof.* Let $\Phi$ be the nonisolated type we are to omit, for notational simplicity a 1-type. Let $C = \{c_i : i < \omega\}$ be a set of new pairwise distinct constants and $\{\psi_n : n < \omega\}$ an enumeration of all sentences from $L(C)$ (note, the latter is still countable). The aim is to find a model $\mathcal{M}$ of $T$ and interpretations $a_i$ of the $c_i$ therein satisfying the following two conditions.
(a)   Every $a_i$ does not realize some formula from $\Phi$.
(b)   The set $\{a_i : i < \omega\}$ satisfies the Tarski-Vaught condition (TV). This guarantees that $\{a_i : i < \omega\}$ forms an elementary substructure of $\mathcal{M}$ (hence a model of $T$) omitting $\Phi$.

We establish this by a chain argument, but this time we successively build an ascending chain of consistent sets of $L(C)$-sentences $T = T_0 \subseteq T_1 \subseteq \ldots \subseteq T_n \subseteq \ldots$ meeting the following four requirements for all $n < \omega$.
($1_n$)   There is $\varphi \in \Phi$ such that $\neg\varphi(c_n) \in T_{n+1}$.

$(2_n)$   If $\psi_n \in T_{n+1}$ has the form $\exists x\, \psi(x)$ for some $\psi$, then $\psi(c_k) \in T_{n+1}$ for some $k < \omega$.

$(3_n)$   $\psi_n \in T_{n+1}$ or $\neg\psi_n \in T_{n+1}$.

$(4_n)$   $T_{n+1} \setminus T$ is finite.

Set $T_\omega = \bigcup_{n < \omega} T_n$. Then condition (1) (more exactly, $(1_n)$ for all $n < \omega$) ensures that every model of $T_\omega$ have the above property (a). Condition (2) guarantees that all models of $T_\omega$ satisfy the Tarski-Vaught condition as in (b), provided that every $L(C)$-sentence $\exists x\, \psi(x)$ is in $T_\omega$, whenever it is true in some model of $T_\omega$. This proviso, in turn, is satisfied if $T_\omega$ is a complete $L(C)$-theory, which is taken care of by condition (3). Condition (4) serves to yield, based on the nonisolation of $\Phi$, in every step of the induction a formula as required by (1)—as we will see shortly.

It thus suffices to find such consistent $T_n$, for then, as mentioned, the set of interpretations of the new constants in any model $\mathcal{M}_\omega$ of $T_\omega$ forms a countable elementary substructure of the $L$-reduct of $\mathcal{M}_\omega$ (hence a countably infinite model of $T$) that omits $\Phi$.

Start with $T = T_0$ and assume $T = T_0 \subseteq T_1 \subseteq \ldots \subseteq T_n$ have been already constructed as to satisfy $(1_k) - (4_k)$ for all $k < n$. We claim,

$(0_n)$   $T_n \cup \{\neg\varphi(c_n)\}$ is consistent for a certain $\varphi \in \Phi$.

Otherwise we would have $T_n \models \varphi(c_n)$ for all $\varphi \in \Phi$. Since the set $T_n \setminus T$ is finite by $(4_{n-1})$, we may form its conjunction, $\theta_n(\bar{c}, c_n)$ say, where $\theta_n(\bar{x}, x) \in L$ and $\bar{c}$ is a tuple from $C$ not containing $c_n$ and $x$ is not contained in $\bar{x}$. then we would have $T \models \theta_n(\bar{c}, c_n) \to \varphi(c_n)$ and hence, by the lemma on new constants (3.3.2), also $T \models \forall x\, (\exists \bar{x}\, \theta_n(\bar{x}, x) \to \varphi(x))$, for all $\varphi \in \Phi$. But the latter is equivalent to $\exists \bar{x}\, \theta_n(\bar{x}, x) \leq_T \Phi$, contradicting the fact that $\Phi$ is not isolated. This proves condition $(0_n)$.

We now describe how to choose $T_{n+1}$. According to $(0_n)$, pick some $\varphi \in \Phi$ such that $T_n \cup \{\neg\varphi(c_n)\}$ is consistent. If $\psi_n$ is inconsistent with $T_n \cup \{\neg\varphi(c_n)\}$, we put $T_{n+1} = T_n \cup \{\neg\varphi(c_n), \neg\psi_n\}$. Otherwise there are two cases to consider. If $\psi_n$ is not of the form $\exists x\, \psi(x)$ for some $\psi$, we set $T_{n+1} = T_n \cup \{\neg\varphi(c_n), \psi_n\}$. If on the other hand $\psi_n$ does have the form $\exists x\, \psi(x)$, we set $T_{n+1} = T_n \cup \{\neg\varphi(c_n), \psi_n, \psi(c_k)\}$ instead, where $k$ is the least index such that $c_k$ does not occur in $T_n \cup \{\neg\varphi(c_n), \psi_n\}$. In any case $T_{n+1}$ is consistent and satisfies, as is easily seen, conditions $(1_n) - (4_n)$. $\square$

**Example.** Showing that the countability $|L| = \aleph_0$ is essential.

Let $A$ be a countably infinite set and $B$ an uncountable set. Consider the (uncountable) language $L = L_=(A \cup B)$ and the $L$-theory $T$ with the axioms $\{b \neq b' : b, b' \in B \text{ and } b \neq b'\}$. Every model of $T$ is clearly uncountable and hence realizes the type $\Phi =_{\mathrm{def}} \{x \neq a : a \in A\}$. Thus this type of $T$

(over the empty set) can in no model of $T$ be omitted. Nevertheless, $\Phi$ is not isolated, as we see next. Every $L$-formula is of the form $\varphi(x, \bar{c})$, where $\varphi(x, \bar{y}) \in L_{=}$ and $\bar{c}$ is a tuple from $A \cup B$. By the quantifier elimination of $T^{\infty}_{=}$ we may assume that $\varphi(x, \bar{y})$ is a boolean combination of equations. If there were such a formula $\varphi(x, \bar{c})$ consistent with $T$ and isolating $\Phi$, we could do already with one disjunct, i. e. with a formula $\varphi(x, \bar{c})$ which is a conjunction of equations and inequations. Consider the conjunction of all inequations occurring in $\varphi(x, \bar{c})$. Informally this conjunction says nothing more than '$x \notin C$' for a certain (finite) set $C$ of entries of $\bar{c}$. If this is all there is in $\varphi(x, \bar{c})$, this formula does not isolate $\Phi$, for it is satisfied in $A$ (in any model). So let's assume, an equation $x = b$ occurs (where $b \in A \cup B$). Since the formula should be satisfiable, we may as well assume that this is the only equation in $\varphi(x, \bar{c})$, and hence this formula is equivalent to $x = b \wedge$ '$x \notin C$'. If it is to isolate $\Phi$, the parameter $b$ cannot be in $A$, whence it must be in $B$. Still, it cannot isolate the type $\Phi$, as there is a model $\mathcal{M}$ of $T \cup \{\exists x \varphi(x, \bar{c})\}$ in which $b$ is interpreted as some $a \in A$, i. e. in which $b^{\mathcal{M}} = a^{\mathcal{M}}$ for some $a \in M$ (in fact, for every $a \in A \smallsetminus C$ there is such a model).

Consequently, no formula $\varphi(x, \bar{c})$ can isolate $\Phi$ in $T$, i. e., this type is nonisolated, and can still not be omitted.

**Exercise 13.1.1.** Let $\mathcal{G}$ be a group in which every nonempty subset definable by a (1-place) formula without parameters contains an element of finite order.

Prove that $\mathcal{G}$ is elementarily equivalent to a periodic group.

**Exercise 13.1.2.** Find an example of a (necessarily incomplete) theory that omits a certain isolated type.

**Exercise 13.1.3.** The counterexample above is incomplete. Complete ones are harder to come by. Find one!

**Exercise 13.1.4.** Use omitting types to obtain, for any characteristic, a purely algebraic[3] algebraically closed field (the field of algebraic numbers).

**Exercise 13.1.5.** Show that every countable model $\mathcal{M}$ of Peano arithmetic has an elementary **end-extension**, i. e. an elementary extension $\mathcal{N}$ such that for every $a \in M$ and $b \in N \smallsetminus M$ we have $a < b$.

## 13.2   Prime models

*Throughout the remainder of this chapter $T$ is a complete theory without finite models[4] (in the countable language $L$).*

---

[3]cf. Exercise 12.2.3.

[4]Therefore we do not have to distinguish between 'countable' and 'countably infinite' models here.

Omitting types is the key to our next structural result.

**Theorem 13.2.1.** (Vaught)
$\mathcal{M}$ *is an elementarily prime model of* $T$ *if and only if* $\mathcal{M}$ *is a countable atomic model of* $T$.

*Proof.* $\Longleftarrow$ follows from 'thinness' (for arbitrary languages $L$, cf. Theorem 12.2.1(1)).
$\Longrightarrow$. Let $\mathcal{M}$ be an elementarily prime model of $T$. It is clear (from the countability of the language) that $\mathcal{M}$ must be countable. To show that $\mathcal{M}$ is atomic, consider any nonisolated type $\Phi \in S_n(T)$. All we have to do is show that $\mathcal{M}$ omits it. But the omitting types theorem yields a model $\mathcal{N}$ of $T$ omitting this type. Then, since $\mathcal{M}$ is prime, it can be elementarily embedded into $\mathcal{N}$, i. e. $\mathcal{M} \overset{\equiv}{\hookrightarrow} \mathcal{N}$. Therefore $\mathcal{M}$ omits $\Phi$ too.  $\square$

(Notice, the direction from right to left did not require the language to be countable.) Together with Theorem 12.2.1(2) we obtain the uniqueness of elementarily prime models.

**Corollary 13.2.2.** $T$ *has (up to isomorphism) at most one elementarily prime model.*  $\square$

Using a stronger theorem about omitting simultaneously countably many types, for which we refer to the cited model-theoretic literature, one can show that theories with a countable saturated model (i. e. small theories) also have a countable atomic model, hence also an elementarily prime model. The converse is not true, as Peano arithmetic shows (for no completion of PA is small, cf. §12.1 and the exercises below).
One can further prove that $T$ *has a countable atomic (hence an elementarily prime) model if and only if the isolated types are dense in* $S_n(T)$ *for all* $n < \omega$ *(in the usual topological sense)*.

*Proof.* The isolated types are dense in $S_n(T)$ if and only if every $\varphi \in L_n$ consistent with $T$ is contained in an isolated type from $S_n(T)$. As every such $\varphi$ is satisfied in any model, this density holds if there is an atomic model. Assume, for the converse, density. For every $n < \omega$ consider the type $\Phi(x_0, \dots, x_{n-1}) =_{\text{def}} \{\neg\varphi : \varphi \in L$ and $\varphi$ is consistent with $T$ and isolates some type from $S_n(T)\}$. These types are not isolated and, by the generalized omitting type theorem, can be simultaneously omitted in a certain countable model of $T$, which is thus atomic.  $\square$

**Exercise 13.2.1.** If a theory (countable or not) has an elementarily prime model and a minimal model (in the sense of Exercise 12.2.8), then these two models are isomorphic.

**Exercise 13.2.2.** If $T$ has an elementarily prime model which is not minimal, then $T$ has an atomic model of power $\aleph_1$.

**Exercise 13.2.3.** Prove that the standard model $\mathbb{N}$ of Peano arithmetic is an elementarily prime and minimal model of true arithmetic $\mathrm{Th}\,\mathbb{N}$.

**Exercise 13.2.4.** Show that the additive group of integers is a minimal model of its own complete theory. (It will be shown in the last chapter that its theory does not have, however, an elementarily prime model.)

# 13.3   Countably categorical theories

Here and there we have looked at countable $\aleph_0$-categorical theories already. Now we are prepared to investigate them systematically. First define the **spectrum function** for any theory $T$ as follows.

Given $\lambda \in \mathbf{Cn}$, the number of isomorphism types of models of $T$ of power $\lambda$ is denoted by $\mathrm{I}(\lambda, T)$. In other words, a theory $T$ has up to isomorphism exactly $\mathrm{I}(\lambda, T)$ many models of power $\lambda$.

Thus $T$ is $\lambda$-categorical if and only if $\mathrm{I}(\lambda, T) = 1$. Further, for theories $T$ as in this chapter, i. e. countable complete ones without finite models, we have $\mathrm{I}(\lambda, T) = 0$ iff $\lambda < \omega$.

Note that $T$ is small, whenever $\mathrm{I}(\aleph_0, T) \leq \aleph_0$ (in particular, if $T$ is $\aleph_0$-categorical), for in countably many countably sets there are only countably many tuples.

**Theorem 13.3.1.** (Engeler, Ryll-Nardzewski, Svenonius)
*The following are equivalent (for any countable complete theory $T$ having infinite models).*
(i)    $T$ is $\aleph_0$-categorical (i. e. $\mathrm{I}(\aleph_0, T) = 1$).
(ii)    $T$ has a countable atomic and saturated model.
(ii$'$)    $T$ has a saturated elementarily prime model.
(iii)    All types in all $S_n(T)$ are isolated.
(iv)    All $S_n(T)$ are finite.
(v)    All Lindenbaum-Tarski algebras $\mathcal{B}_n(T)$ of $T$ are finite, i. e., up to $T$-equivalence there are, for all $n < \omega$, only finitely many formulas in $L_n$.
(vi)    All models of $T$ are atomic.
(vii)    All countable models of $T$ are atomic.

This theorem is often called *Ryll-Nardzewski's theorem*. To be fair it should be mentioned that it was proved by three people independently, see

[Engeler, E. : A characterization of theories with isomorphic denumerable models, Notices Am. Math. Soc. **6** (1959) 161],

[Ryll-Nardzewski, C. : On the categoricity in power $\aleph_0$, Bulletin Acad. Polon. Sci. Sér. Sci. Math. Astronom. Phys. **7** (1959) 545 – 548], and

[Svenonius, L.: $\aleph_0$-categoricity in first-order predicate calculus, Theoria (Lund) **25** (1959) 82 – 94].

*Proof.* (i) $\Longrightarrow$ (ii'). Since every model has a countable elementary substructure and there is only one such up to isomorphism, 'the' countable model is an elementarily prime model. Further, as mentioned before, $T$ is small (if it is $\aleph_0$-categorical). Thus it has a countable saturated model, which must be (isomorphic to) the elementarily prime model.

(ii') $\Longleftrightarrow$ (ii). See Theorem 13.2.1.

(ii) $\Longrightarrow$ (iii). As all types from $S_n(T)$ are realized in every saturated model, they are all realized in an atomic model. Hence they are all isolated.

(iii) $\Longrightarrow$ (iv). Infinite compact spaces contain nonisolated points; then refer to Remark (3) in §11.4.

(iv) $\Longrightarrow$ (v). Formulas of $L_n$ that lie in exactly the same types from $S_n(T)$ must be $T$-equivalent. Hence there are no more formulas (up to $T$-equivalence) than subsets of $S_n(T)$. Consequently, $|\mathcal{B}_n(T)| \leq 2^{|S_n(T)|} < \aleph_0$.

(v) $\Longrightarrow$ (iii). $\mathcal{B}_n(T)$ is then atomic (below every formula, after finitely many steps down one arrives at an atom).

(iii) $\Longrightarrow$ (vi). If all types of $T$ are isolated, all models of $T$ are atomic.

(vi) $\Longrightarrow$ (vii) is trivial.

(vii) $\Longrightarrow$ (i) follows from the uniqueness of countable atomic models. $\quad\Box$

**Corollary 13.3.2.** *$T$ is $\aleph_0$-categorical if and only if for all (or for some) $\mathcal{M} \models T$ and all (or some) tuple(s) $\bar{a}$ from $\mathcal{M}$, the theory $\mathrm{Th}(\mathcal{M}, \bar{a})$ is $\aleph_0$-categorical.*

*Proof.* $\Longrightarrow$. If $\mathcal{M} \models T$, $\bar{a}$ is from $\mathcal{M}$, and $\mathcal{B}_{n+m}(T)$ is finite, then also $\mathcal{B}_n(\mathrm{Th}(\mathcal{M}, \bar{a}))$ is finite.

$\Longleftarrow$. Let $\mathcal{M}$ and $\bar{a}$ be as before. Assume $\mathrm{Th}(\mathcal{M}, \bar{a})$ is $\aleph_0$-categorical. Since $T$ is complete, for all $\varphi, \psi \in L_n$ we have $\varphi \sim_T \psi$ iff $\varphi(\mathcal{M}) = \psi(\mathcal{M})$ iff $\varphi \sim_{\mathrm{Th}(\mathcal{M}, \bar{a})} \psi$. Hence $|\mathcal{B}_n(T)| \leq |\mathcal{B}_n(\mathrm{Th}(\mathcal{M}, \bar{a}))|$, and the assertion follows. $\quad\Box$

Let $\mathcal{M}$ be an arbitrary structure and $n < \omega$. The **orbit** of $\bar{a} \in M^n$ under $\mathrm{Aut}\,\mathcal{M}$ is the set $\{f[\bar{a}] : f \in \mathrm{Aut}\,\mathcal{M}\}$. It is obvious that the orbits of all $n$-tuples of $M$ under $\mathrm{Aut}\,\mathcal{M}$ partition the set $M^n$ (in other words, to be in the same orbit is an equivalence relation).

**Exercise 13.3.1.** Prove that tuples from the same orbit of $M$ have the same type over $\emptyset$; if $\mathcal{M}$ is countable and saturated, the converse holds.

**Exercise 13.3.2.** Show that $T$ is $\aleph_0$-categorical if and only if every countable model $\mathcal{M}$ of $T$ has finitely many orbits of $n$-tuples (under $\operatorname{Aut}\mathcal{M}$) for every $n$.

**Exercise 13.3.3.** Prove that an elementarily prime model of a countable $\aleph_1$-categorical, but not $\aleph_0$-categorical theory is minimal.

**Exercise 13.3.4.** Refute the preceding assertion for totally categorical theories.

## 13.4  Finitely many countable models

In the previous section we investigated theories that have only one countable model. We conclude this chapter with an example of a complete theory with exactly 3 countable models, which can be modified, for every $n > 3$, as to possess precisely $n$ countable models. All the more surprising is the following

**Theorem 13.4.1.** (Vaught)
$I(\aleph_0, T) \neq 2$ for every (countable complete) theory $T$.

*Proof.* Assume, every countable model of $T$ is isomorphic to one of the countable models $\mathcal{M}_0$ and $\mathcal{M}_1$. We have to show that the latter two are isomorphic. As mentioned before the previous theorem, $T$ is small, and thus has a countable saturated model. Suppose this model is $\mathcal{M}_1$. By the universality of saturated structures, we have $\mathcal{M}_0 \overset{\equiv}{\hookrightarrow} \mathcal{M}_1$. Every $\mathcal{N} \models T$ has, by Löwenheim-Skolem, a countable elementary substructure, hence $\mathcal{M}_0 \overset{\equiv}{\hookrightarrow} \mathcal{N}$ or $\mathcal{M}_1 \overset{\equiv}{\hookrightarrow} \mathcal{N}$. In any case this yields $\mathcal{M}_0 \overset{\equiv}{\hookrightarrow} \mathcal{N}$, hence $\mathcal{M}_0$ is an elementarily prime model of $T$, hence atomic. If also $\mathcal{M}_1$ is atomic, then, by uniqueness, $\mathcal{M}_0 \cong \mathcal{M}_1$.

If, on the other hand, $\mathcal{M}_1$ were not atomic, it would realize a nonisolated type $\Phi \in S_n(T)$, say by $\bar{a}$. Consider $T' = \operatorname{Th}(\mathcal{M}_1, \bar{a})$. As $T \cup \Phi(\bar{a}) \subseteq T'$, every model of $T'$ has the form $(\mathcal{N}, \bar{c})$, where $\mathcal{N} \models T$ and $\bar{c} \models \Phi$ in $\mathcal{N}$.

It suffices to show that $T'$ is $\aleph_0$-categorical, for then, in view of the 'for some' version of Corollary 13.3.2, also $T$ would be $\aleph_0$-categorical, hence $\mathcal{M}_1$ would have to be atomic, in contrast to the assumption.

All countable models of $T'$ have the form $(\mathcal{M}_i, \bar{c})$, where $\bar{c} \models \Phi$ in $M_i$ ($i < 2$). As $\mathcal{M}_0$ is atomic and thus omits $\Phi$, we must have $i = 1$. So let $(\mathcal{M}_1, \bar{c})$ and $(\mathcal{M}_1, \bar{c}')$ be countable models of $T'$. We show that these are isomorphic. First of all, we have $\bar{c} \models \Phi$ and $\bar{c}' \models \Phi$, hence by the completeness of $\Phi$ also $\operatorname{tp}^{\mathcal{M}_1}(\bar{c}) = \operatorname{tp}^{\mathcal{M}_1}(\bar{c}')$. The homogeneity of saturated structures now yields $f \in \operatorname{Aut}\mathcal{M}_1$ such that $f[\bar{c}] = \bar{c}'$. But then $f : (\mathcal{M}_1, \bar{c}) \cong (\mathcal{M}_1, \bar{c}')$, and $T'$ is $\aleph_0$-categorical.                                                                    $\square$

We conclude with an example of a complete theory with exactly 3 countable models. For every $n > 3$, by adding $n - 2$ additional predicates one can obtain an example from this with exactly $n$ countable models.

Ehrenfeucht's **Example** of a complete theory $T_3$ with $I(\aleph_0, T_3) = 3$.

In the language $L_<$, consider the model $\eta = (\mathbb{Q}; <) \models \text{DLO}_{--}$. Adding new constants $\underline{n}$ for all $n < \omega$ we obtain the expansion $(\eta; \omega)$. Set $T_3 = \text{Th}(\eta; \omega)$. This is an $L_<(\omega)$-theory. Consider the following three models of this theory, the standard model $\mathcal{M}_1 = (\eta; \omega)$, and two $L_<(\omega)$-expansions of $\eta$ in which the $\underline{n}$ have an upper bound. Interpret the $\underline{n}$ so that in one of these, $\mathcal{M}_2$, they have even a least upper bound, while in the other, $\mathcal{M}_3$, they don't. Then these three models are countable and pairwise nonisomorphic. We are going to show that every countable model of $T_3$ is isomorphic to one of these.

To this end let $\mathcal{N}' \models T_3$. Then $\mathcal{N}' = (\mathcal{N}, \{a_i : i < \omega\})$, where $a_0 < a_1 < a_2 < \ldots$. Put $A_{\mathcal{N}'} = \{a \in N : a \le a_n \text{ for some } n < \omega\}$ and $B_{\mathcal{N}'} = N \setminus A_{\mathcal{N}'}$. Notice, the latter contains no interpretations of new constants and is thus either empty or merely a dense linear ordering; moreover, if it is so it must be a model of either $\text{DLO}_{--}$ or $\text{DLO}_{+-}$ (it can't have a right endpoint!).

The orderings $(-\infty, a_0)$ and $(a_{n-1}, a_n)$ are isomorphic, by Cantor's theorem. Patching the corresponding isomorphisms together we obtain an isomorphism from $A_{\mathcal{N}'}$ onto $(\eta; \omega)$.

If $B_{\mathcal{N}'} = \emptyset$, then clearly $\mathcal{N}' \cong \mathcal{M}_1$. If not, there are two cases to distinguish.

*Case 1.* $B_{\mathcal{N}'}$ has a least element $e$. Then $B_{\mathcal{N}'} \setminus \{e\}$ is a model of $\text{DLO}_{--}$, hence isomorphic to $\eta$. Then $\mathcal{N}' \cong \mathcal{M}_2$.

*Case 2.* $B_{\mathcal{N}'}$ has no least element. Then $B_{\mathcal{N}'} \models \text{DLO}_{--}$ and $B_{\mathcal{N}'} \cong \eta$. In this case $\mathcal{N}' \cong \mathcal{M}_3$.

This completes the proof of $I(\aleph_0, T_3) = 3$. To show the completeness of $T_3$, note that the algebraically prime model $\mathcal{M}_1$ is, in fact, an elementarily prime model (exercise!). Hence $T_3$ is complete. □

**Exercise 13.4.1.** Prove that $\mathcal{M}_1$ is an elementarily prime model of $T_3$.

**Exercise 13.4.2.** For all $\alpha$ with $3 < \alpha \le \aleph_0$, find a complete theory $T_\alpha$ such that $I(\aleph_0, T_\alpha) = \alpha$.

**Exercise 13.4.3.** Prove $I(\aleph_0, \text{PA}) = 2^{\aleph_0}$.

**Exercise 13.4.4.** Prove the same for the ($L_{\mathbb{Z}}$-theory of) the additive group of the integers.

It is not known if there is a countable complete theory $T$ for which $I(\aleph_0, T)$ assumes an uncountable value different from (and hence, by Lemma 15.7.1, necessarily smaller than) $2^{\aleph_0}$. The assertion that this is not the case is known as **Vaught's conjecture**. A discussion of the latter can be found e. g. in Hodges' model theory books.

# Part V

# Two applications

We conclude with two independent applications of our entire material, the so-called strongly minimal theories whose dimension theory generalizes and yields as a special case that of Ernst Steinitz for algebraically closed fields (Chapter 14), and the complete theory of the abelian group $\mathbb{Z}$ of the integers (Chapter 15), whose behaviour gives us the opportunity to draw the reader's attention to concepts from stability theory (§§15.7–8), a possible and desirable subject of further reading.

*Unless stated otherwise, theories are arbitrary again, in a fixed, but arbitrary signature $\sigma$ (with $L = L(\sigma)$, the corresponding language). Further, $\mathcal{M}$ and $\mathcal{N}$ are arbitrary L-structures.*

# Chapter 14

# Strongly minimal theories

Our concern here are theories that are minimal with respect to the sets that are definable in their models. Namely, a complete theory with infinite models is said to be **strongly minimal** if every parametrically definable subset of each of its models is either finite or cofinite.[1] This is the least possible amount of sets definable by 1-place formulas with parameters, for—in any structure—finite sets are definable by disjunctions of equations, while cofinite sets are definable by negations of such. These theories were introduced by William Marsh in

[Marsh, W. E. : On $\omega_1$-categorical and not $\omega$-categorical theories, Dissertation, Dartmouth College 1966, unpublished]

where their basic properties were derived, including their most interesting feature, a dimension theory like the one for vector spaces or that for algebraically closed fields. We present Marsh' theory in the first five sections. And in §14.4 we show that this general theory has as particular instances the paradigmatic algebraic examples. Namely, we derive Steinitz' theorem as a consequence of the more general theory and indicate in the exercises the same for the dimension theory of vector spaces.

As a final application of the dimension theory, we obtain, among other things, the impossibility of a finite axiomatization of countably categorical strongly minimal theories, which was first proved by Johann A. Makowsky and independently by Robert Vaught.

---

[1]Recall, a set is cofinite if its complement is finite.

## 14.1   Basic properties

Given a finite subset $A = \{a_0, \ldots, a_{n-1}\}$ of a structure $\mathcal{M}$, the formula $\bigvee_{i<n} x = a_i$ defines the subset $A$ in $\mathcal{M}$, and its negation defines its complement (in both definitions we used the parameters $a_0, \ldots, a_{n-1}$). Therefore, as mentioned, finite and cofinite sets are *always* parametrically definable. Thus, in a finite structure *every* subset is (parametrically) definable. To exclude this vacuous case we required a strongly minimal theory to be complete without finite models. Note that such a complete theory without finite models is strongly minimal if and only if every 1-place formula with parameters from a model of that theory is either itself algebraic or else so is its negation.

**Examples.** (1)   $T_{\equiv}^{\infty}$ is strongly minimal, due to quantifier elimination (to be shown in Exercise 9.2.6), for in every model $\mathcal{M}$ all 1-place formulas are equivalent to $x = x$, $x \neq x$, or to a boolean combination of equations $x = a$, where $a \in M$. A formula of the latter kind is algebraic if and only if every disjunct contains an (unnegated) equation of the form $x = a$, where $a \in M$. Otherwise its negation must be algebraic.

(2)   We leave as an exercise that, analogously, the theory $T_{\mathcal{K}}^{\infty}$ of all infinite vector spaces over a division ring $\mathcal{K}$ is strongly minimal (cf. Exercise 9.2.7).

(3)   Similarly, recalling the discussion before Lemma 11.4.3 it is clear that $\mathrm{ACF}_q$ is strongly minimal for every characteristic $q$.

(4)   The (complete theories of the) Prüfer groups are strongly minimal (exercise!).

(5)   The simplest example of a complete theory with infinite models which is not strongly minimal is an infinite set with a predicate defining an infinite subset whose complement is also infinite.

The diversity of these examples indicates that strongly minimal theories can be quite complicated. And the example of $\mathrm{ACF}_0$ (where algebraic geometry takes place!) may serve as a warning to think *all* definable sets are simple: only the definable *sub*sets are, i. e. the sets definable by 1-place formulas (with parameters), while already 2-place formulas may define complicated curves.[2]

Nevertheless, the structure of the class of models of a strongly minimal theory is as simple as it can get—it forms an elementary chain, as we will

---

[2]Algebraic geometry deals with these latter. Recent developments show that model theory has something to say about this, even in a much more general context (which is part of *geometric stability theory*, a powerful theory created by Boris Zilber).

see in due course. We start with much simpler things.

**Lemma 14.1.1.** *Let* Th $\mathcal{M}$ *be strongly minimal.*
*The infinite algebraically closed subsets of* $\mathcal{M}$ *are exactly the (universes of) elementary substructures of* $\mathcal{M}$.

*Proof.* The direction from right to left is clear, as by Lemma 11.5.2(1) every elementary substructure is algebraically closed (and infinite, as so is $\mathcal{M}$).

For the converse, suppose $A \subseteq M$ is infinite and algebraically closed. To apply Tarski-Vaught (in the form of Corollary 8.3.3), let $\varphi \in L_1(A)$ be satisfied in $\mathcal{M}$. We have to show that $\varphi(\mathcal{M}) \cap A$ is not empty. If $\varphi$ is algebraic, we even have $\varphi(\mathcal{M}) \subseteq \mathrm{acl}_{\mathcal{M}} A \subseteq A$. If not, $\varphi(\mathcal{M})$ is cofinite and thus has a nonempty intersection with the (infinite!) set $A$, as desired. $\square$

**Proposition 14.1.2.** *If* $\mathcal{K} \models \mathrm{ACF}$ *and* $A \subseteq K$, *then* $\mathrm{acl}_{\mathcal{K}} A \preccurlyeq \mathcal{K}$; *in particular, all algebraically closed subsets of algebraically closed fields are themselves algebraically closed fields. Further, given an algebraically closed field* $\mathcal{K}$ *of any characteristic* $q$, *the field* $\mathrm{acl}_{\mathcal{K}} \emptyset$ *is an elementarily prime model of* $\mathrm{ACF}_q$.

*Proof.* In view of the preceding lemma, for the first part we only have to verify that all algebraically closed subsets of algebraically closed fields are infinite. But this follows by the same argument as that showing that algebraically closed fields are infinite (see Remark at the beginning of §9.4).

For the assertion about the prime model it suffices to recall from Lemma 11.5.2(3) that $\mathrm{acl}_{\mathcal{M}} A = \mathrm{acl}_{\mathcal{M}'} A$, whenever $\mathcal{M} \equiv \mathcal{M}'$ and $A \subseteq M \cap M'$. $\square$

**Proposition 14.1.3.** *The following are equivalent for any structure* $\mathcal{M}$.
(i) Th $\mathcal{M}$ *is strongly minimal.*
(ii) *For every* $A \subseteq M$ *there is exactly one nonalgebraic type in* $\mathrm{S}_1^{\mathcal{M}}(A)$.
(iii) $\mathcal{M}$ *is infinite and for all* $\mathcal{N} \succcurlyeq \mathcal{M}$ *and* $A \subseteq M$ *all elements of* $N \setminus \mathrm{acl}\, A$ *have the same type over* $A$ *(even over* $\mathrm{acl}\, A$).

*Proof.* Lemma 11.4.3(4) says that the existence of nonalgebraic types is equivalent to $\mathcal{M}$ being infinite.

Assume, over some set $A \subseteq M$ there are two distinct nonalgebraic complete 1-types $\Phi$ and $\Psi$. Then $\Phi$ contains a formula $\varphi$ from $L_1(A)$ whose negation is in $\Psi$. Then neither $\varphi$ nor $\neg\varphi$ are algebraic, hence Th $\mathcal{M}$ is not strongly minimal.

If, for the converse, there is a nonalgebraic formula $\varphi \in L_1(A)$ whose negation is not algebraic either, then Lemma 11.4.3(3) yields a nonalgebraic type in $S_1^{\mathcal{M}}(A)$ that contains $\varphi$ and one that contains $\neg\varphi$, i. e. at least two nonalgebraic types in $S_1^{\mathcal{M}}(A)$.

This proves the first equivalence. The second follows from the easy observation that the set of elements of $N$ with an algebraic type over $A$ (or, equivalently, over $\mathrm{acl}_N A$) is exactly $\mathrm{acl}_N A = \mathrm{acl}_{\mathcal{M}} A$.         $\square$

We conclude with a technical result about isolation of types, which we need in the next section.

**Lemma 14.1.4.** *Let* $\mathrm{Th}\,\mathcal{M}$ *be strongly minimal and* $A \subseteq M$.

*If* $\mathrm{acl}_{\mathcal{M}} A$ *is finite, then*

(1)   *the set* $\mathrm{acl}_{\mathcal{M}} A$ *is definable in* $\mathcal{M}$ *with parameters from* $A$,

(2)   *every type from* $S_1^{\mathcal{M}}(A)$ *is isolated (and exactly one of them nonalgebraic), and*

(3)   *if* $a \in M$, *then every tuple from* $\mathrm{acl}_{\mathcal{M}}(A \cup \{a\})$ *has an isolated type in* $\mathcal{M}$.

*Proof.* Ad (1). If $\mathrm{acl}\, A$ is finite, it is a finite union of (finite) sets definable with parameters from $A$, hence itself definable with parameters from $A$ in $\mathcal{M}$.

Ad (2). The algebraic types in $S_1^{\mathcal{M}}(A)$ (i. e. the types of elements of $\mathrm{acl}\, A$) are isolated anyway. By (1), the complement $M \smallsetminus \mathrm{acl}\, A$ is definable with parameters from $A$, by a formula $\varphi \in L_1(A)$, say. If we can show that (the equivalence class of) $\varphi$ is an atom in the Lindenbaum-Tarski algebra $\mathcal{B}_1(\mathrm{Th}(\mathcal{M}, A))$, then there is only one 1-type over $A$ containing this formula, which is thus isolated. So let $\psi \in L_1(A)$ be such that $\psi(\mathcal{M}) \subseteq \varphi(\mathcal{M})$. By strong minimality, $\psi(\mathcal{M})$ or $\varphi(\mathcal{M}) \smallsetminus \psi(\mathcal{M})$ is finite, hence contained in $\mathrm{acl}\, A$. But $\mathrm{acl}\, A$ and $\varphi(\mathcal{M})$ are disjoint, so one of these sets is in fact empty. In other words, $\psi$ is either $\mathcal{M}$-equivalent to $\varphi$ or to $\bot$, i. e., $\varphi$ is an atom, and we are done.

Ad (3). Let $A = \{a_0, \ldots, a_{n-1}\}$. Set $a_n = a$ and $\bar{a} = (a_0, \ldots, a_n)$. Every tuple of $\mathrm{acl}(A \cup \{a\}) = \mathrm{acl}\, \bar{a}$ is contained in a tuple of the form $\bar{a}{}^\frown \bar{c}$, where $\bar{c}$ is a tuple from $\mathrm{acl}(A \cup \{a\}) \smallsetminus (A \cup \{a\})$. Therefore (by the easy direction of the transitivity of isolation), we may assume without loss that the tuple in question is $\bar{a}{}^\frown \bar{c}$. Now $\mathrm{tp}(\bar{c}/\bar{a})$ is algebraic, hence isolated by Lemma 11.5.2(2). Hence, by the aforementioned transitivity again, we are left with showing that $\mathrm{tp}(\bar{a})$ is isolated. By transitivity, in turn, this reduces to the verification of the isolation of the types $\mathrm{tp}(a_i/\{a_0, \ldots, a_{i-1}\})$ for all $i \leq n$. But this follows from (2), as the sets $\mathrm{acl}\{a_0, \ldots, a_{i-1}\}$ are finite for all $i \leq n$.         $\square$

**Exercise 14.1.1.** Prove that the Prüfer groups have a strongly minimal complete theory.

**Exercise 14.1.2.** Show that for any division ring $\mathcal{K}$ the theory $T_{\mathcal{K}}^{\infty}$ of all infinite $\mathcal{K}$-vector spaces is strongly minimal.

**Exercise 14.1.3.** Prove that a countable strongly minimal theory has a countable saturated model.

**Exercise 14.1.4.** Given an arbitrary 1-place formula $\psi(x, \bar{y})$ in a strongly minimal theory $T$, show that there is a finite bound on the power of the *finite* sets among the sets of the form $\psi(\mathcal{M}, \bar{a})$, where $\mathcal{M} \models T$ and $\bar{a}$ is in $M$.

A formula $\varphi \in L_n(M)$ is said to be **minimal** in an infinite $L$-structure $\mathcal{M}$ if for every formula $\psi \in L_n(M)$, either $\varphi(\mathcal{M}) \cap \psi(\mathcal{M})$ or $\varphi(\mathcal{M}) \smallsetminus \psi(\mathcal{M})$ is finite. (So, to be minimal in a structure is really a property of (parametrically) definable sets.) A formula is said to be **strongly minimal** in $T$ if it is minimal in every model of $T$. Note that a complete theory with infinite models is strongly minimal if and only if $x = x$ is strongly minimal in $T$.

**Exercise 14.1.5.** Let $T$ be a complete theory with infinite models.
Prove that $T$ is strongly minimal if and only if it has an $\omega$-saturated model in which $x = x$ is minimal.

**Exercise 14.1.6.** Find an infinite structure in which $x = x$ is minimal, but *not* strongly minimal.

**Exercise 14.1.7.** Show that every commutative ring in which $x = x$ is minimal is a field.

**Exercise 14.1.8.** Consider the example $\mathcal{A} = (A, g)_{g \in G}$ from Exercise 9.2.8. Given $g \in G$, set $\text{fix}_A\, g = \{a \in A : g(a) = a\}$.
Prove that Th $\mathcal{A}$ is strongly minimal if and only if $x = x$ is minimal in $\mathcal{A}$ if and only if all fixed sets $\text{fix}_A\, g$, where $g$ runs over $G$, are finite or cofinite in $A$.

In the next exercise, let $\mathcal{M}$ be a structure whose language contains the language of groups, $L' = L(1; \cdot, {}^{-1})$, and assume that its $L'$-reduct, $\mathcal{M}'$, is a group. Further, suppose $x = x$ is minimal in $\mathcal{M}$.

**Exercise 14.1.9.** (1) Show that every proper subgroup of (the $L'$-reduct of) $\mathcal{M}$ which is, as a subset of $M$, definable (possibly with parameters) in $\mathcal{M}$, is finite.
(2) Let $f$ be a group homomorphism from $\mathcal{M}'$ to itself whose graph is definable in $\mathcal{M}$.
Prove that $f$ is onto whenever its image is infinite.
(3) For $f$ as before, prove that it is the zero map, whenever its kernel is infinite.

In the remaining exercises we present a recent result of Frank Wagner confirming a conjecture of Klaus-Peter Podewski, in case of positive characteristic, saying that a field in which the formula $x = x$ is minimal must be algebraically closed, and hence strongly minimal. (Note that the characteristic 0 case is still open.)

[Wagner, F. : Minimal fields, J. Symbolic Logic, to appear]

We work in a field $\mathcal{K}$ of characteristic $p > 0$. Since we do not yet know that it is algebraically closed, we do not know either if algebraic in the field-theoretic sense is the same as algebraic in the model-theoretic sense. Therefore, in the remaining exercises, we reserve the word *algebraic* for the field-theoretic term, and denote by $\mathcal{K}_0$ the field-theoretic closure of $\emptyset$ in $\mathcal{K}$ (i. e. the union of the solution sets of polynomials over the prime field $\mathbb{F}_p$), while, as usual, $\mathrm{acl}_{\mathcal{K}}\emptyset$ denotes the model-theoretic closure of $\emptyset$ in $\mathcal{K}$. Recall that $\mathcal{K}_0 \subseteq \mathrm{acl}_{\mathcal{K}}\emptyset$. One of the things to prove will be the inverse inclusion.

The first exercise is for algebraic preparation. Recall that a **finite extension** of a field $\mathcal{K}$ is an extension of $\mathcal{K}$ which is, as a vector space over $\mathcal{K}$, finite-dimensional.

**Exercise 14.1.10.** Let $a, b \in \mathcal{K}$.
(1)  Verify $(a + b)^p = a^p + b^p$ and $(a - b)^p = a^p - b^p$.
(2)  Derive that $p$th roots are unique in $\mathcal{K}$.
(3)  Derive that the map given by $\varphi : x \mapsto x^p$ is an embedding of the field $\mathcal{K}$ into itself, the so-called **Frobenius** homomorphism.
(4)  Prove that, if $\mathcal{K}$ contains one root of the polynomial $x^p - x - a$, then it contains all of them (i. e., $\mathcal{K}$ is, so to speak, *normal* with respect to polynomials of the form $x^p - x - a$).
(5)  Show that every finite extension of a finite field—in particular, every extension of a finite field obtained by adjoining finitely many (absolutely) algebraic elements—is a finite field.
(6)  Derive that every element algebraic over a finite field is a root of unity (i. e., for any such element, $c$, there is a natural number $n > 0$ such that $c^n = 1$).
(7)  Derive that the subfield $\mathcal{K}_0$ of $\mathcal{K}$ is a locally finite field (i. e., every finite subsets of $\mathcal{K}_0$ generates (or is contained in) a finite subfield).
(8)  Prove that any element algebraic over a locally finite field is a root of unity.

The following fact uses Galois theory and may be extracted from

[Macintyre, A. : On $\omega_1$-categorical theories of fields, Fundamenta Math. **71** (1971) 1 – 25]

**Fact.** *Let $n > 0$ be a natural number. Consider $\mathcal{L}$, the extension of $\mathcal{K}$ obtained by adjoining all $n$th roots of unity (in other words, $\mathcal{L} = \mathcal{K}(e)$, where $e$ is a primitive $n$th root of unity). Then $\mathcal{L}$ can be obtained by successively adjoining $m$th roots (of arbitrary elements from preceding extensions), where $m < n$, and roots of the polynomials $x^p - x - a$ over the preceding extensions (i. e., where $a$ is from a preceding extension).*

*Therefore, if $\mathcal{K}$ contains, along with all of its elements, also all of their $m$th roots, where $0 < m < n$, and all roots of the polynomials $x^p - x - a$, where $a \in \mathcal{K}$, then $\mathcal{K}$ contains all $n$th roots of unity.*

Let, from now on, $x = x$ be minimal in $\mathcal{K}$.

**Exercise 14.1.11.** (1)  Show that every polynomial $x^p - x - a$ (where $a \in \mathcal{K}$) has a root in $\mathcal{K}$.

(2)  Derive that $\mathcal{K}$ contains *all* roots of polynomials of the form $x^p - x - a$, where $a \in \mathcal{K}$.

(3)  Show that every element in $\mathcal{K}$ has an $n$th root in $\mathcal{K}$ (for every natural number $n > 0$).

(4)  Using the above fact, derive that $\mathcal{K}$ is closed under taking $n$th roots of elements, for all natural numbers $n > 0$. In particular, $\mathcal{K}$ contains *all* roots of unity.

(5)  Derive further that the Frobenius homomorphism $\varphi$ is an automorphism of $\mathcal{K}$.

**Exercise 14.1.12.** Let, as before, $\varphi$ be the Frobenius automorphism of $\mathcal{K}$.
(1)  Show that the fixed points of the powers $\varphi^n$, where $n > 0$, are in $\mathcal{K}_0$.
(2)  Derive that if $a \in \mathcal{K} \smallsetminus \mathcal{K}_0$, then $\{\varphi^n(a) : 0 < n < \omega\}$ is infinite.
(3)  Derive that $\mathcal{K}_0 = \mathrm{acl}_{\mathcal{K}} \emptyset$.

**Exercise 14.1.13.** (1)  Show that every element that is (field-theoretically) algebraic over $\mathcal{K}_0$ is a root of unity.
(2)  Derive that $\mathcal{K}_0$ is algebraically closed.
(3)  Verify that $\mathcal{K}_0$ is an elementary substructure of $\mathcal{K}$.
(4)  Derive that $\mathcal{K}$ is algebraically closed and hence strongly minimal.

It should be noted that a similar result for groups has been known since [Reineke, J. : Minimale Gruppen, Z. Math. Logik Grundlagen Math. **21** (1975) 357 – 359].

## 14.2  Categoricity, saturated and atomic models

First we show that 'most' of the models of a strongly minimal theory are saturated.

**Lemma 14.2.1.** *Let $\lambda$ be an infinite cardinal.*

*A model $\mathcal{M}$ of a strongly minimal theory is $\lambda$-saturated if and only if, for all subsets $A \subseteq M$ with $|A| < \lambda$, we have $\mathrm{acl}_{\mathcal{M}} A \neq M$.*

*Consequently, a model $\mathcal{M}$ of a strongly minimal theory is saturated if and only if, for all subsets $A \subseteq M$ with $|A| < |M|$, we have $\mathrm{acl}_{\mathcal{M}} A \neq M$.*

*Proof.* The condition is certainly necessary for saturation (for acl $A$ realizes only algebraic types, while $\mathcal{M}$ has also nonalgebraic types over $A$, cf. Remark (2) in §12.1). To see that it is sufficient, assume acl $A \neq M$. Then all 1-types over $A$ are realized in $\mathcal{M}$: for the algebraic ones this is clear anyway, while the only nonalgebraic one (cf. Proposition 14.1.3) is realized by *every* element of $M \smallsetminus \mathrm{acl}\, A$. $\qquad\square$

**Theorem 14.2.2.** *Let $T$ be a strongly minimal theory.*
(1)   *All models of $T$ of power $> |T|$ are saturated.*
(2)   *$T$ is categorical in every $\lambda > |T|$.*

*Proof.* Ad (1). If $|M| > |T|$, then $|\operatorname{acl} A| \leq |A| + |T| < |M|$ (see Lemma 11.5.4), hence $\operatorname{acl} A \neq M$, provided $A \subseteq M$ and $|A| < |M|$. The assertion is now clear in view of the lemma.

(2) follows from (1) and the uniqueness of saturated models (we are using here the generalization of Theorem 12.1.1(2) asked for in Exercise 12.1.2).                                                                        □

Once we have the dimension theory at hand, we can prove that even more models are saturated (see Theorem 14.5.1 below).

**Corollary 14.2.3.** (Steinitz) *For any characteristic $q$, the theory $\operatorname{ACF}_q$ is categorical in all uncountable powers.*                                                □

In Exercise 14.1.3 it was to be shown that countable strongly minimal theories have a countable saturated model (in particular, $\operatorname{ACF}_q$ does). Next we exhibit a concrete one.

**Proposition 14.2.4.** *Suppose $\mathcal{M}$ is an uncountable model of a countable strongly minimal theory.*
    *If $a_i \in M \smallsetminus \operatorname{acl}\{a_j : j < i\}$ for all $i < \omega$, then $\operatorname{acl}\{a_i : i < \omega\}$ is a countable and saturated elementary substructure of $\mathcal{M}$.*

*Proof.* Consider $M_\omega =_{\mathrm{def}} \operatorname{acl}\{a_i : i < \omega\}$. This set is infinite and algebraically closed and thus the universe of an elementary substructure $\mathcal{M}_\omega$ of $\mathcal{M}$. In view of the countability of the language we further have $|\mathcal{M}_\omega| \leq |\operatorname{acl}\{a_i : i < \omega\}| + \aleph_0 = \aleph_0$.
    To prove saturation let $A \subseteq M$ be finite. We have to prove $\operatorname{acl} A \neq M$ (Lemma 14.2.1). Each of the finitely many elements of $A$ is contained in the algebraic closure of finitely many $a_i$. Hence the same is true of $A$, i. e., $A \subseteq \operatorname{acl}\{a_i : i < n\}$ for some $n < \omega$. Then, by hypothesis, e. g. $a_n$ is not contained in $\operatorname{acl} A$, as desired.                                                          □

This result holds true, mutatis mutandis, for uncountable strongly minimal theories. The proof of this, however, makes use of the dimension theory to be developed in the next section and will be thus postponed. We turn to the description of atomic models instead.

**Proposition 14.2.5.** *Let $\mathcal{M}$ be a model of a strongly minimal theory.*

*Then there is a countable (finite or infinite) well-ordered subset $A$ of $M$ whose algebraic closure in $\mathcal{M}$ is the universe of an atomic structure $\mathcal{M}_0 \preccurlyeq \mathcal{M}$ such that $\operatorname{acl} B$ is finite for all sections $B$ of $A$.*

*Proof.* If $\operatorname{acl} \emptyset$ is infinite, set $A = \emptyset$. Otherwise pick any $a_0$ in $M \smallsetminus \operatorname{acl} \emptyset$. If now $\operatorname{acl}\{a_0\}$ is infinite, set $A = \{a_0\}$. Otherwise pick any $a_1$ in $M \smallsetminus \operatorname{acl}\{a_0\}$ and so forth. Then we can either continue our choice of the $a_i$ infinitely often without making the algebraic closures of the finite sections infinite—in which case we set $A = \{a_i : i < \omega\}$—or else, after finitely many steps, we obtain an infinite algebraic closure, say $\operatorname{acl}\{a_0, \ldots, a_n\}$, while $\operatorname{acl}\{a_0, \ldots, a_{n-1}\}$ is still finite. In this latter case we set $A = \operatorname{acl}\{a_0, \ldots, a_n\}$. In both cases $\operatorname{acl} A$ is (the universe of) an elementary substructure $\mathcal{M}_0$ of $\mathcal{M}$ (Lemma 14.1.1).

We are left with showing that $\mathcal{M}_0$ is atomic. In the finite case this is immediate from Lemma 14.1.4(2). In the infinite case, on the other hand, $\mathcal{M}_0$ is the union of the algebraic closures of the finite sections, which contain, by that same lemma, only tuples realizing isolated types. But this means that $\mathcal{M}_0$ is atomic. □

**Remarks.**

(1)  In this proposition, if the strongly minimal theory is countable, then so is $\mathcal{M}_0$. Hence it is an elementarily prime model by Theorem 12.2.1(1).

(2)  This yields another proof of the fact proved in Proposition 14.1.2 that the field $\operatorname{acl}_{\mathbb{C}} \emptyset$ of all algebraic numbers is an elementarily prime model of $\mathrm{ACF}_0$.

(3)  Similarly for positive characteristics $p$.

**Exercise 14.2.1.** Generalize the above Remark (1) to arbitrary cardinalities.

**Exercise 14.2.2.** Let $\mathcal{A} = (A, g)_{g \in G}$ as in Exercise 14.1.8 have a strongly minimal theory. Exhibit a saturated model of this theory.

**Exercise 14.2.3.** Exhibit all the saturated models of the complete theory of the Prüfer group.

# 14.3   Dimension

We now develop the aforementioned dimension theory for models of strongly minimal theories—modelled after that for vector spaces or algebraically closed fields.

*From now on, unless stated otherwise, all sets are subsets of (the universe of) $\mathcal{M}$, a fixed $L$-structure, and we write $\operatorname{acl}$ instead of $\operatorname{acl}_{\mathcal{M}}$. After the* *following general definitions, we assume, in addition, that* $\mathrm{Th}\,\mathcal{M}$ *be strongly minimal.*

The algebraic closure operator gives rise to a dependence relation between elements and subsets of $M$. Namely, given $x \in M$ and $Y \subseteq M$, we say that $x$ **depends (algebraically)** on $Y$ if $x \in \operatorname{acl} Y$. Further, a set $Y$ is called **(algebraically) independent** if none of its elements depends on the rest, i. e., if $y \notin \operatorname{acl}(Y \smallsetminus \{y\})$ for all $y \in Y$. An independent subset $Y$ of a subset $X$ of $M$ is called a **basis** of $X$ (in $\mathcal{M}$), if $X \subseteq \operatorname{acl} Y$. A basis of $M$ in $\mathcal{M}$ is called a **basis** of the structure $\mathcal{M}$. We say that a set $X$ **has dimension** if it has a basis and all its bases have the same cardinality, which is called the **dimension** of $X$ (in $\mathcal{M}$), in symbols $\dim X$. Otherwise, by definition, $X$ has **no dimension**. If $X$ is all of $M$, its dimension (in $\mathcal{M}$) is called the **dimension** of the structure $\mathcal{M}$ and denoted by $\dim \mathcal{M}$.

Note that the dimension $\dim X$ depends on the ambient structure $\mathcal{M}$, and we indicate it occasionally by writing $\dim_{\mathcal{M}} X$. E. g. $\dim \mathcal{M}$ is, more exactly, $\dim_{\mathcal{M}} M$.

**Remark.** It follows from the corresponding discussion of algebraicity in §11.5 (exercise!) that, in the context of vector spaces or algebraically closed fields, dependence is linear dependence or algebraic dependence of fields, respectively. Hence bases, as defined above, are, in those contexts, the usual bases of vector spaces or the **transcendence bases** of fields, respectively. And dimension, as defined above, turns into the usual dimension of vector spaces or the **transcendence degree** of fields, respectively.

We are going to show that all models, moreover, all subsets of models, of strongly minimal theories have a dimension: one also says that the dimension is **well-defined**.

Then, for the following trivial (cardinality) reasons, sets of power bigger than that of the language have a (well-defined) dimension—provided they have a basis at all (whose existence we will prove only after the exchange lemma below).

**Lemma 14.3.1.** (1)  *If $B$ is a basis of $A$, then $|B| \leq |A| \leq |B| + |L|$.*
(2)  *Any basis of a set of power $\lambda > |L|$ has power $\lambda$.*
(3)  *If a set of power $\lambda > |L|$ has a basis, it has (a well-defined) dimension.*

*Proof.* The cardinality estimates for algebraic closures (Lemma 11.5.4) used in the preceding section yield similar estimates for dimensions: the powers of an independent set and its algebraic closure can differ by at most $|L|$, which proves (1). The rest is immediate.  □

We have to work only a little more to show the same under the weaker hypothesis that the set in question have an infinite basis. A cardinality argument still suffices, as we show next—this time involving, in addition, the finite character of algebraic closure (see Proposition 11.5.3(1)).

**Lemma 14.3.2.** *If a subset of $M$ has an infinite basis, it has dimension.*

*Proof.* As a basis of $A \subseteq M$ is also a basis of $\operatorname{acl} A$, we may confine ourselves to algebraically closed subsets of $M$. So consider $\operatorname{acl} X$, where $X$ is an infinite independent set. We have to show that every basis of $\operatorname{acl} X$ has the same power as $X$.

Let $Y$ be such a basis. Then $X \subseteq \operatorname{acl} Y$ and every $y \in Y$ is algebraic over a finite set $X_y \Subset X$. Then $X \subseteq \operatorname{acl} \bigcup_{y \in Y} X_y$. Since, being independent, $X$ cannot be algebraic over a proper subset, we have $X = \bigcup_{y \in Y} X_y$. Then $Y$ must be infinite too and $|X| = |\bigcup_{y \in Y} X_y| \leq |Y| \cdot \aleph_0$ (by Lemma 7.6.6), hence $|X| \leq |Y|$ (as $Y$ is infinite).

We have proved that any basis $Y$ has a power not smaller than the power of the given infinite basis $X$. Interchanging the roles of $X$ and $Y$, this implies that all bases have the same power, as desired. □

Finally we have to deal with the most complicated case of (algebraically closed) sets with a *finite* basis (all of whose bases are, by the preceding lemma, finite). As in the case of vector spaces or fields, mere cardinality arguments no longer suffice: this requires the exchange property.

**Lemma 14.3.3.** (The Exchange Lemma)[3] *Let $\mathcal{M}$ be a model of a strongly minimal theory, $A \subseteq M$, and $b, c \in M$.*

*If $c \in \operatorname{acl}(A \cup \{b\}) \smallsetminus \operatorname{acl} A$, then $b \in \operatorname{acl}(A \cup \{c\})$. Consequently, if $B \subseteq M$ is an independent set and $b \in M \smallsetminus \operatorname{acl} B$, then $B \cup \{b\}$ is independent too.*

*Proof.* The second assertion follows from the first on taking $A = B \smallsetminus \{c\}$ and $c$ arbitrary in $B$.

By the premise, there is a formula $\eta(x, y)$ in $L(A)$ such that the set $\eta(b, \mathcal{M})$ is finite and contains $c$. Let $|\eta(b, \mathcal{M})| = m < \omega$. Denote the $L(A)$-formula $\exists^{=m} y \, \eta(x, y)$ by $\varphi(x)$. As this formula is satisfied by $b$, it cannot be algebraic, for otherwise, by transitivity of algebraicity, $\operatorname{acl}(A \cup \{b\}) = \operatorname{acl} A$, contradicting the assumption on $c$. So $\varphi(\mathcal{M})$ must be infinite and hence cofinite in $M$.

If (the $L(A)$-formula!) $\eta(x, c)$ is algebraic, $b \in \operatorname{acl}(A \cup \{c\})$, and we are done. If, on the other hand, this formula is not algebraic, strong minimality

---

[3] Also called Steinitz' Lemma, as it was publicised by Steinitz's seminal work in the context of fields, however, it had been known, in various contexts, before.

yields that its negation is, i. e., $|\mathcal{M} \smallsetminus \eta(\mathcal{M}, c)| = n$ for some $n < \omega$. We claim that then the formula $\exists^{=n} x \, \neg\eta(x, y)$—which we denote by $\psi(y)$—cannot be algebraic. For this formula has parameters only from $A$ and is satisfied by $c$, so if it were algebraic, $c$ would be algebraic over $A$, which it is not.

So we are left with the case that $\psi(y)$ is not algebraic. Choose $m + 1$ elements $c_0, \ldots, c_m$ satisfying it. Then the sets $\eta(\mathcal{M}, c_i)$ are all cofinite, hence so is their intersection with the cofinite set $\varphi(\mathcal{M})$. In particular—$\mathcal{M}$ being infinite—this intersection is not empty. Pick any of its elements, $a$. Then the set $\eta(a, \mathcal{M})$ contains all $m + 1$ elements chosen before. But $\varphi$ says that this set has exactly $m$ elements, contradiction.                $\square$

It is easy to see that the exchange lemma—in fact, everything in this section—holds under the weaker hypothesis that $x = x$ be minimal in $\mathcal{M}$ (in the sense of the end of §14.1).

From the exchange lemma one can easily derive that the bases of a given set $X$ are exactly the maximal independent subsets of $X$ (exercise!). Thus the existence of bases is guaranteed by Zorn's lemma. Moreover, it follows that every independent subset can be extended to a basis. The second half of the lemma suggests to construct bases by successively adjoining nonalgebraic elements—as known from vector spaces and algebraically closed fields.

To this end, we define a **Morley sequence** of length $\lambda$ in $\mathcal{M}$ to be a sequence $A = \{a_i : i < \lambda\}$ of elements of $M$ such that $a_i \notin \operatorname{acl} A_i$ for all $i < \lambda$ (where, as in Chapter 7, $A_i$ denotes the section $\{a_j : j < i\}$ of $A$). Clearly, an independent set is a Morley sequence. Next we see that the exchange lemma implies the converse. Therefore we should try and find bases by building Morley sequences, which we also do in the next lemma.

**Proposition 14.3.4.** *Let $\lambda$ be an infinite cardinal.*

(1)  *Every Morley sequence is independent.*

(2)  *Every subset of $\mathcal{M}$ has a basis. Moreover, every independent subset of a given subset can be extended to a basis of that subset.*

(3)  *If $\mathcal{M}$ is $\lambda$-saturated, there is a Morley sequence of length $\lambda$ in every infinite (parametrically) definable subset of $\mathcal{M}$. In particular, if $\mathcal{M}$ is saturated, every infinite definable subset of $\mathcal{M}$ has a basis of power $|M|$.*

*Proof.* Ad (1). Assume, $A = \{a_i : i < \lambda\}$ is not independent. Choose a minimal $i < \lambda$ such that $a_i \in \operatorname{acl}(A \smallsetminus \{a_i\})$. Then there is a finite subset $B$ in $A \smallsetminus \{a_i\}$ such that $a_i \in \operatorname{acl} B$. Choose $B$ minimal such.

Let $j$ be the maximal index of elements occurring in $B$. Note that $a_i \in \operatorname{acl} B \smallsetminus \operatorname{acl}(B \smallsetminus \{a_j\})$ by the minimal choice of $B$. Clearly, $i \neq j$. If

$j < i$, $A$ is not a Morley sequence. If $i < j$, invoking the exchange lemma, we get $a_j \in \mathrm{acl}(B \cup \{a_i\} \smallsetminus \{a_j\}) \subseteq A_j$. Hence, in this case, $A$ is not a Morley sequence either.

Ad (2). Let $C \subseteq M$. Starting with $A_0$, the independent subset of $C$ to be extended, and taking unions at limit stages, we obtain a Morley sequence inside $C$ by successively adjoining elements from $C$ that are not in the algebraic closure of what has been constructed so far. More exactly, suppose the Morley sequence $A_i$ has already been constructed (where $i$ is an ordinal). If its closure contains all of $C$, it is a basis of $C$. If not, pick an element $a_i$ in $C$ that is not algebraic over $A_i$—making $A_{i+1} = A_i \cup \{a_i\}$ a Morley sequence. We can continue until we reach some $A_j$ whose algebraic closure contains $C$ and which is thus the desired basis.

Ad (3). Let $D = \varphi(\mathcal{M}, \bar{a})$ be the definable set. Starting with $A_0 = \emptyset$ and taking unions at limit stages as before, we successively realize a (the only) non-algebraic type containing $\varphi(x, \bar{a})$ over $\bar{a}$ and what has been constructed so far. More exactly, suppose $A_i \subseteq D$ (of power $< \lambda$) has already been constructed, where $i < \lambda$. Pick an element $a_i$ in $D$ not algebraic over $A_i$ (which exists by Remark (2) in §12.1). Set $A_{i+1} = A_i \cup \{a_i\}$ and continue until $A_\lambda = \bigcup_{i < \lambda} A_i$ is reached, which is a Morley sequence of length $\lambda$. $\square$

For obvious reasons we call the sequence constructed in the proof of (2) above a **Morley sequence over** $A_0$. We are now ready to prove the main result of this section saying that dimension is well-behaved in $\mathcal{M}$.

**Theorem 14.3.5.** *Every subset of $\mathcal{M}$ has a dimension.*

*Proof.* The existence of bases was proved in the preceding proposition. So we are left with showing that any two bases of a given subset have the same power. Two lemmas ago this was proved for subsets having an infinite basis (and we mentioned that it suffices to consider algebraically closed subsets). So consider a finite independent set $X = \{x_i : i < m\}$ and any other basis $Y$ of $\mathrm{acl}\, X$. We have to show $|Y| = m$. We know already that $Y$ must be finite. Write $Y = \{y_i : i < n\}$ and assume, without loss of generality, that $n \geq m$.

Inductively on $j < n$—replacing some element from $X$ by $y_j$—we are going to construct sets $X(j)$ of power $m$ containing $Y_j = \{y_k : k < j\}$ such that $\mathrm{acl}\, X(j) = \mathrm{acl}\, X$. Then, in the final step $j = n$, we will have $Y \subseteq X(n)$, hence $n \leq m$, and hence also $m = n$, which proves the assertion.

We start with $X(0) = X$, and for each $j < n$ we replace, if necessary, some element from the already constructed $X(j)$ by $y_j$ as follows. If $X(j)$ contains $y_j$ already, there's nothing to do, and we set $X(j + 1) = X(j)$.

Otherwise, $y_j$ is not in $X(j)$, hence (by the independence of $Y$) not algebraic over, $X(j) \cap Y$. Since $y_j \in \mathrm{acl}\, X = \mathrm{acl}\, X(j)$, we may choose a minimal subset $X'(j)$ of $X(j)$ over which $y_j$ is algebraic. Since $y_j$ is not algebraic over $X(j) \cap Y$, this chosen subset $X'(j)$ is not entirely contained in $Y$. So there is $x \in X'(j) \smallsetminus Y$. Then $y_j$ is algebraic over $X'(j)$, but not over $X'(j) \smallsetminus \{x\}$. By the exchange lemma, $x$ is algebraic over $\{y_j\} \cup (X'(j) \smallsetminus \{x\})$. Set $X(j+1) = \{y_j\} \cup (X(j) \smallsetminus \{x\})$ (i. e., replace $x \in X(j)$ by $y_j$). Clearly, $x$ is algebraic also over the latter. Consequently, $\mathrm{acl}\, X(j) \subseteq \mathrm{acl}\, X(j+1)$. As the inverse inclusion holds too (for $y_j$ is in $\mathrm{acl}\, X = \mathrm{acl}\, X(j)$), we have $\mathrm{acl}\, X(j+1) = \mathrm{acl}\, X$. Finally, it is clear from the construction that $X(j+1)$ has the same power as $X(j)$. This completes the proof.                    $\square$

**Remark.** In model theory one can relativize many concepts to a corresponding concept 'over a subset' of a given model by adding constants for the elements of this subset to the language. One example was that of Morley sequence *over $A_0$* above. In the same way one obtains independence, dimension etc. *over subsets*. (Since strong minimality was defined in terms of *parametrically* definable subsets, this does not affect the theory's being strongly minimal.) In the case of fields, this yields the usual concepts of transcendence degree over another field etc., where transcendence degree over the empty set (or, equivalently, over the prime field) is usually called **absolute transcendence degree**.

**Exercise 14.3.1.** Show that the dimension of the field of complex numbers is $2^{\aleph_0}$ (the power of the continuum).

**Exercise 14.3.2.** Prove that in vector spaces (in)dependence is the usual linear (in)dependence , and hence dimension, as defined in this section, is the usual vector space dimension.

**Exercise 14.3.3.** Prove that in algebraically closed fields (in)dependence is the usual algebraic (in)dependence, and hence dimension, as defined in this section, is the transcendence degree of fields.

**Exercise 14.3.4.** Suppose $\mathcal{M}$ is a model of a strongly minimal theory and $B \subseteq A \subseteq M$. Prove the equivalence of the following three conditions.
(i)   $B$ is a basis of $A$.
(ii)  $B$ is a maximal independent subset of $A$.
(iii) $B$ is a minimal subset of $A$ with $A \subseteq \mathrm{acl}\, B$.

**Exercise 14.3.5.** Let $A$ be an infinite subset of $M$ definable in $\mathcal{M}$ without parameters.
    Prove that any basis of $A$ is a basis of $\mathcal{M}$.

**Exercise 14.3.6.** What is the dimension of the Prüfer group (regarded as a model of a strongly minimal theory)? What are the bases of the other models of this theory?

The remaining exercises deal with unpublished joint material of Martin Ziegler and the author concerning a particular sort of basis in a general, not necessarily strongly minimal context. So let $\mathcal{M}$ be any structure and $C \subseteq M$. We define a **low sequence** of length $\lambda$ in $C$ to be a sequence $A = \{a_i : i < \lambda\}$ of elements of $C$ such that, for all $i < \lambda$, the element $a_i$ is not algebraic over $A_i$ (where, as in Chapter 7, $A_i$ denotes the section $\{a_j : j < i\}$ of $A$) *and* $\mathrm{acl}_{\mathcal{M}}(A_i \cup \{a_i\})$ is minimal (with respect to set inclusion) among the sets of the form $\mathrm{acl}_{\mathcal{M}}(A_i \cup \{c\})$, where $c$ runs over $C$. A **low basis** of $C$ (in $\mathcal{M}$) is a low sequence $A$ in $C$ such that $C \subseteq \mathrm{acl}_{\mathcal{M}} A$. The first exercise is asking for an example showing that these need not exist. However, finite sets clearly always have low bases.

**Exercise 14.3.7.** Find a structure in which there are infinite descending chains of algebraic closures of singletons (so that the structure itself has no low sequence, not even of length 1).

**Exercise 14.3.8.** Suppose $\mathcal{M}$ is a countable saturated structure and $B$ and $C$ are finite subsets.

Show that there is an automorphism $f$ of $\mathcal{M}$ such that $f[C] \cap B \subseteq \mathrm{acl}_{\mathcal{M}} \emptyset$.

**Exercise 14.3.9.** Prove the same for infinite sets $B$ and $C$ in a saturated structure of power bigger than that of $B$ and $C$.

# 14.4   Steinitz' Theorem—categoricity revisited

In §14.2 we proved that strongly minimal theories are categorical in all powers bigger than that of the language. As a consequence we obtained that the theories of algebraically closed fields of fixed characteristic are categorical in all uncountable powers. Our proof was based on the uniqueness of saturated models, i. e. on a very general back-and-forth construction. Steinitz' original proof used what he called *irreducible systems of elements* and what was later named *transcendence bases*. (In our terminology, these are just the *bases*, which follows from Proposition 11.5.1, cf. Exercise 14.3.3 above). We are going to generalize his proof to (models of) strongly minimal theories, and thus obtain a more concrete proof of categoricity (avoiding the uncountable generalization of the uniqueness of saturated models asked for in Exercise 12.1.2). Moreover, as a special case, we get a full proof of (the full statement of) Steinitz' theorem saying that the isomorphism type of an algebraically closed field is uniquely determined by its characteristic and its *transcendence degree*. Note that (by Proposition 11.5.1, cf. Exercise 14.3.3, again) this transcendence degree is nothing but the dimension as defined in the preceding section.

Recall from the exercises in §11.5 that a **partial elementary map** from an $L$-structure $\mathcal{M}$ to an $L$-structure $\mathcal{N}$ is a map $f$ from a subset of $\mathcal{M}$ to $N$ such that $\mathcal{M} \models \varphi(\bar{a})$ if and only if $\mathcal{N} \models \varphi(f[\bar{a}])$, for all $\varphi$ from $L$ and all matching tuples $\bar{a}$ from dom $f$. This is the same as saying $(\mathcal{M}, \text{dom } f) \equiv (\mathcal{N}, f[\text{dom } f])$ (cf. Lemma 8.2.1 about *total* elementary maps). In particular, a partial elementary map is an isomorphism between its domain and its image.

*For the remainder of this section, let $\mathcal{M}$ and $\mathcal{N}$ be models of a given strongly minimal theory.*

**Lemma 14.4.1.** *Let $A$ be an independent subset of $\mathcal{M}$ and $B$ an independent subset of $\mathcal{N}$.*

*Any injective map of $A$ into $B$ (which exists only if $|A| \leq |B|$) is a partial elementary map from $\mathcal{M}$ to $\mathcal{N}$.*

*Proof.* Let $f$ be such a map and $A = \{a_i : i < \lambda\}$. Given $i < \lambda$, we use $A_i$ to denote the subset $\{a_j : j < i\}$. Inductively on $i < \lambda$ we show that $f_i = f \restriction A_i$ is a partial elementary map. This is true for $i = 0$, as $f_0$ is empty. Assuming $f_i$ is (partial) elementary, we show that also $f_{i+1}$ is. (The limit stage of the induction, including the final statement that $f$ is elementary, is trivial, since the union of a chain of partial elementary maps is obviously again (partial) elementary.)

Let $\varphi$ be a sentence over $A_{i+1}$. Letting $\bar{a}$ be the parameters occurring in $\varphi$ which are different from $a_i$, we have $\bar{a} \subseteq A_i$ and may write $\varphi$ as $\varphi(\bar{a}, a_i)$. We have to show that

$$\mathcal{M} \models \varphi(\bar{a}, a_i) \Longleftrightarrow \mathcal{N} \models \varphi(f[\bar{a}], f(a_i)).$$

To show the implication from left to right, assume $\mathcal{M} \models \varphi(\bar{a}, a_i)$.

As $A$ is independent, $a_i \notin \text{acl}\,\bar{a}$, hence $\varphi(\bar{a}, x)$ is not algebraic. Then the induction hypothesis that $f_i$ is elementary yields

$$\mathcal{M} \models \exists^{>n} x\, \varphi(\bar{a}, x) \Longleftrightarrow \mathcal{N} \models \exists^{>n} x\, \varphi(f[\bar{a}], x)$$

for all natural numbers $n$. Hence $\varphi(f[\bar{a}], x)$ is not algebraic either. Then, by strong minimality, $\neg\varphi(f[\bar{a}], x)$ is algebraic. As $f$ is injective, $f(a_i) \notin f[\bar{a}]$, hence $f(a_i) \notin \text{acl}\, f[\bar{a}]$ by the independence of $B$. So $f(a_i)$ cannot satisfy $\neg\varphi(f[\bar{a}], x)$. Hence it must satisfy $\varphi(f[\bar{a}], x)$.

We have proved the implication

$$\mathcal{M} \models \varphi(\bar{a}, a_i) \Longrightarrow \mathcal{N} \models \varphi(f[\bar{a}], f(a_i)),$$

where, in fact, $\varphi$ was arbitrary. Thus we have this implication also for $\neg\varphi$. As a consequence we obtain the converse of this implication for $\varphi$.     □

The proof of the next lemma was asked for in Exercise 11.5.4.

**Lemma 14.4.2.** *Let $A$ be a subset of $\mathcal{M}$ and $B$ a subset of $\mathcal{N}$.*
*Any partial elementary map from $\mathcal{M}$ to $\mathcal{N}$ that maps $A$ onto $B$ (which exists only if $|A| = |B|$) can be extended to a partial elementary map from $\mathcal{M}$ to $\mathcal{N}$ mapping $\mathrm{acl}_{\mathcal{M}}\, A$ onto $\mathrm{acl}_{\mathcal{N}}\, B$.*     □

**Theorem 14.4.3.**
(1) *Given independent sets $A$ in $\mathcal{M}$ and $B$ in $\mathcal{N}$, there is a bijection between $\mathrm{acl}\, A$ and $\mathrm{acl}\, B$ that is a partial elementary map from $\mathcal{M}$ to $\mathcal{N}$ if and only if $|A| = |B|$.*
(2) *$\mathcal{M}$ and $\mathcal{N}$ are isomorphic if and only if they have the same dimension.*
(3) *$\dim \mathcal{M} = \dim \mathcal{N}$ if and only if any bijection between independent sets $A \subseteq M$ and $B \subseteq N$ such that $|A| < \dim \mathcal{M}$ and $|B| < \dim \mathcal{N}$ can be extended to an isomorphism between $\mathcal{M}$ and $\mathcal{N}$.*

*Proof.* Ad (1). As in the proof of Lemma 14.4.1, partial elementary maps preserve (in)dependence. Thus, if $f$ is a partial elementary bijection between $\mathrm{acl}\, A$ and $\mathrm{acl}\, B$, then $f(A)$ is a basis of $\mathrm{acl}\, B$, hence has, as dimension is well-defined, the same power as $B$. Then $|B| = |f(A)| = |A|$.

Conversely, if $|A| = |B|$, there is a bijection between these bases, which is (partial) elementary by Lemma 14.4.1 and can be extended to an elementary bijection between $\mathrm{acl}\, A$ and $\mathrm{acl}\, B$, by Lemma 14.4.2.

Ad (2). Let $A$ and $B$ be bases of $\mathcal{M}$ and $\mathcal{N}$, respectively. The result just proved says in this case that $|A| = |B|$ iff there is an elementary bijection between $\mathcal{M}$ and $\mathcal{N}$. But an elementary bijection between models is an isomorphism (check!).

Ad (3). For the non-trivial direction, let $f$ be a bijection between $A$ and $B$. Extend each of $A$ and $B$ to bases $A'$ and $B'$ of $\mathcal{M}$ and $\mathcal{N}$, respectively. Since $\dim \mathcal{M} = \dim \mathcal{N} > |A| = |B|$, we have $|A' \smallsetminus A| = |B' \smallsetminus B|$. Hence we can find a bijection between $A'$ and $B'$ extending $f$. Now proceed as before.     □

**Corollary 14.4.4.**
*$\mathcal{M}$ is elementarily embeddable in $\mathcal{N}$ if and only if $\dim \mathcal{M} \leq \dim \mathcal{N}$.*     □

The theorem yields another proof of the categoricity of a strongly minimal theory $T$ in all cardinals bigger than $|T|$. It also shows (in fact already the preceding lemma does) that the isomorphism type of the algebraic closure of a substructure, or even just a subset, $A$, of a given model $\mathcal{M}$ of $T$ depends only on the 'type' of $A$ in $\mathcal{M}$: if $\mathcal{N}$ is another model of $T$ containing $A$ such that $A$ is 'sitting inside $\mathcal{N}$ the same way' as it is in $\mathcal{M}$, which should mean that the identity map on $A$ viewed as a partial map from $\mathcal{M}$ to $\mathcal{N}$ is elementary, then $\mathrm{acl}_{\mathcal{M}} A$ and $\mathrm{acl}_{\mathcal{N}} A$ are $A$-isomorphic. An equivalent hypothesis on $A$ implying this uniqueness of its algebraic closure would be that $A$ has the same dimension, whether considered as being in $\mathcal{M}$ or in $\mathcal{N}$. In case $T$ has quantifier elimination, this is clear for substructures $\mathcal{A}$: this can be seen either from elementary amalgamation over $\mathcal{A}$, as given by Remark (5) in §9.2, or directly from the completeness of $T \cup \mathrm{D}(\mathcal{A})$ (as $L(A)$-theory), which implies that $(\mathcal{M}, A)$ and $(\mathcal{N}, A)$ are elementarily equivalent, whence the identity map on $A$ is as required before. It is easy to extend this to the case of an arbitrary isomorphism between $\mathcal{A}$ and another substructure of some model of $T$. This allows us to define *the* **dimension** of any substructure of a model of $T$ without reference to that model. In the case of fields, this dimension is just the transcendence degree. Let us summarize.

**Corollary 14.4.5.** *Let $T$ be a strongly minimal theory with quantifier elimination.*

*Then the algebraic closure of a substructure $\mathcal{A}$ of some model of $T$ is unique up to isomorphism over $A$ (even when computed in another model of $T$ containing $\mathcal{A}$ as a substructure). More generally, every isomorphism between substructures of models of $T$ can be extended to an isomorphism of their algebraic closures.*

*If $\mathcal{N}$ is a model of $T$, then $\mathcal{A}$ is embeddable in $\mathcal{N}$ if and only if $\dim \mathcal{A} \le \dim \mathcal{N}$.* □

We may now formulate Steinitz' theorem in its full strength.

**Corollary 14.4.6.** (Steinitz' Theorem)

*Every field $\mathcal{K}$ has an algebraic closure, which is unique up to isomorphism over $\mathcal{K}$, and a uniquely determined dimension—its transcendence degree. Moreover, every isomorphism between two fields can be extended to an isomorphism of their algebraic closures.*

*Algebraically closed fields are isomorphic if and only if they have the same characteristic and the same transcendence degree. (In particular, for a given characteristic, $q$, the theory $\mathrm{ACF}_q$ is categorical in all uncountable powers.)*

*A field $\mathcal{K}$ is embeddable in an algebraically closed field $\mathcal{N}$ if and only if $\dim \mathcal{K} \le \dim \mathcal{N}$.* □

[Steinitz, E. : Algebraische Theorie der Körper, J. f. Reine u. Angewandte Math. **137** (1910) 167 – 309, Reprint, R. Baer, H. Hesse (eds.), Chelsea, N. Y. 1950, 176 pp.]

**Remark.** Note that this rests on the fact formulated at the beginning of §9.5 and (to be) proved in Exercise 12.1.4 that every field is contained in an algebraically closed field (which is a special case of the more general fact to be proved in Exercise 12.1.3 that every model of an inductive theory is contained in an existentially closed model).

**Exercise 14.4.1.** Find another, more elegant, proof of Lemma 14.4.1 using Proposition 14.1.3 and the conjugation of types by the partial elementary maps $f_i$ already constructed.

**Exercise 14.4.2.** Given a model $\mathcal{M}$ of a strongly minimal theory, prove the following assertions.
(1) $\mathcal{M}$ has at least $2^{\dim \mathcal{M}}$ automorphisms.
(2) If $\mathcal{M}$ is countable and saturated, then it has precisely continuum many automorphisms.
(3) The field of complex numbers has exactly $2^{2^{\aleph_0}}$ automorphisms.

A subset $A$ of $M$ is said to be **indiscernible** (in $\mathcal{M}$) if any two tuples of the same length from $A$ have the same type in $\mathcal{M}$.

**Exercise 14.4.3.** Prove that independent subsets of $\mathcal{M}$ are indiscernible.

**Exercise 14.4.4.** Show that every partial elementary map preserves dimension, i. e., if $f$ is such a map and $A \subseteq \operatorname{dom} f$, then $\dim A = \dim f[A]$.

## 14.5 The chain of models

We already know that in higher powers (more precisely, in powers bigger than the cardinality of the theory) a strongly minimal theory has, up to isomorphism, exactly one model—a saturated one. And the universality of saturated models (to be shown, for uncountable ones, in Exercise 12.1.2) implies that, by moving to isomorphic copies if necessary, these can be arranged to form an elementary chain. In this section, using the dimension theory, we prove that *all* models of a strongly minimal theory (can be arranged to) form an elementary chain.

We begin with a simple characterization of saturation. The dimension theory allows us now to improve Theorem 14.2.2(1) considerably.

**Theorem 14.5.1.** *Let $\lambda$ be an infinite cardinal.*
*$\mathcal{M}$ is $\lambda$-saturated if and only if $\dim \mathcal{M} \geq \lambda$.*
*In particular, $\mathcal{M}$ is saturated if and only if $\dim \mathcal{M} = |\mathcal{M}|$.*

*Proof.* In view of Lemma 14.2.1 it suffices to show that $\dim \mathcal{M} \geq \lambda$ if and only if, for all subsets $A \subseteq M$ with $|A| < \lambda$, we have $\operatorname{acl} A \neq M$. If $\dim \mathcal{M} \geq \lambda$, no set of power smaller than $\lambda$ can span the whole model $\mathcal{M}$, hence $\operatorname{acl} A \neq M$ for all such $A$, as desired. For the converse, let $B$ be a basis of $\mathcal{M}$, i. e., $M = \operatorname{acl} B$ and $\dim \mathcal{M} = |B|$, and assume that $\operatorname{acl} A \neq M$ for all $A \subseteq M$ with $|A| < \lambda$. Then $|B|$ must be at least $\lambda$, i. e. $\dim \mathcal{M} \geq \lambda$.  $\square$

A strongly minimal theory $T$ has two important cardinal invariants. Define $\iota(T)$ to be the smallest cardinal $\iota$ such that some (hence, by the proof of Theorem 14.4.3, every) algebraically closed set of dimension $\iota$ (in any model of $T$) is infinite, and define $\lambda(T)$ to be the smallest cardinal $\lambda \geq \iota(T)$ such that some (hence every) algebraically closed set of dimension $\lambda$ has power $\lambda$.

Obviously, $\iota(T)$ is either a natural number or $\aleph_0$, depending on whether $T$ has a finite-dimensional model (of dimension $\iota(T)$) or all finite-dimensional algebraically closed sets are finite. In the latter case we say, $T$ is **locally finite.** (Notice, this notion says more than just that every model of $T$ is·a locally finite structure in the sense of §6.3: not only does a substructuré of a model generated by a finite set have to be finite, but even its algebraic closure.)

**Remark.** All algebraically closed sets of dimension $\geq \iota(T)$, and only these, are infinite, hence models of $T$ (by 14.1.1 above). Since $\lambda(T) \geq \iota(T)$, every algebraically closed subset of dimension $\lambda(T)$ embeds one of dimension $\iota(T)$, hence it must be infinite. Consequently, $\lambda(T)$ is always infinite. On the other hand, the computation of Proposition 11.5.4 for the cardinality of algebraic closures shows that algebraic closures of sets of power $\lambda \geq |T|$ have power $\lambda$, hence $\lambda(T) \leq |T|$. Further, the power of every algebraically closed set of dimension $\geq \lambda(T)$ is equal to its dimension. Hence all such algebraically closed sets, and only these, form saturated models.

Recall that a chain of sets indexed by ordinals is said to be *continuous* if the entry with index a limit ordinal is the union of the preceding ones.

**Theorem 14.5.2.** *The models of any strongly minimal theory $T$ can be arranged to form a continuous elementary chain containing exactly one of each dimension $\lambda \geq \iota(T)$. More precisely, $T$ has, for every $\lambda \geq \iota(T)$, a model $\mathcal{M}_\lambda$ of dimension $\lambda$ such that the following hold.*
(1)  $\mathcal{M}_{\iota(T)}$ *is an elementarily prime model of $T$ (of dimension $\iota(T)$).*
(2)  $\mathcal{M}_\lambda \preccurlyeq \mathcal{M}_\kappa$ *if and only if $\lambda \leq \kappa$.*
(3)  $\mathcal{M}_\lambda$ *is saturated if and only if $\lambda \geq \lambda(T)$.*

(4) *Every model of $T$ is isomorphic to exactly one of the $\mathcal{M}_\lambda$.*

(5) *Exactly one of the following two cases occurs.*

(i) *$T$ is totally categorical, $\iota(T) = \aleph_0$, hence $\mathcal{M}_{\aleph_0}$ is elementarily prime and countable, and*

$$|\mathcal{M}_\lambda| = \lambda \text{ for all } \lambda \geq \aleph_0.$$

(ii) *$T$ is not totally categorical, $\iota(T) < \aleph_0$, hence $\mathcal{M}_n$ is elementarily prime for some natural number $n$,*

$$|\mathcal{M}_\lambda| = |\mathcal{M}_{\aleph_0}| + \lambda \text{ for all } \lambda \geq \aleph_0,$$

*and either*

$$|\mathcal{M}_m| = |\mathcal{M}_{\aleph_0}| = \aleph_0, \text{ whenever } n \leq m \leq \aleph_0,$$

*or else, for some $\lambda > \aleph_0$, we have*

$$|\mathcal{M}_{\aleph_0}| = |\mathcal{M}_\kappa| = \lambda, \text{ whenever } \aleph_0 \leq \kappa \leq \lambda.$$

*Proof.* We first choose the $\mathcal{M}_\lambda$ by induction on $\lambda \geq \iota(T)$. Let $\mathcal{M}$ be any model of $T$. Since $\dim \mathcal{M}$ must be at least $\iota(T)$, we can find an algebraically closed subset of dimension $\iota(T)$ (take any subset of power $\iota(T)$ of a basis of $\mathcal{M}$ and then the algebraic closure thereof). Note that the same argument (invoking Löwenheim-Skolem upward to make $\mathcal{M}$ arbitrarily large) provides us with models of arbitrary dimension $\geq \iota(T)$, however, we want to choose them so that they form an elementary chain. To this end, let now $\lambda > \iota(T)$ and assume, an elementary chain of models $\mathcal{M}_\kappa$ of dimension $\kappa$ ($\kappa < \lambda$) has already been chosen. If $\lambda$ is a limit cardinal, let $\mathcal{M}_\lambda$ simply be their union. If, on the other hand, $\lambda = \kappa^+$, choose $\mathcal{M}_\lambda$ as follows. Consider a basis $B$ of $\mathcal{M}_\kappa$, and find, in an elementary extension $\mathcal{M}$ of $\mathcal{M}_\kappa$ of a power bigger than $|L| + \kappa$, a maximal Morley sequence $C$ over $B$ (as in the proof of Proposition 14.3.4(2)). Then $C$ is a basis of $\mathcal{M}$ and has, by Lemma 14.3.1, power $|\mathcal{M}|$. Since, by choice, $|\mathcal{M}| \geq \lambda$, we can find an initial segment, $A$, of $C$ of power $\lambda$. Letting $\mathcal{M}_\lambda$ be the algebraic closure of $A$ in $\mathcal{M}$, it is easily seen that $\dim \mathcal{M}_\lambda = \lambda$ and $\mathcal{M}_\kappa \preccurlyeq \mathcal{M}_\lambda \preccurlyeq \mathcal{M}$.

Assertion (3) was already mentioned (it is immediate from Theorem 14.5.1), while assertion (2) is true by construction.

Further, by the uniqueness of dimension, the $\mathcal{M}_\lambda$ are pairwise not isomorphic. Since the dimension of any model of $T$ is at least $\iota(T)$, Theorem 14.4.3(2) tells us that every model is isomorphic to some of the chosen $\mathcal{M}_\lambda$. This proves (4), which in turn implies (1).

We are left with (5). There are two cases to consider.

*Case 1, $\iota(T) = \aleph_0$.* Then $T$ is locally finite. Given $\lambda \geq \iota(T)$, let $B$ be a basis of $\mathcal{M}_\lambda$. Then the elementarily prime model $\mathcal{M}_{\iota(T)} = \mathcal{M}_{\aleph_0}$ is the algebraic closure of the countable initial segment $B_\omega$ of $B$ and, being the union of the finite sets $\operatorname{acl} B_n$ ($n$ running over the natural numbers), it is countable. Moreover, $\mathcal{M}_\lambda$ is equal to $\operatorname{acl} B = \bigcup_{C \in B} \operatorname{acl} C$, hence a union of $\lambda$ finite sets (by local finiteness). Consequently, $|\mathcal{M}_\lambda| = \lambda$ for all $\lambda \geq \aleph_0$. Then all $\mathcal{M}_\lambda$ have different cardinality, which shows that $T$ is totally categorical.

*Case 2, $\iota(T) = n < \aleph_0$.* Let $\lambda$ be infinite. As $\mathcal{M}_\lambda$ contains $\mathcal{M}_{\aleph_0}$ as well as a basis of power $\lambda$, we have $|\mathcal{M}_\lambda| \geq |\mathcal{M}_{\aleph_0}| + \lambda$. On the other hand, as in the first case, $\mathcal{M}_\lambda$ is the union of $\lambda$ algebraically closed subsets of finite dimension. This time the bound on the size of these, however, is only $|\mathcal{M}_{\aleph_0}|$. Still, $|\mathcal{M}_\lambda| \leq \lambda \cdot |\mathcal{M}_{\aleph_0}| = |\mathcal{M}_{\aleph_0}| + \lambda$, hence $|\mathcal{M}_\lambda| = \lambda$, whenever $\lambda \geq |\mathcal{M}_{\aleph_0}|$.

Finally we prove that one of the two cardinality assertions must hold in this case. Then it follows that $T$ is not totally categorical, for either we have two (even infinitely many) countable models of different (countable) dimension, or else we have, in an uncountable power $\lambda$, at least two models of different dimension ($\leq \lambda$).

If the model $\mathcal{M}_{\aleph_0}$ is countable, we get the first assertion, as then, being infinite, the finite-dimensional models must be countable too. If, on the other hand, $|\mathcal{M}_{\aleph_0}| = \lambda > \aleph_0$, then, by the above cardinality computations, all models of infinite dimension at most $\lambda$ have power $\lambda$. This completes the proof of the theorem. □

This allows us to slightly improve Theorem 14.2.2(2) about the categoricity spectrum of strongly minimal theories.

**Corollary 14.5.3.** *If $T$ is strongly minimal, then $\lambda(T) = |\mathcal{M}_{\aleph_0}|$ and $T$ is categorical in all $\lambda > \lambda(T)$. Further, either $\lambda(T) = \aleph_0 = \iota(T)$ and $T$ is totally categorical, or else $\lambda(T) \geq \aleph_0 > \iota(T)$ and $T$ has infinitely many nonisomorphic models of power $\lambda(T)$.*

*Proof.* [4] The first statement is immediate from the theorem (and the definition of $\lambda(T)$). The first half of the second one then follows if (5i) of the theorem takes place, and we are left with case (5ii). In this case, as is easily seen, $\lambda(T) \geq \aleph_0 > \iota(T)$. Further, if $\lambda(T)$ is countable, the models $\mathcal{M}_m$ are countable whenever $\iota(T) \leq m \leq \aleph_0$, hence we have infinitely many nonisomorphic models of power $\lambda(T)$. So assume, $\lambda(T)$ is uncountable. Then, by (5ii), there are at least as many nonisomorphic models of power $\lambda(T)$ as there

---

[4]Some of the arguments in this proof stem from Laskowski's work cited below.

are infinite dimensions $\leq \lambda(T)$, i. e. as there are infinite cardinals $\leq \lambda(T)$. So, if $\lambda(T) = \aleph_\alpha$ and $\alpha \geq \omega$, there are infinitely many such. Thus we are left with the case that $\alpha$ is finite. Then $\lambda(T) = \aleph_\alpha$ is regular, by Exercise 7.6.4. Hence $\mathcal{M}_{\aleph_0}$ is of regular and uncountable cardinality. But this model is the union of the countably many models $\mathcal{M}_m$, where $m < \omega$. Hence, by regularity, one of the latter must have cardinality $\lambda(T)$. As $|\mathcal{M}_m| \leq |\mathcal{M}_k| \leq |\mathcal{M}_{\aleph_0}|$ for all $m \leq k \leq \aleph_0$, from some $m$ on all these models have power $\lambda(T)$. Thus, also in this case there are infinitely many nonisomorphic ones. □

In the countable case we obtain

**Corollary 14.5.4.** *A countable strongly minimal theory $T$ is either $\aleph_0$-categorical, i. e. has (up to isomorphism) exactly one countable model, or else there is an elementary chain of pairwise nonisomorphic countable models*

$$\mathcal{N}_0 \preccurlyeq \mathcal{N}_1 \preccurlyeq \mathcal{N}_2 \preccurlyeq \ldots \preccurlyeq \mathcal{N}_\omega$$

*such that the following hold.*
*(1) $\mathcal{N}_0$ is an elementarily prime model.*
*(2) $\mathcal{N}_\omega$ is the only countable saturated model of $T$.*
*(3) Every countable model of $T$ is isomorphic to one of the $\mathcal{N}_\nu$.*
*(4) $\iota(T)$ is finite and $\dim \mathcal{N}_\nu = \iota(T) + \nu$ for all $\nu \leq \omega$.*
*Consequently, a countable strongly minimal theory has, up to isomorphism, either just 1 or $\aleph_0$ countable models.*

*Proof.* If $\iota(T) = \aleph_0$, we are in case (5i) of the theorem. Then all $\mathcal{M}_\lambda$ with uncountable $\lambda$ are uncountable, while $\mathcal{M}_{\aleph_0}$ is countable. So $\mathcal{M}_{\aleph_0}$ is the only countable model.

If $\iota(T) = n < \aleph_0$, we are in case (5ii). As the language is countable, every model of countable dimension is countable (Proposition 11.5.4). Hence $\mathcal{M}_{\aleph_0}$ is countable and so $\mathcal{M}_\lambda$ is countable if and only if $\lambda \leq \aleph_0$. Now set $\mathcal{N}_\nu = \mathcal{M}_{n+\nu}$. As $n + \nu$ is countable if and only if $\nu$ is, the corollary is proved. □

It should be mentioned that in

[Baldwin, J. T. and Lachlan, A. H. : On strongly minimal sets, J. Symbolic Logic **36** (1971) 79 – 96]

John Baldwin and Alistair Lachlan publicised and considerably generalised Marsh's work. They worked in the much more general setting of arbitrary theories categorical in an uncountable power, albeit only in the case of a countable language (we will say a few words about the uncountable case below). Baldwin and Lachlan showed that the above corollary holds in this much more general situation, a result that is generally known as the Baldwin-Lachlan theorem and whose proof is beyond the scope of this book. It had been shown in Morley's seminal paper that such theories are categorical in *all* uncountable powers, thus confirming what was known as Łoś' conjecture, cf. p. 123. And the main motive for Baldwin and Lachlan seems to have been the confirmation of Vaught's conjecture for such theories, saying that a countable complete theory has either countably or continuum many countable models (cf. end of previous chapter). For uncountably categorical theories (in a countable language), they obtained the much stronger result that the possible numbers are just 1 and $\aleph_0$, which we—following Marsh—proved for the strongly minimal case only, where the proof is sufficiently simple. The only ingredient in the strongly minimal case was dimension theory, which allowed us to argue without any restrictions on the power of the language. It is interesting to note that the general categorical case is much harder for uncountable theories. In fact, only after a good deal of *geometric stability theory* had been developed, was it possible to generalize the Baldwin-Lachlan theorem to the uncountable case. This was done by Chris Laskowski in

[Laskowski, M. C. : Uncountable theories categorical in a higher power, J. Symbolic Logic **53** (1988) 512 – 530].

**Exercise 14.5.1.** Find examples of vector spaces for the phenomena formulated in (5) of the above theorem.

**Exercise 14.5.2.** Which of the cases formulated in (5) of the above theorem takes place in algebraically closed fields, which in the theory of pure identity, $T_{=}^{\infty}$?

**Exercise 14.5.3.** Show that the elementarily prime model of a strongly minimal theory is a minimal model (in the sense of p. 193 at the end of §12.2) if and only if it has finite dimension.

**Exercise 14.5.4.** Show that a complete theory has a locally finite model in which the formula $x = x$ is minimal (in the sense of Exercise 14.1.5) if and only if it is strongly minimal and totally categorical.

**Exercise 14.5.5.** Prove that, if $\aleph_\alpha \geq \iota(T)$, then the isomorphism types of models of the strongly minimal theory $T$ of power at most $\aleph_\alpha$ form a chain of order type $\alpha + 1$ or $\omega + \alpha + 1$ depending on whether $\iota(T)$ is infinite or not.

# 14.6   Homogeneity and total categoricity

We are going to show that a totally categorical strongly minimal theory can have no finite axiomatization. The proof of this is based on a certain saturation property (which we consider first) and some homogeneity behaviour of the finite-dimensional models of arbitrary strongly minimal theories, similar to that of saturated and atomic structures (as considered in the main theorems of Chapter 12).

**Lemma 14.6.1.** *Let $T$ be strongly minimal and $n$ a natural number.*
*All types from $S_n(T)$ are realized in every algebraically closed subset of dimension at least $n$ of any model of $T$.*

*Proof.* Suppose $\Phi \in S_n(T)$ and $\bar{a}$ be an independent sequence of length $n$ of the model $\mathcal{M}$ of $T$. We show that $\Phi$ is realized in $\mathcal{M}$ by some ($n$-) tuple from $\operatorname{acl} \bar{a}$.

Consider a realization $\bar{b}$ of $\Phi$ in some model $\mathcal{N}$ of $T$, and let $\bar{e}$ be a basis of $\bar{b}$ (regarded as the set of its entries). Since $\mathrm{l}(\bar{e}) \leq \mathrm{l}(\bar{b}) = n$, there is an injective map from $\bar{e}$ into $\operatorname{acl} \bar{a}$, which is a partial elementary map from $\mathcal{N}$ to $\mathcal{M}$, by Lemma 14.4.1. By Lemma 14.4.2, this map is extendable to a partial elementary map $f$ from $\mathcal{N}$ to $\mathcal{M}$ mapping $\operatorname{acl} \bar{e}$ into $\operatorname{acl} \bar{a}$. Since $\bar{b} \subseteq \operatorname{acl} \bar{e}$, we see that $f[\bar{b}]$ is in $\operatorname{acl} \bar{a}$. But, as $f$ is elementary, this latter tuple also realizes $\Phi$ (in $\mathcal{M}$). $\square$

We say that a subset $C$ of a model $\mathcal{M}$ is ∞-**homogeneous** (with respect to $\mathcal{M}$) if every partial elementary map from $\mathcal{M}$ to $\mathcal{M}$ whose domain and range are contained in $C$ can, for every given $a \in C$, be extended to a partial elementary map from $\mathcal{M}$ to $\mathcal{M}$ whose domain contains also $a$ and whose range is still contained in $C$. In other words, given $A \subseteq C$, $B \subseteq C$, and $a \in C$, if $(\mathcal{M}, A) \equiv (\mathcal{M}, B)$, there is $b \in C$ such that $(\mathcal{M}, A \cup \{a\}) \equiv (\mathcal{M}, B \cup \{b\})$. For the sake of notational simplicity, we write $(\mathcal{M}, Aa) \equiv (\mathcal{M}, Bb)$ for the latter. More generally and more precisely, given two sequences (finite or infinite), $\bar{a}$ and $\bar{b}$, we simply use $\bar{a}\bar{b}$ to denote their concatenation, i. e. the sequence whose entries are first all the entries of $\bar{a}$ (in the given order) and then all the entries of $\bar{b}$. And we may 'concatenate' sets with tuples and even single elements—as above—so that expressions like $Ab\bar{a}$ make sense, where the order written suggests the order of concatenation. This allows us to write partial elementary maps very efficiently: e. g., the expression $(\mathcal{M}, Ac\bar{a}) \equiv (\mathcal{M}, Bd\bar{b})$ means that there is a partial elementary map from $\mathcal{M}$ to $\mathcal{N}$ mapping the set $A \subseteq M$ to the set $B \subseteq N$, the element $c \in M$ to the element $d \in N$, and the tuple $\bar{a}$ from $M$ to the tuple $\bar{b}$ from $N$.

**Lemma 14.6.2.** *Every finite-dimensional algebraically closed subset of a model of a strongly minimal theory is ∞-homogeneous.*

*Proof.* Consider a subset $C$ of dimension $n$ of a model $\mathcal{M}$, and let $A, B \subseteq C$ be such that $(M, A) \equiv (M, B)$ and $a \in C$. We have to find $b \in C$ such that $(\mathcal{M}, Aa) \equiv (\mathcal{M}, Bb)$.

Clearly, $A$ and $B$ are of dimension $\leq n$, and, by Exercise 14.4.4, their dimension is equal, to $m \leq n$ say. Since we can extend elementary maps between arbitrary sets to their algebraic closures, we may assume, $A$ and $B$ are algebraically closed. Let $\bar{a}$ and $\bar{b}$ be their respective bases (hence $m$-tuples).

If $a \in \text{acl}\,\bar{a}$, then $a \in A$ and we trivially have an image $b \in B$ such that $(M, Aa) \equiv (M, Bb)$. If, on the other hand, $a \notin \text{acl}\,\bar{a}$, then $\bar{a}a$ is independent. As the latter is contained in $C$, we must have $m < n$. But then $C \smallsetminus B$ cannot be empty, so we can find $b \in C \smallsetminus B$. Then also $\bar{b}b$ is independent, and, by indiscernibility (see Exercise 14.4.3), $\bar{a}a$ and $\bar{b}b$ have the same type, i. e. $(M, \bar{a}a) \equiv (M, \bar{b}b)$. Extending to algebraic closures again (and then restricting appropriately), we obtain $(M, Aa) \equiv (M, Bb)$.          $\square$

We now turn to the totally categorical case. We give three equivalent conditions for this, the first two equivalences of which are immediate from Theorem 14.5.2, while the last one is known to us—for countable theories—from the theorem in §13.3. In the strongly minimal case there is an easier proof, avoiding omitting types and thus allowing to drop the countability assumption.

**Theorem 14.6.3.** *The following are equivalent for any strongly minimal theory $T$.*
(i)    $T$ *is totally categorical.*
(ii)   $\iota(T) = \aleph_0$.
(iii)  $T$ *is locally finite.*
(iv)   $S_n(T)$ *is finite for all $n$.*

*Proof.* By Theorem 14.5.2, a strongly minimal theory is totally categorical if and only if $\iota(T)$ is infinite if and only if $T$ is locally finite. This proves the first two equivalences.

Lemma 14.6.1 showed that every type in $S_n(T)$ can be realized in a fixed $n$-dimensional algebraically closed subset $A$ of some model $M$ of $T$. But in a locally finite theory, finite-dimensional sets are finite. So $A$ is finite and consequently so is $S_n(T)$. This proves (iii) $\Longrightarrow$ (iv).

For the converse, fix a natural number $n$, and let $\bar{a} \in M^n$ and $M \models T$. We have to show, $A = \text{acl}\,\bar{a}$ is finite. Assuming the contrary, we are going to find infinitely many $n + 1$-types over $\emptyset$. Let $a_0$ be any element in $A$. Then $a_0$ satisfies some algebraic formula $\eta_0$ over $\bar{a}$. As $\eta_0(M)$ is finite, we can find $a_1 \in A \smallsetminus \eta(M)$. Then $tp(\bar{a}a_0) \neq tp(\bar{a}a_1)$. Further, we have another algebraic formula $\eta_1$ over $A$ algebraizing $a_1$. Again we can find an element $a_2 \in A \smallsetminus (\eta_0(M) \cup \eta_1(M))$, and so forth. In this way we obtain infinitely many $n + 1$-tuples $\bar{a}a_i$ in $A$, all having a different type, contradicting the finiteness of $S_{n+1}(T)$.      □

We are now ready to prove that a totally categorical, strongly minimal theory cannot be finitely axiomatized. Recall from §3.4 that this means, such a theory cannot be axiomatized by a single sentence.

**Theorem 14.6.4.** *Any sentence that is consistent with a totally categorical, strongly minimal theory has a finite model.*

*Proof.* Let $\varphi$ be a sentence consistent with (and, by completeness, true in every model of) the totally categorical, strongly minimal theory $T$. Write $\varphi$ in prenex normal form as $Q_1 x_1 \ldots Q_n x_n \psi$, where $\psi = \psi(x_1, \ldots, x_n)$ is quantifier-free and, for each $i$, the symbol $Q_i$ stands for one of the quantifiers. In some big enough model $\mathcal{M}$ of $T$, consider an independent set $B$ of power $n$ and let $A$ be its algebraic closure. By the previous theorem, $A$ is finite. Consider the substructure $\mathcal{A}$ of $\mathcal{M}$ with universe $A$ (which exists since $A$ is algebraically closed, cf. Remark (2), §11.5).

We are going to show that $\mathcal{A} \models \varphi$. By adding dummy variables if necessary, we may assume that $n$ is even and $Q_{2i+1} = \forall$ and $Q_{2i+2} = \exists$. For all $i \leq \frac{n}{2}$, denote the formula

$$\forall x_{n-2i+1} \exists x_{n-2i+2} \ldots \forall x_{n-1} \exists x_n \psi(x_1, \ldots, x_n)$$

by $\theta_i(x_1, \ldots, x_{n-2i})$. Then $\theta_{\frac{n}{2}}$ is $\varphi$, hence we are done if we can prove the following claim.

**Claim.** Given $0 \leq i \leq \frac{n}{2}$ and $\bar{a} \in A^{n-2i}$, if $\mathcal{M} \models \theta_i(\bar{a})$, then $\mathcal{A} \models \theta_i(\bar{a})$.

We proceed by induction on $i$. The initial step being trivial—as $\theta_0$ is the quantifier-free formula $\psi$ (and $\mathcal{A}$ is a substructure of $\mathcal{M}$)—let us prove the claim for $i + 1 \leq \frac{n}{2}$, assuming that it is true for $i$.

So let $\bar{a}$ be an $(n - 2i - 2)$-tuple in $A$ and $\mathcal{M} \models \theta_{i+1}(\bar{a})$, i. e. ,

$$\mathcal{M} \models \forall x_{n-2i-1} \exists x_{n-2i} \theta_i(\bar{a}, x_{n-2i-1}, x_{n-2i}).$$

If we can find, for every $a_{n-2i-1} \in A$, some $a_{n-2i} \in A$ such that

$$\mathcal{M} \models \theta_i(\bar{a}, a_{n-2i-1}, a_{n-2i}),$$

hence, by induction hypothesis,

$$\mathcal{A} \models \theta_i(\bar{a}, a_{n-2i-1}, a_{n-2i}),$$

then we also have $\mathcal{A} \models \theta_{i+1}(\bar{a})$.[5]

The proof of this is based on the homogeneity of $A$ (as given by Lemma 14.6.2) and on the fact (proved in Lemma 14.6.1) that every $n$-type (of $\mathcal{M}$) over the empty set is realized in $A$. Namely, given $a_{n-2i-1} \in A$, the hypothesis of the claim yields $b \in M$ such that $\mathcal{M} \models \theta_i(\bar{a}, a_{n-2i-1}, b)$. Realize the $(n - 2i)$-type of $\bar{a}a_{n-2i-1}b$ (over the empty set) by some $(2i + 1)$-tuple $\bar{a}'a'_{n-2i-1}b'$ from $A$. Then $(\mathcal{M}, \bar{a}a_{n-2i-1}b) \equiv (\mathcal{M}, \bar{a}'a'_{n-2i-1}b')$, in particular, $(\mathcal{M}, \bar{a}a_{n-2i-1}) \equiv (\mathcal{M}, \bar{a}'a'_{n-2i-1})$. Homogeneity now yields $a_{n-2i} \in A$ satisfying $(\mathcal{M}, \bar{a}a_{n-2i-1}a_{n-2i}) \equiv (\mathcal{M}, \bar{a}'a'_{n-2i-1}b')$, hence also $(\mathcal{M}, \bar{a}a_{n-2i-1}a_{n-2i}) \equiv (\mathcal{M}, \bar{a}a_{n-2i-1}b)$.

Consequently, $\mathcal{M} \models \theta_i(\bar{a}, a_{n-2i-1}, b)$ implies $\mathcal{M} \models \theta_i(\bar{a}, a_{n-2i-1}, a_{n-2i})$, as desired. $\qquad\qquad\qquad\qquad\qquad\qquad\qquad\qquad\qquad\qquad\qquad\qquad$ □

Since strongly minimal theories have no finite models, this immediately yields

**Corollary 14.6.5.** *No totally categorical, strongly minimal theory is finitely axiomatizable.* $\qquad\qquad\qquad\qquad\qquad\qquad\qquad\qquad\qquad\qquad\qquad\qquad\qquad$ □

This was proved in Makowsky's 1971 diploma thesis

[Makowsky, J. A. : Kategorizität und endliche Axiomatisierbarkeit, Diplomarbeit, ETH Zürich, 1971, mimeographed]

and later published in

[Makowsky, J. A. : On some conjectures connected with complete sentences, Fundamenta Math. **81** (1974) 193 – 202],

where it was noted that Vaught had proved it independently.

The proof of the corollary via finite models as in the theorem raises the question whether the only obstruction to finite axiomatizability is that being infinite is not finitely axiomatizable in this context. So Lachlan asked if, nevertheless, these theories could be axiomatized by the infinite set $T_=^{\infty}$, saying all models are infinite, plus one more sentence, i. e., if they were finitely axiomatizable modulo the (infinite) axiom scheme $T_=^{\infty}$. Such theories were called **quasi-finitely axiomatizable**. Note that a theory $T$ is quasi-finitely axiomatizable if and only if $T^{\infty}$ is finitely axiomatizable. Lachlan's question was answered positively by Gisela Ahlbrandt and Martin Ziegler in

[Ahlbrandt, G. and Ziegler, M. : Quasi-finitely axiomatizable totally categorical theories, Annals of Pure and Applied Logic **30** (1986) 63 – 82].

Note the interesting consequence that there are only countably many totally categorical strongly minimal theories.

---

[5]The point here is to find $a_{n-2i}$ in $A$, as opposed to just in $M$.

It should be noted that Makowsky, as well as Ahlbrandt and Ziegler, worked in the more general context of **almost strongly minimal theories**, that is theories $T$ having a strongly minimal 1-place formula $D$ (without parameters) such that every model $\mathcal{M}$ of $T$ is algebraic over $D(\mathcal{M})$ (in $\mathcal{M}$).[6] Later both results were generalized to arbitrary totally categorical countable theories (the context for which the problems were stated originally)—the non-finite axiomatizability by Boris Zilber in

[Зильбер, Б. И. (Zilber, B. I.) : О решении проблемы конечной аксиоматизируемости для теорий, категоричных во всех бесконечных мощностях, in Исследования по теоретическому программированию, Kazakh. Gos. University, Alma-Ata 1981, pp. 69 – 74. (English translation as *On the solution of the problem of finite axiomatizability for theories categorical in all infinite powers* in American Mathematical Society Translations (2) **135** (1987) 13 – 17.)],

the quasi-finite axiomatizability by Ehud Hrushovski in

[Hrushovski, E. : Totally categorical structures, Transactions of the American Mathematical Society **313** 131 – 159].

All this is presented in Pillay's *Geometric Stability Theory*.

The first two exercises deal with a few different concepts of homogeneity. We have met three already, $\infty$-homogeneity, ultrahomogeneity, and the homogeneity of saturated or atomic structures in Chapter 12. The latter two were defined via automorphisms, while the definition of the former went through partial elementary maps. Let us unify all this a little. To this end fix an infinite cardinal $\lambda$ and a structure $\mathcal{M}$. We say $\mathcal{M}$ is $\lambda$-**homogeneous** if, for all sequences $A \subseteq M$, $B \subseteq M$ of length smaller than $\lambda$ and such that $(\mathcal{M}, A) \equiv (\mathcal{M}, B)$, and every $a \in M$, there is $b \in M$ such that $(\mathcal{M}, Aa) \equiv (\mathcal{M}, Bb)$. We say $\mathcal{M}$ is **strongly $\lambda$-homogeneous** if, for all sequences $A \subseteq M$, $B \subseteq M$ of length smaller than $\lambda$ and such that $(\mathcal{M}, A) \equiv (\mathcal{M}, B)$, there is an automorphism of $\mathcal{M}$ taking $A$ to $B$. (Note that ultrahomogeneity is somewhat like strong $\omega$-homogeneity, only that the cardinality restriction applies to a set of generators rather than to the set itself.) The structure $\mathcal{M}$ is said to be (**strongly**) **homogeneous**, if it is (strongly) $|M|$-homogeneous.

**Exercise 14.6.1.** Verify: $\mathcal{M}$ is $\infty$-homogeneous if and only if it is $\lambda$-homogeneous for all infinite cardinals $\lambda$, if and only if it is $|M|^+$-homogeneous.

**Exercise 14.6.2.** Show that strongly $\lambda$-homogeneous implies $\lambda$-homogeneous and that in case $\lambda = |M|$ the two concepts coincide, i. e., $\mathcal{M}$ is homogeneous (i. e. $|M|$-homogeneous) if and only if it is strongly homogeneous (i. e. strongly $|M|$-homogeneous). (For simplicity, assume $\mathcal{M}$ is countable first.)

**Exercise 14.6.3.** Prove that almost strongly minimal theories (as defined above) are categorical in all powers bigger than the language.

---

[6]The correct definition is a little more general, but the one adopted here serves our purposes sufficiently.

## 14.7   Tiny models

We conclude this chapter with a closer look at the finite-dimensional models of a strongly minimal theory $T$. We established, in §14.5, that $T$ has such models only if the invariant $\iota(T)$ is finite. *So, for the rest of this section we assume $\iota(T) = n < \omega$.*

Our concern is the power of the finite-dimensional models. Since they are all countable (and infinite) if the theory itself is countable, *only uncountable theories interest us here.* We saw that the cardinality of the model of dimension $\aleph_0$ plays a crucial role in the classification of the models of $T$, and we derived $\lambda(T) = |\mathcal{M}_{\aleph_0}|$. Further, from the fact that $\mathcal{M}_{\aleph_0}$ is the union of the finite-dimensional models we obtained that in case $\lambda(T)$ is regular, all but finitely many of the finite-dimensional models have power $\lambda(T)$. What about the other ones, and what about the possible cardinalities of them? We want to discuss this question briefly, mostly in order to point out that it is largely open and thus provoke some interest in further reading. Let's first give these models a name: following Jim Loveys, we call a model of $T$ **tiny** if it has power smaller than $\lambda(T)$. So, only finite-dimensional models can be tiny.

As any tiny model has to be finite-dimensional, if $T$ has one, it cannot be totally categorical. However, there need not be one at all. In general it is not known how many tiny models a strongly minimal theory can have (not even whether there is one with *two* nonisomorphic tiny models). But there are strongly minimal theories with one tiny model (see the exercise below) and there is a structural condition—so-called triviality—under which we can prove that there cannot be a second one.

$T$ is said to be **trivial** if the closure operator acl is trivial in the sense that $\operatorname{acl} A = \bigcup_{a \in A} \operatorname{acl} a$ for all sets $A$ (in any model of $T$). (Obviously it suffices to require this for finite sets $A$.) This is equivalent to the property that every pairwise independent set is independent (check!). The easiest example of such a theory is $T_=$. Another one occurs in the last exercise below. Many algebraic examples are not trivial (exercise!).

**Proposition 14.7.1.** *Let $T$ be strongly minimal and trivial.*

*If $\mathcal{N} \models T$ is tiny, then $\mathcal{N}$ is the elementarily prime model of $T$; moreover, $\mathcal{N} = \operatorname{acl} \emptyset$ (i.e. $\mathcal{N} = \mathcal{M}_0$ in the notation of Theorem 14.5.2).*

*Proof.* As mentioned, being tiny, $\mathcal{N}$ must be finite-dimensional, $\mathcal{N} = \mathcal{M}_n$ say. All we have to do is show that $n = 0$. Choose a basis $\bar{a} = (a_0, \ldots, a_{n-1})$ of $\mathcal{N}$. Then $\mathcal{N} = \operatorname{acl}(\bar{a})$. But $T$ is trivial, so $N = \bigcup_{i<n} \operatorname{acl} a_i$.

Remember, all the $a_i$ have the same type (by the indiscernibility of bases). Since $\mathcal{N}$ is an elementary substructure of the saturated model $\mathcal{M}_{\lambda(T)}$, the homogeneity of saturated models (cf. §12.1 Remark (5)) then shows that the $a_i$ can be mapped onto each other by an automorphism of that saturated extension. Since then the same is true for the algebraic closures of these basis elements, we see that $|\operatorname{acl} a_i|$ is equal to a fixed cardinal $\mu$ for all $i$. Then $|N| = n\mu$, hence $\mu$ must be infinite (as $N$ is) and $|N| = n\mu = \mu$.

We claim $|\mathcal{M}_{\aleph_0}| = \mu$. To this end, using triviality, write also $M_{\aleph_0}$ as the union of algebraic closures of basis elements $a_i$ ($i < \omega$). By the same argument, all these closures of basis elements have power $\mu$, hence $|M_{\aleph_0}| = \aleph_0 \cdot \mu = \aleph_0 + \mu = \mu$, for $\mu$ is infinite.

If we now had $n \neq 0$, we would get $|N| = \mu = |M_{\aleph_0}|$, hence $\mathcal{N}$ would not be tiny. Thus $n = 0$ and so $N = \operatorname{acl} \emptyset$. □

Another question is what the power of a tiny model can be. In all known examples of strongly minimal theories with tiny models only the elementarily prime model is tiny and it is always countable. So the question arises whether any tiny model of a strongly minimal theory has to be countable. Here the situation is even more surprising, namely, this question turns out to be independent from set theory. A proof of this would go far beyond the scope of this text, but the interested reader is referred to

[Laskowski, M. C., Pillay, A., and Rothmaler, Ph. : Tiny models of categorical theories, Archive for Math. Logic **31** (1992) 385 – 396],

(where the above proposition appeared in more generality). Many more interesting questions and results around strong minimality, especially those connected with the surprising discoveries of Ehud Hrushovski, are discussed in the literature mentioned in the appendix. To get an idea of how the theory of strong minimality is embedded in the general model theory, the reader is referred to the introduction of Pillay's *Geometric Stability Theory*.

**Exercise 14.7.1.** Show that complete theories of vector spaces or algebraically closed fields are not trivial.

In the last exercise we discuss an example, due to Loveys, of a strongly minimal theory having a tiny model. (There are more algebraic examples, like the (countable) Prüfer group as a module over its (uncountable) endomorphism ring. To verify that they are, though, is less easy.)

Consider sequences of 0's and 1's of order type $\omega$. We define the sum of two such sequences of the same length componentwise (modulo 2). Then $G$, the set of all such sequences, forms an abelian group of the power of the continuum (the Nth direct power of $\mathbb{Z}_2$, the additive group with 2 elements). Given $g \in G$ and a natural number $n$, use $g \upharpoonright n$ to denote the initial segment of $g$ of length $n$. Let $A$ be the (countable!) set of all these finite $0, 1$-sequences.

Given $g \in G$ and $a \in A$ of length $n$, we define $g + a = (g \upharpoonright n) + a$. This provides us with an action of $g$ on $A$ as a permutation, and a structure $\mathcal{A} = (A, g)_{g \in G}$ as in Exercise 14.1.8. Hence its theory is strongly minimal.

**Exercise 14.7.2.** Verify the following statements about the above example.
(1)  $A = \operatorname{acl} \emptyset$, i. e., $\mathcal{A}$ has dimension 0.
(2)  Every elementary extension of $\mathcal{A}$ has at least the power of the continuum.
(3)  $\mathcal{A}$ is a tiny model of $\operatorname{Th} \mathcal{A}$.
(4)  $\operatorname{Th} \mathcal{A}$ is trivial.

# Chapter 15

# $\mathbb{Z}$

This final chapter is wholly devoted to the complete theory of the abelian group $(\mathbb{Z}; 0; +, -)$ of the integers. In passing, whenever it seems appropriate, we mention some general model-theoretic terms that play an important role in the contemporary theory.

From now on we simply write $\mathbb{Z}$ for the group of integers, as well as $\mathbb{Q}$ for that of the rationals—and in general we treat the distinction between structures and their universes quite sloppily.

*In this chapter **group** means **abelian group**.*

## 15.1 Axiomatization, pure maps and ultrahomogeneous structures

Let us first recall some old and introduce some new terminology and notation.

As before, $L_\mathbb{Z}$ is the language of the signature $(0; +, -)$. Given $n < \omega$ and a(n abelian) group $\mathcal{A}$, we use both $n\mathcal{A}$ and $nA$ to denote the subgroup $\{na : a \in A\}$ of $\mathcal{A}$. The cyclic group $\mathbb{Z}/n\mathbb{Z}$ of order $n$ is denoted by $\mathbb{Z}_n$. For the *trivial group* $\{0\}$ we simply write $0$ (hence $0\mathcal{A} = \mathcal{A}0 = 0$).

Further, $(m_0, \ldots, m_{n-1})$ is used to denote the greatest common divisor (**gcd**) of $\{m_i : i < n\} \subseteq \mathbb{Z}$, while $[m_0, \ldots, m_{n-1}]$ stands for their least common multiple (**lcm**). $n \mid x$ is an abbreviation of the $L_\mathbb{Z}$-formula $\exists y \, (ny = x)$, and $n \nmid x$ is one for $\neg(n \mid x)$. AG denotes the $L_\mathbb{Z}$-theory of all (abelian) groups, and $\text{AG}_{\text{tf}}$ that of the torsionfree ones, cf. §5.2. As mentioned in the introduction, all groups are *abelian* in this chapter.

The following will turn out to be an axiomatization of the complete theory of ℤ. Let $T^{\mathbb{Z}}$ denote the $L_{\mathbb{Z}}$-theory with the set of axioms

$$\mathrm{AG_{tf}} \cup \left\{ \exists u \left( \bigwedge_{0<m<n} n \nmid mu \wedge \forall x \bigvee_{m<n} n \mid (x - mu) \right) : 1 < n < \omega \right\}.$$

Here the extra axiom says that there is *one* generating element $u$ of order $n$ in $\mathcal{M}/n\mathcal{M}$. Hence the models of $T^{\mathbb{Z}}$ are exactly the torsionfree groups $\mathcal{M}$ for which $\mathcal{M}/n\mathcal{M}$ is cyclic of order $n$, i. e. isomorphic to $\mathbb{Z}_n$, for all $n > 1$.

Using a little divisibility theory one can show that, modulo $\mathrm{AG_{tf}}$, it suffices to require this for prime numbers $n$.

Being nontrivial torsionfree groups, all models of $T^{\mathbb{Z}}$ are infinite. Hence so is every subgroup $n\mathcal{M}$ of every model $\mathcal{M} \models T^{\mathbb{Z}}$ for every $0 < n < \omega$.

$\mathbb{Z}$ itself, as well as all subgroups $n\mathbb{Z}$ for $0 < n < \omega$ (which are isomorphic to $\mathbb{Z}$) are models of $T^{\mathbb{Z}}$. Being a subgroup (i. e. a substructure), $n\mathbb{Z}$ satisfies the same quantifier-free $L_{\mathbb{Z}}(n\mathbb{Z})$-sentences as $\mathbb{Z}$. But $\mathbb{Z} \models n \mid n$, while $n\mathbb{Z} \models n \nmid n$, hence the formula $n \mid x$ cannot be equivalent to a quantifier-free one. Thus $T^{\mathbb{Z}}$ does not admit quantifier elimination. Moreover, $T^{\mathbb{Z}}$ is not even model-complete, for, as mentioned, the model $n\mathbb{Z}$ is a subgroup of the model $\mathbb{Z}$, but not an elementary one. This also implies that $\mathbb{Z}$ is a minimal model of $T^{\mathbb{Z}}$ in the sense of Exercise 12.2.8.

In the next section we will show that $T^{\mathbb{Z}}$ admits elimination up to atomic formulas and formulas of the form $n \mid t(\bar{x})$ (where $t$ is an arbitrary $L_{\mathbb{Z}}$-term). For this reason we extend our language $L_{\mathbb{Z}}$ by a predicate $D_n$ for every $n > 1$ (with the intended meaning of $n \mid x$), and denote the resulting language by $L^{\mathrm{D}}$ ('D' stands for 'divisibility'). Further, D is used to denote the set of all atomic $L^{\mathrm{D}}$-formulas of the form $D_n(t(\bar{x}))$, where $1 < n < \omega$ and $t$ is an $L_{\mathbb{Z}}$-term, i. e. of the form $\sum k_i x_i$ for some $k_i \in \mathbb{Z}$. We call the formulas from D simply D-**formulas**. Correspondingly, a D-**literal** is, by definition, a D-formula or a negation of a D-formula. Given an $L_{\mathbb{Z}}$-theory $\Sigma$, we use $\Sigma^{\mathrm{D}}$ to denote the $L^{\mathrm{D}}$-theory obtained from $\Sigma$ by adding the axioms $\forall x \, (D_n(x) \leftrightarrow n \mid x)$ for all $n > 1$. We reserve the symbol $T^{\mathrm{D}}$ for the $L^{\mathrm{D}}$-theory $(T^{\mathbb{Z}})^{\mathrm{D}}$.

$\Sigma^{\mathrm{D}}$ is what is also called a **definitional expansion** of $\Sigma$, because the new syntactic elements can be expressed (*defined*) by the old ones. As a consequence, every model $\mathcal{M}$ of $\Sigma$ can be expanded in a unique way to a model $\mathcal{M}^{\mathrm{D}}$ of $\Sigma^{\mathrm{D}}$ (namely, by setting $D_n(\mathcal{M}) = n\mathcal{M}$), and every model of $\Sigma^{\mathrm{D}}$ is of the form $\mathcal{M}^{\mathrm{D}}$ for some $\mathcal{M} \models \Sigma$, i. e. the $L^{\mathrm{D}}$-expansion of a model of $\Sigma$.

We use $^{\mathrm{D}} : \mathrm{Mod\,AG} \to \mathrm{Mod\,AG^{\mathrm{D}}}$ to denote this map bringing a group to its canonical $L^{\mathrm{D}}$-expansion. The expression $\mathcal{M}^{\mathrm{D}}$ thus always includes that

$\mathcal{M}$ is a group.

For any $L^D$-substructure $\mathcal{A}^D$ of $\mathcal{M}^D \models AG^D$, we clearly have $D_n(\mathcal{A}^D) = \mathcal{A} \cap n\mathcal{M}$. However, this does not mean that $D_n(\mathcal{A}^D) = n\mathcal{A}$, because from the divisibility of $a \in A$ in $\mathcal{M}$ one can in general not derive its divisibility in $\mathcal{A}$. One can do so only if $\mathcal{A}^D$ is itself a model of $AG^D$. This leads us to the important notion of pure subgroup.

A group homomorphism $f : \mathcal{A} \to \mathcal{B}$ is said to be **pure**, if $\mathcal{B} \models n \mid f(a)$ implies $\mathcal{A} \models n \mid a$ for every $a \in A$ and $n < \omega$. We write $f : \mathcal{A} \hookrightarrow_{rd} \mathcal{B}$ for this. (To justify the hook notation, note that every pure homomorphism is a monomorphism: consider $n = 0$). A **pure** subgroup of a group $\mathcal{B}$ is a subgroup $\mathcal{A}$ of $\mathcal{B}$ such that the inclusion map of $\mathcal{A}$ in $\mathcal{B}$ is pure. $\mathcal{A}$ is also just said to be **pure** in $\mathcal{B}$, in symbols $\mathcal{A} \subseteq_{rd} \mathcal{B}$. (Here 'rd' stands for 'relatively divisible', a synonym for 'pure'.)

Thus a subgroup $\mathcal{A}$ of a group $\mathcal{B}$ is pure in $\mathcal{B}$ if and only if $n\mathcal{B} \cap \mathcal{A} = n\mathcal{A}$ for all $n < \omega$. Every direct summand of a group is trivially a pure subgroup (for $\mathcal{B} = \mathcal{A} \oplus \mathcal{C}$ implies $n\mathcal{A} = n\mathcal{B} \cap \mathcal{A}$). But there are other pure maps. E. g., every elementary homomorphism is pure (for the definition of 'pure' is like that of 'elementary', but with far less formulas to consider). Looking at any finite direct summand of an infinite group, we see that the converse is not true (for the simple reason that, one being finite and the other not, these groups can't be elementarily equivalent).

We are interested in pure maps inasmuch as, restricted to maps between models of $T^{\mathbb{Z}}$, they are exactly the elementary ones, as we will see in §15.3. Note, for the time being, that every embedding between models of $AG^D$ is, regarded as an $L_{\mathbb{Z}}$-map between their $L_{\mathbb{Z}}$-reducts, a pure homomorphism.

Let us summarize.

**Remarks.**

(1)  Let $f : \mathcal{A}^D \hookrightarrow \mathcal{M}^D$ be an embedding of an $L^D$-structure $\mathcal{A}^D$ in a model $\mathcal{M}^D \models AG^D$.
    Then $f : \mathcal{A} \hookrightarrow_{rd} \mathcal{M}$ iff $\mathcal{A}^D \models AG^D$.

(2)  As $n0 = 0$, the trivial group $0$ is pure in every group. Its $L^D$-expansion $0^D$ (in which $D_n(0)$ holds for all $n$) is thus a model, hence an algebraically prime model of $AG^D$.

We conclude this section with a lemma on ultrahomogeneous structures needed in the next section. Recall that a structure is ultrahomogeneous if every isomorphism between finitely generated substructures can be lifted to an automorphism of the entire structure (cf. §9.2).

**Lemma 15.1.1.** *Let $\mathcal{M}$ be an ultrahomogeneous structure.*

*For every finitely generated substructure $\mathcal{A} \subseteq \mathcal{M}$ and any embedding $h$ of $\mathcal{A}$ in an isomorphic structure $\mathcal{N} \cong \mathcal{M}$, we have $(\mathcal{M}, A) \equiv (\mathcal{N}, h[A])$.*

*Proof.* Let $f : \mathcal{N} \cong \mathcal{M}$. Then the restriction of $fh$ onto $A$ is an isomorphism of $\mathcal{A}$ onto $fh[A] \subseteq \mathcal{M}$, hence extendible to an automorphism $g \in \text{Aut}\,\mathcal{M}$. Then $\mathcal{M} \models \varphi(\bar{c})$ iff $\mathcal{M} \models \varphi(g[\bar{c}])$, for all formulas $\varphi$ and all matching tuples $\bar{c}$ from $M$. In particular, $\mathcal{M} \models \varphi(\bar{a})$ iff $\mathcal{M} \models \varphi(fh[\bar{a}])$, for all tuples $\bar{a}$ from $A$. But this means $(\mathcal{M}, A) \equiv (\mathcal{M}, fh[A])$. On the other hand, $f : \mathcal{N} \cong \mathcal{M}$ yields $(\mathcal{M}, fh[A]) \equiv (\mathcal{N}, h[A])$. $\square$

**Exercise 15.1.1.** Show that any group $\mathcal{A}$ has the same definable sets as its $L^{\mathrm{D}}$-expansion $\mathcal{A}^{\mathrm{D}}$.

**Exercise 15.1.2.** Prove that a cyclic group is ultrahomogeneous if and only if it is finite.

**Exercise 15.1.3.** Find a finite (abelian) group that is not ultrahomogeneous.

**Exercise 15.1.4.** Let $\mathcal{M} \models T^{\mathbb{Z}}$ and $\mathcal{D}$ a torsionfree divisible group (cf. §11.1). Prove that also $\mathcal{M} \oplus \mathcal{D}$ is a model of $T^{\mathbb{Z}}$.

## 15.2   Quantifier elimination and completeness

It is fundamental for the investigation of $T^{\mathbb{Z}}$ that $T^{\mathrm{D}}$, i. e. $(T^{\mathbb{Z}})^{\mathrm{D}}$, has quantifier elimination. In order to apply the criterion of Theorem 9.2.2(iii), we have to consider simply primitive $L^{\mathrm{D}}$-formulas, i. e. formulas $\varphi = \varphi(\bar{y})$ of the form $\exists x\, \psi(x, \bar{y})$, where $\psi = \psi(x, \bar{y})$ is a conjunction of D-literals and ($L_{\mathbb{Z}}$-) term equations and inequations in the variables $x, \bar{y}$.

The D-formulas are the only atomic $L^{\mathrm{D}}$-formulas not already in $L_{\mathbb{Z}}$. We first consider the case where the quantifier-free matrix $\psi$ of $\varphi$ contains only D-literals (and no term equations or inequations). Obviously, every D-formula $D_n(s(x, \bar{y}))$ (where $s$ is an $L_{\mathbb{Z}}$-term) is AG-equivalent to a formula $D_n(mx - t(\bar{y}))$, where $m \in \mathbb{Z}$ and $t$ is a term in $\bar{y}$. In torsionfree groups this formula is clearly equivalent to $D_{nk}(kmx - kt(\bar{y}))$ for arbitrary $k \neq 0$. Thus, modulo $\text{AG}_{\mathrm{tf}}^{\mathrm{D}}$, every finite collection of D-literals has a 'common denominator,' and we may therefore assume that our $\psi$ (in the case without term equations or inequations) has the form

$$\bigwedge_{i<k} D_n(m_i x - t_i(\bar{y})) \wedge \bigwedge_{k \le i < j} \neg D_n(m_i x - t_i(\bar{y})).$$

This latter formula is $\text{AG}^{\mathrm{D}}$-equivalent to the $L_{\mathbb{Z}}$-formula

$$\bigwedge_{i<k} n \mid (m_i x - t_i(\bar{y})) \wedge \bigwedge_{k \le i < j} n \nmid (m_i x - t_i(\bar{y})).$$

Let us stick with this $n$ for a while and fix some more temporary notation. Given an element $a$ in a group $\mathcal{A}$, let $a_A$ denote the coset $a + n\mathcal{A}$. Further, given a tuple $\bar{a}$ in $A$, let $\bar{a}_A$ denote the tuple of the corresponding cosets, and for any subset $X \subseteq A$ let $X_A$ denote the set $\{a_A : a \in X\}$. Then $\mathcal{A}_A$ is the factor group $\mathcal{A}/n\mathcal{A}$ (as an $L_\mathbb{Z}$-structure). Let $\psi^*(x, \bar{y})$ be the $L_\mathbb{Z}$-formula

$$\bigwedge_{i<k} m_i x = t_i(\bar{y}) \wedge \bigwedge_{k \leq i < j} m_i x \neq t_i(\bar{y}).$$

Then, for every matching tuple $\bar{a}$ and every element $b$ of a torsionfree group $\mathcal{M}$ we have $\mathcal{M}^D \models \psi(b, \bar{a})$ iff $\mathcal{M}_M \models \psi^*(b_M, \bar{a}_M)$.

This implies $\psi(\mathcal{M}, \bar{a}) = \bigcup\{b + nM : b_M \in \psi^*(\mathcal{M}_M, \bar{a}_M)\}$, since $b'_M = b_M$ for all $b' \in b + nM$. Clearly, the same is true for any other model $\mathcal{N}^D$ of $AG_{tf}^D$.

If now $\mathcal{M}^D$ and $\mathcal{N}^D$ are two models of $T^D$ ($\supseteq AG_{tf}^D$) with a joint ($L^D$-) substructure $\mathcal{A}$, then $\mathcal{M}^D \models D_n(a)$ iff $\mathcal{A}^D \models D_n(a)$ iff $\mathcal{N}^D \models D_n(a)$, hence $a_M = 0$ (in $\mathcal{M}_M$) iff $a_N = 0$ (in $\mathcal{N}_N$), for all $a \in A$.

As $(a - b)_M = a_M - b_M$ and similarly for $N$, the assignment $a_M \mapsto a_N$ thus defines a bijection $h$ between $A_M$ and $A_N$. But $\mathcal{A}$ is a subgroup of $\mathcal{M}$, hence $\mathcal{A}_M$ is a subgroup of $\mathcal{M}_M$, and similarly $\mathcal{A}_N$ is a subgroup of $\mathcal{N}_N$. Obviously, $h$ is an isomorphism from $\mathcal{A}_M$ onto $\mathcal{A}_N$. In particular, $h$ is an embedding of $\mathcal{A}_M$ in $\mathcal{N}_N$. The theory $T^D$ further guarantees $\mathcal{N}_N \cong \mathcal{M}_M \cong \mathbb{Z}_n$. By Exercise 15.1.2, these latter are ultrahomogeneous, hence Lemma 15.1.1 yields $(\mathcal{M}_M, A_M) \equiv (\mathcal{N}_N, h[A_M])$. Together with the equivalences already proved, $\mathcal{M}^D \models \varphi(\bar{a})$ iff $\mathcal{M}_M \models \exists x\, \psi^*(x, \bar{a}_M)$, and $\mathcal{N}^D \models \varphi(\bar{a})$ iff $\mathcal{N}_N \models \exists x\, \psi^*(x, \bar{a}_N)$, this proves $\mathcal{M}^D \models \varphi(\bar{a})$ iff $\mathcal{N}^D \models \varphi(\bar{a})$, as $h[\bar{a}_M] = \bar{a}_N$.

In the case where $\varphi$ is a conjunction of D-literals, the criterion of Theorem 9.2.2(iii) is thus satisfied. Moreover, either $\psi(\mathcal{M}, \bar{a}) = \psi(\mathcal{N}, \bar{a}) = \emptyset$ or else $\psi(\mathcal{M}, \bar{a})$ as well as $\psi(\mathcal{N}, \bar{a})$ are infinite, since along with every element $b$ these sets contain the entire cosets $b + nM$ or $b + nN$, respectively. This allows us to admit (in the quantifier-free matrix of $\varphi$) also term inequations $k_i x \neq s_i(\bar{y})$, for if $\varphi$ has the form $\exists x\, (\psi(x, \bar{y}) \wedge \bigwedge_{i<m} k_i x \neq s_i(\bar{y}))$, then $\mathcal{M}^D \models \varphi(\bar{a})$ clearly implies that the set $\psi(\mathcal{M}, \bar{a})$ is not empty, whence both, $\psi(\mathcal{M}, \bar{a})$ and $\psi(\mathcal{N}, \bar{a})$ are infinite. Then neither is the set $\psi(\mathcal{N}, \bar{a}) \setminus \{b \in N : k_i b = s_i(\bar{a}), i < m\}$ empty, as every term equation $k_i x = s_i(\bar{a})$ has at most one solution, by torsionfreeness, and thus the set to be subtracted is finite. Hence also in this case $\mathcal{M}^D \models \varphi(\bar{a})$ implies $\mathcal{N}^D \models \varphi(\bar{a})$. For reasons of symmetry the converse follows too.

We are left with the case, where the matrix of $\varphi$ contains a term equation $kx = t(\bar{y})$. In this case we can directly eliminate the quantifier in

$\varphi$ as follows. As before, we can multiply all other conjuncts of the matrix of $\varphi$ by $k$ in order to obtain an $AG_{tf}^D$-equivalent formula having all coefficients at $x$ divisible by $k$. In other words, the matrix of $\varphi$ is $AG_{tf}^D$-equivalent to a formula of the form $kx = t(\bar{y}) \wedge \psi(kx, \bar{y})$, where $\psi(z, \bar{y})$ is a conjunction of D-literals and term equations and inequations. Then $\varphi$, that is $\exists x \, (kx = t(\bar{y}) \wedge \psi(kx, \bar{y}))$, is $AG_{tf}^D$-equivalent to the quantifier-free $L^D$-formula $D_k(t(\bar{y})) \wedge \psi(t(\bar{y}), \bar{y})$, as is easily seen by direct inspection. Since quantifier-free formulas are preserved in substructures and extensions, we obtain again $\mathcal{M}^D \models \varphi(\bar{a})$ iff $\mathcal{A}^D \models D_k(t(\bar{a})) \wedge \psi(t(\bar{a}), \bar{a})$ iff $\mathcal{N}^D \models \varphi(\bar{a})$, i. e., also in this case clause (iii) of Theorem 9.2.2 holds.

We have thus proved

**Theorem 15.2.1.** $T^D$ has quantifier elimination. $\qquad\qquad\qquad\Box$

**Corollary 15.2.2.** $T^D$ and $T^{\mathbb{Z}}$ are complete (hence $T^{\mathbb{Z}} = \mathrm{Th}\,\mathbb{Z}$).

*Proof.* In the previous section we remarked that $0^D$ is a prime structure for $\mathrm{Mod}\,T^D$ (Remark (2)). Hence the completeness of $T^D$ follows from Remark (1) of §9.2. Then all models of $T^D$ are elementarily equivalent (Proposition 8.1.2), which is clearly inherited by their $L_{\mathbb{Z}}$-reducts. But all models of $T^{\mathbb{Z}}$ arise in such a way (remember, $T^D$ was a definitional expansion of $T^{\mathbb{Z}}$), hence all models of $T^{\mathbb{Z}}$ are elementarily equivalent too, i. e. the latter theory is complete. Since $\mathbb{Z}$ is a model of $T^{\mathbb{Z}}$, we finally see that $\mathrm{Th}\,\mathbb{Z}$ and $T^{\mathbb{Z}}$ must be equal. $\qquad\qquad\qquad\Box$

Now we translate everything we know about $T^D$ back into $L_{\mathbb{Z}}$. From now on we use the symbol $\Delta$ exclusively for the set of all $L_{\mathbb{Z}}$-formulas of the form $n \mid t(\bar{x})$ (where $n < \omega$ and $t$ is a term). Call these formulas, correspondingly, $\Delta$-**formulas**. We single out the ones that 'truly' need a quantifier and let $\Delta^+$ stand for the set of all $\Delta$-formulas of the form $n \mid t(\bar{x})$ where $n > 1$. Further, $\widetilde{\Delta}$ is used to denote the set of all boolean combinations of $\Delta$-formulas.

All $\Delta^+$-formulas are $AG^D$-equivalent to D-formulas and vice versa. (The only difference between a $\Delta^+$-formula and an equivalent D-formula is that the former is a (nonatomic) $L_{\mathbb{Z}}$-formula, while the latter is an atomic $L^D$-formulas). Any term equation $t(\bar{x}) = 0$ is AG-equivalent to the $\Delta$-formula $0 \mid t(\bar{x})$. Therefore, up to AG-equivalence, all quantifier-free $L_{\mathbb{Z}}$-formulas are contained in $\widetilde{\Delta}$. We are ready to draw another conclusion from the above quantifier elimination.

**Corollary 15.2.3.** $T^{\mathbb{Z}}$ has $\widetilde{\Delta}$-elimination.

*Proof.* By the preceding discussion and the above theorem, every $L_\mathbb{Z}$-formula is $T^D$-equivalent to a formula from $\tilde{\Delta}$. As $\tilde{\Delta} \subseteq L_\mathbb{Z}$, the remarks in §15.1 about $T^D$ being a definitional expansion of $T^\mathbb{Z}$ show that this equivalence holds in fact modulo $T^\mathbb{Z}$.   □

As all constant $L_\mathbb{Z}$-terms are AG-equivalent to the term 0, every $\Delta$-*sentence* is AG-equivalent to a sentence of the form $n \mid 0$, hence AG-equivalent to $\top$. Consequently, every $L_\mathbb{Z}$-sentence is $T^\mathbb{Z}$-equivalent to $\top$ or $\bot$, this yielding another argument for the completeness of $T^\mathbb{Z}$.

For later purposes, we conclude with some straightforward simplifications of certain $\Delta$-formulas.

**Remarks.**
(1)   Every $\Delta$-formula in the free variable $x$ is AG-equivalent to a formula $n \mid mx$, where $n < \omega$ and $m \in \mathbb{Z}$. As these are AG-equivalent to $n \mid -mx$, we may always assume that also $m$ is a natural number.
(2)   If $m = 0$ or $n = 1$, then $(n \mid mx) \sim_{\mathrm{AG}} (x = x)$.
(3)   If $m > 0$ and $n = 0$, then $(n \mid mx) \sim_{\mathrm{AG}} (mx = 0) \sim_{\mathrm{AG_{tf}}} (x = 0)$.
(4)   If $m > 0$ and $n > 1$ are relatively prime, then $(n \mid mx) \sim_{\mathrm{AG}} (n \mid x)$.
(5)   If $m > 0$ and $n > 1$, then $(n \mid mx) \sim_{\mathrm{AG}} (n \mid (n,m)x) \sim_{\mathrm{AG_{tf}}} (\frac{n}{(n,m)} \mid x)$.

*Proof.* (1) through (3) are clear, and (4) is immediate from (5) on letting $(n,m) = 1$.

Ad (5). First of all, $(n,m)$ is contained in $n\mathbb{Z} + m\mathbb{Z}$ and can be represented as $(n,m) = nk + ml$ for some $k, l \in \mathbb{Z}$. Then $(n,m)x = nkx + mlx$ holds in any group, hence $n \mid mx$ implies $n \mid (n,m)x$. The converse being trivial, we have the first equivalence. The second, for torsionfree groups, has been mentioned a few times already.   □

The proof of the following simple lemma is left to the reader.

**Lemma 15.2.4.** *Suppose* $\delta = \delta(x, \bar{y})$ *is the $\Delta$-formula* $n \mid (mx - t(\bar{y}))$, $\mathcal{A}$ *is a group,* $\bar{a} \in A^{l(\bar{y})}$, *and* $\bar{0}$ *is the $l(\bar{y})$-tuple with all entries 0.*
(1)   $\delta(\mathcal{A}, \bar{0})$ *is the subgroup defined by* $n \mid mx$ *in* $\mathcal{A}$ *(i. e. , in case of a torsionfree $\mathcal{A}$ and $m > 0$ and $n > 1$, the subgroup $\frac{n}{(n,m)}\mathcal{A}$).*
(2)   $\delta(\mathcal{A}, \bar{a}) = a + \delta(\mathcal{A}, \bar{0})$ *for all* $a \in \delta(\mathcal{A}, \bar{a})$ *(i. e., $\delta(x, \bar{a})$ and $\delta(x - a, \bar{0})$ define the same coset of $\delta(\mathcal{A}, \bar{0})$, provided there is such an a).*

**Exercise 15.2.1.** Prove the above lemma.

**Exercise 15.2.2.** By Corollary 9.2.3 and Exercise 15.1.2, the (theories of the) finite cyclic groups have quantifier elimination. Give a proof of this by directly finding a quantifier-free formula for every simply primitive formula.

**Exercise 15.2.3.** Find a direct quantifier elimination for $T^{\mathrm{D}}$.

**Exercise 15.2.4.** The arguments for the preceding two exercises simplify considerably if we ad a new constant for the 1 in $\mathbb{Z}_n$ or $\mathbb{Z}$.

Formulate a corresponding axiomatization $T_1^{\mathbb{Z}}$ of $(\mathbb{Z}; 1)$ (or, more exactly, of $(\mathbb{Z}; 0, 1; +, -)$), eliminate the quantifiers in $\mathrm{Th}(\mathbb{Z}_n; 1)$ and $(T_1^{\mathbb{Z}})^{\mathrm{D}}$, and derive the completeness of $T_1^{\mathbb{Z}}$.

## 15.3   Elementary maps and prime models

Since every nonzero element in a torsionfree group generates a group isomorphic to $\mathbb{Z}$, the latter is an algebraically prime model of $T^{\mathbb{Z}}$. We know further that for $n > 1$ the inclusion $n\mathbb{Z} \subseteq \mathbb{Z}$ is not elementary and that therefore $T^{\mathbb{Z}}$ is not model-complete. The question arises if there always is, nevertheless, an elementary embedding of $\mathbb{Z}$ into every model (for instance, in the case of $n\mathbb{Z}$ the isomorphism $\mathbb{Z} \cong n\mathbb{Z}$ is trivially such an elementary embedding). However, in general this is not so.

**Theorem 15.3.1.** $T^{\mathbb{Z}}$ has no elementarily prime model.

[Baldwin, J. T., Blass, A. R., Glass, A. M. W., and Kueker, D. W. : A 'natural' theory without a prime model, algebra universalis **3** (1973) 152 – 155]

*Proof.* If there were such a model, it would be embeddable in $\mathbb{Z}$, hence isomorphic to $\mathbb{Z}$. We are going to build a model of $T^{\mathbb{Z}}$, however, which embeds $\mathbb{Z}$, but does not so elementarily.

For every prime $p$, consider the group $\mathbb{Z}_{(p)}$ of rational numbers whose denominator is not divisible by $p$. Clearly, the factors $p^k\mathbb{Z}_{(p)}/p^{k+1}\mathbb{Z}_{(p)}$ of the chain $\mathbb{Z}_{(p)} \supseteq p\mathbb{Z}_{(p)} \supseteq p^2\mathbb{Z}_{(p)} \supseteq \ldots$ are isomorphic to $\mathbb{Z}_p$ and hence $\mathbb{Z}_{(p)}/p^k\mathbb{Z}_{(p)}$ is isomorphic to $\mathbb{Z}_{p^k}$. Every element of $\mathbb{Z}_{(p)}$ is divisible by every other prime, hence $\mathbb{Z}_{(p)} = m\mathbb{Z}_{(p)}$ for every $m$ not divisible by $p$. In general, we thus have the isomorphism $\mathbb{Z}_{(p)}/m\mathbb{Z}_{(p)} \cong \mathbb{Z}_{p^n}$ for arbitrary natural numbers $m = m'p^n$ with $p \nmid m'$.

Consider the direct sum $\mathcal{M} = \bigoplus_{p \in \mathbb{P}} \mathbb{Z}_{(p)}$. We claim, this is a model of $T^{\mathbb{Z}}$. Clearly, $\mathcal{M}$ is a torsionfree group, and it remains to show that $\mathcal{M}/m\mathcal{M}$ is isomorphic to $\mathbb{Z}_m$ for all $m > 0$.

Given a decomposition into prime factors, $m = \prod_{i<n} p_i^{k_i}$, we now have $m\mathcal{M} = \bigoplus_{p \in \mathbb{P}} m\mathbb{Z}_{(p)}$, where $m\mathbb{Z}_{(p)} = \mathbb{Z}_{(p)}$ for all $p \neq p_i$, and $m\mathbb{Z}_{(p_i)} = p_i^{k_i}\mathbb{Z}_{(p_i)}$ for all $i < n$. Thus in $\mathcal{M}/m\mathcal{M} \cong \bigoplus_{p \in \mathbb{P}} \mathbb{Z}_{(p)}/m\mathbb{Z}_{(p)}$ only the $n$ summands $\mathbb{Z}_{(p_i)}/p_i^{k_i}\mathbb{Z}_{(p_i)} \cong \mathbb{Z}_{p_i^{k_i}}$ are different from 0. Hence $\mathcal{M}/m\mathcal{M}$ is isomorphic to $\bigoplus_{i<n} \mathbb{Z}_{p_i^{k_i}}$, which in turn is isomorphic to $\mathbb{Z}_m$. Consequently, $\mathcal{M}$ is a model of $T^{\mathbb{Z}}$.

Every element of $\mathcal{M}$ has finite support (cf. §4.1) and is therefore already contained in a subgroup that is a direct sum of finitely many $\mathbb{Z}_{(p)}$. This means that such an element is divisible by all other (infinitely many!) primes. Hence every element of $\mathcal{M}$ is divisible by all but finitely many primes. Consequently, $\mathbb{Z}$ cannot be elementarily embedded in $\mathcal{M}$: e. g., $1 \in \mathbb{Z}$ is divisible by no prime. □

Next we describe the elementary maps between models of $T^{\mathbb{Z}}$ in a more algebraic fashion.

**Lemma 15.3.2.** *A homomorphism between models of $T^{\mathbb{Z}}$ is elementary if and only if it is pure.*

*Proof.* Let $\mathcal{M}, \mathcal{N} \models T^{\mathbb{Z}}$ and $f : \mathcal{M} \to \mathcal{N}$.

For the nontrivial direction, assume $f$ is pure. To show it is elementary, consider an $L_{\mathbb{Z}}$-formula $\varphi = \varphi(\bar{x})$ and a matching tuple $\bar{a}$ from $M$. By the $\tilde{\Delta}$-elimination of $T^{\mathbb{Z}}$ (Corollary 15.2.3), $\varphi$ is equivalent to a boolean combination of $\Delta$-formulas. Hence, without loss of generality, we may assume that $\varphi$ itself is in $\Delta$. We have to show $\mathcal{M} \models \varphi(\bar{a})$ iff $\mathcal{N} \models \varphi(f[\bar{a}])$.

The direction from left to right is satisfied by *all* $\Delta$-formulas (under any circumstances), while the converse follows from the purity of $f$ (if $\varphi$ is of the form $n \mid t(\bar{x})$, just consider $a = t(\bar{a}) \in M$ and $n \mid a$, taking into account $f(a) = t(f[\bar{a}])$). □

**Exercise 15.3.1.** Verify that $T^{\mathbb{Z}}$ has no atomic models.

**Exercise 15.3.2.** Show directly that $\mathbb{Z}$ is not atomic.

**Exercise 15.3.3.** What happens with these questions if a constant symbol for 1 *or* a function symbol for multiplication is added to the signature?

## 15.4 Types and stability

Due to the $\tilde{\Delta}$-elimination of $T^{\mathbb{Z}}$ the description of types becomes quite simple. Namely, complete types are determined by the $\Delta$-formulas they contain (cf. Remark (8) in §11.3), i. e. $\mathrm{tp}(\bar{b}/A) = \mathrm{tp}(\bar{c}/A)$ if and only if $\mathrm{tp}_\Delta(\bar{b}/A) = \mathrm{tp}_\Delta(\bar{c}/A)$.

In order to infer $\bar{c} \models \mathrm{tp}(\bar{b}/A)$, however, not only does $\bar{c}$ have to realize $\mathrm{tp}_\Delta(\bar{b}/A)$, but of course also $\{\neg\delta(\bar{x}, \bar{a}) \in \mathrm{tp}(\bar{b}/A) : \delta \in \Delta\}$, the set of all negated instances of $\Delta$-formulas contained in $\mathrm{tp}(\bar{b}/A)$.

The actually needed formulas can be further reduced. Namely, Lemma 15.2.4(2) says that the instance $n \mid t(\bar{x}, \bar{a})$ of the $\Delta$-formula $n \mid t(\bar{x}, \bar{y})$ defines, in any group $\mathcal{M}$, a coset of the subgroup defined by $n \mid t(\bar{x}, \bar{0})$ in $\mathcal{M}$, provided the formula is satisfiable in $\mathcal{M}$. As any two cosets of the same group are either identical or else disjoint (hence inconsistent), this implies that one needs to consider only *one* instance of each formula $\delta \in \Delta$ in every $\Delta$-type $\mathrm{tp}_\Delta(\bar{b}/A)$. Then, due to the countability of $\Delta$, every $\Delta$-type is determined by a countable set of formulas. More precisely, $\mathrm{tp}_\Delta(\bar{b}/A)$ is determined by a map assigning to every $\Delta$-formula $\delta(\bar{x}, \bar{y})$ with $l(\bar{x}) = l(\bar{b})$ either an instance $\delta(\bar{x}, \bar{a})$ (or just the corresponding tuple $\bar{a}$ from $A$) or the answer 'not contained' for the case that $\mathrm{tp}_\Delta(\bar{b}/A)$ contains *no* instance of this formula.

This allows us to count types. There are no more pairwise inconsistent $\Delta$-types over $A$ than maps from $\Delta$ into the set $A^{<\omega}$ of all tuples from $A$ plus one further element (for the negative answer above). If $A$ is infinite, it has the same power as $A^{<\omega}$, and adding this one extra element does not change this situation. Hence in the infinite case the number of maps under consideration is $|{}^\Delta A| = |A|^{|\Delta|} = |A|^{\aleph_0}$.

This leads to an important definition—for any complete theory $T$ in an arbitrary language $L$. Given an infinite cardinal $\lambda$, the theory $T$ is said to be **stable** in $\lambda$, or $\lambda$-**stable**, if $|A| \leq \lambda$ implies $|S_1^\mathcal{M}(A)| \leq \lambda$, for all $\mathcal{M} \models T$ and $A \subseteq M$. The theory $T$ is said to be **stable** if it is stable in some infinite cardinal, and **superstable** if it is stable from some cardinal on, i. e., if there is a cardinal $\kappa$ such that $T$ is stable in every $\lambda \geq \kappa$.

**Remarks.**

(1)  Since increasing the set $A$ cannot decrease the number of types over it, in the above definition one needs to check only 'big' sets $A$ of power $\lambda$. E. g., provided $\lambda \geq |L|$, it suffices to consider the case where $A$ itself is a model, for, by Löwenheim-Skolem, we can always extend $A$ to a model (of the same power) containing $A$. That is, $T$ is stable in some $\lambda \geq |L|$ if and only if $|S_1^\mathcal{M}(M)| = \lambda$ for all models $\mathcal{M} \models T$ of power $\lambda$. (Equality of powers can be assumed, since over $M$ there are at least $|M|$ types anyway, for $x = a$ gives a different one for every $a \in M$.)

(2)  In view of the $\widetilde{\Delta}$-elimination, the above estimate of pairwise inconsistent $\Delta$-types is, in fact, a bound on the number of *all* types. Hence $T^\mathbb{Z}$ is stable in every cardinal of the form $\lambda^{\aleph_0}$, for $|A| \leq \lambda^{\aleph_0}$ implies $|S_1(A)| \leq |A|^{|\Delta|} = |A|^{\aleph_0} \leq (\lambda^{\aleph_0})^{\aleph_0} = \lambda^{\aleph_0 \cdot \aleph_0} = \lambda^{\aleph_0}$.

The last statement can be improved on, as the 1-types can be pinned down even further. Given $\mathcal{M} \models T^\mathbb{Z}$, every instance of a 1-place $\Delta$-formula

$\delta(x, \bar{a})$ over $M$ (i. e. with $\bar{a}$ from $M$) defines the empty set or a coset of $k_\delta M$. For $k_\delta > 0$ there are exactly $k_\delta$ such cosets—the subgroup $k_\delta M$ has **index** $k_\delta$ in $M$. As a consequence, up to $T^\mathbb{Z}$-equivalence, every $\delta \in \Delta$ with $k_\delta > 0$ has only finitely many (namely $k_\delta$) instances $\delta(x, \bar{a})$ (where $\bar{a}$ is from $M$). Then, up to equivalence, there are at most as many $\Delta^+$-1-types of $M$ as there are maps from $\omega$ to $\bigcup_{0<k<\omega} M/kM$ or, equivalently, from $\omega$ to $\bigcup_{0<k<\omega} \mathbb{Z}_k$. As both sets are countable, there are exactly $\aleph_0^{\aleph_0} = 2^{\aleph_0}$ such maps. But $2^{\aleph_0}$ is not only an *upper* bound on the number of pairwise inconsistent $\Delta^+$-1-types. For every coset of $kM$ is partitioned into $m$ cosets of the subgroup $mkM$ (as $kM/mkM \cong mM$). If we now consider e. g. only the subgroups of the form $2^n M$, then every 0-1-sequence of length $\omega$ yields another $\Delta^+$-1-type, hence at least $2^{\aleph_0}$ many pairwise consistent ones.

Thus we have shown that over every model of $T^\mathbb{Z}$ there are exactly $2^{\aleph_0}$ pairwise inconsistent $\Delta^+$-1-types.

A complete 1-type over $M$ is not completely determined by its $\Delta^+$-part. However, the only $\Delta$-formulas that are not $\Delta^+$-formulas are (up to equivalence) the term equations of the form $kx = t(\bar{y})$ where $k > 0$, and there are at most $|M|$ instances of any such formula. By torsionfreeness, every 1-type can contain only one such equation (up to equivalence again, as always), as it has to be realizable. Hence, besides its $\Delta^+$-part the only information missing is, which of the $|M|$ nontrivial term equations (i. e. with $k > 0$) it contains or if it contains none of these altogether. As $M$ is infinite, we have $|M|$ extra possibilities for this, which, together with the bound on the $\Delta^+$-1-types, yields $|S_1(M)| \leq 2^{\aleph_0} \cdot |M| = \max\{2^{\aleph_0}, |M|\}$.

This implies that $T^\mathbb{Z}$ is stable in all $\lambda \geq 2^{\aleph_0}$, hence superstable, for, if $|M| \geq 2^{\aleph_0}$, then $|S_1(M)| \leq |M|$ (and hence $|S_1(M)| = |M|$). We will see shortly that only in these cardinals is $T^\mathbb{Z}$ stable. First we describe the algebraic 1-types in more detail.

**Lemma 15.4.1.** *Suppose $\mathcal{M} \models T^\mathbb{Z}$, $A \subseteq M$, and $c$ is an element of $\mathcal{M}$.*
$\mathrm{tp}^\mathcal{M}(c/A)$ *is algebraic iff* $\mathrm{tp}_\Delta^\mathcal{M}(c/A)$ *is algebraic iff* $\mathrm{tp}^\mathcal{M}(c/A)$ *contains a nontrivial term equation (i. e. one of the form $kx = t(\bar{y})$ with $k > 0$).*

*Proof.* The directions from right to left are trivial or mentioned already (nontrivial term equations have at most one solution in torsionfree groups).

We are left with showing that any algebraic type $\mathrm{tp}^\mathcal{M}(c/A)$ contains a nontrivial term equation. By definition it contains an algebraic formula, which, by $\widetilde{\Delta}$-elimination, can be taken a boolean combination of $\Delta$-formulas. Since then *every* disjunct must be algebraic and *one* of them must be, by completeness, contained in the type, the latter contains an algebraic formula $\psi$ that is a conjunction of $\Delta$-formulas and negations of such. Assume, there were no nontrivial term equations among these. We lead this to a contradiction. Namely, $\psi$ would then be a conjunction of D-literals and term inequations. However, as we saw in the proof of (and before) Theorem 15.2.1, such formulas define infinite sets in all models of $T^D$. This would contradict the algebraicity of $\psi$.                                                □

Thus the type $\mathrm{tp}^{\mathcal{M}}(c/A)$ is not algebraic if and only if (up to trivial formulas equivalent to $x = x$) the types $\mathrm{tp}_{\Delta}^{\mathcal{M}}(c/A)$ and $\mathrm{tp}_{\Delta+}^{\mathcal{M}}(c/A)$ are the same. This leads to a simple criterion for their realization in case $A$ is an elementary substructure of $\mathcal{M}$ (i. e. for complete 1-types over models of $T^{\mathbb{Z}}$). We emphasized in the beginning that in general $b \models \mathrm{tp}_{\Delta}(c/A)$ does not imply $b \models \mathrm{tp}(c/A)$, since for the latter $b$ has to satisfy also the negations $\{\neg \delta \in \mathrm{tp}(c/A) : \delta \in \Delta\}$. But in the case where $A$ is elementary in $\mathcal{M}$ and $\mathrm{tp}^{\mathcal{M}}(c/A)$ is algebraic, these negations are redundant, as then $\mathrm{tp}^{\mathcal{M}}(c/A)$ contains the equation $x = c$ (and $c$ is in $A$), which is contained already in $\mathrm{tp}_{\Delta}^{\mathcal{M}}(c/A)$. Also for the nonalgebraic complete 1-types over models of $T^{\mathbb{Z}}$ we have a similar simplification.

**Lemma 15.4.2.** *Suppose $\Phi \in S_1^{\mathcal{M}}(M)$ and $\mathcal{M} \models T^{\mathbb{Z}}$.*

*$\Phi$ is not algebraic if and only if, for all $\mathcal{N} \succcurlyeq \mathcal{M}$ and $b \in N \setminus M$, $b \models \Phi_{\Delta+}$ already implies $b \models \Phi$.*

*(Here $\Phi_{\Delta+}$ is the $\Delta^+$-part of $\Phi$, hence $\mathrm{tp}_{\Delta+}(c/M)$ if $\Phi = \mathrm{tp}(c/M)$.)*

*Proof.* If $\Phi$ is algebraic, it can't be realized outside $M$, while $\Phi_{\Delta+}$ is.

Let $\Phi = \mathrm{tp}(c/M)$ not be algebraic. Then $\Phi_{\Delta} \sim_{\mathrm{Th}(\mathcal{M},M)} \Phi_{\Delta+}$. Let further $b$ be as in the hypothesis. As $b \notin M$, the type $\mathrm{tp}(b/M)$ is not algebraic, and we have $\mathrm{tp}_{\Delta}(b/M) \sim_{\mathrm{Th}(\mathcal{M},M)} \mathrm{tp}_{\Delta+}(b/M)$. To show $b \models \Phi$ it therefore remains to verify $\mathrm{tp}_{\Delta+}(b/M) \sim_{\mathrm{Th}(\mathcal{M},M)} \Phi_{\Delta+}$ (this uses $\tilde{\Delta}$-elimination). To this end, let $\{a_i : i < n\} \subseteq M$ be a set of representatives of the partition of $M$ into cosets of $nM$ ($n > 0$).

$\mathcal{M} \models \forall x \bigvee_{i<n} n \mid (x - a_i)$ and the completeness of $\Phi$ implies that one of the formulas $n \mid (x - a_i)$ must be in $\Phi$ ($\Phi$ decides every coset of $nM$ in $M$). Since $\mathcal{M} \models \forall x (n \mid (x - a_i) \to \bigwedge_{i \neq j < n} n \nmid (x - a_j))$, this makes the negations of $\Delta^+$-formulas redundant. In other words, every realization of $\Phi_{\Delta+}$ realizes all negations of instances of $\Delta^+$-formulas contained in $\Phi$. Hence $b \models \Phi_{\Delta+}$ implies $\mathrm{tp}_{\Delta+}(b/M) = \Phi_{\Delta+}$, as desired.                     □

Now that we have seen that the negations of nonalgebraic (instances of) $\Delta$-formulas are redundant in complete 1-types over models, we are going to reduce the parameters required for this phenomenon. Consider two types without parameters which play an important role in this.

Let $\mathbb{1}$ denote the set $\{n \nmid x : 1 \neq n < \omega\}$, and
let $\mathbb{1}_*$ denote the set $\{n \mid x : 0 < n < \omega\} \cup \{x \neq 0\}$.

Notice, up to AG-equivalence also $\mathbb{1}$ contains the formula $x \neq 0$ (which is equivalent to $0 \nmid x$). Both, $\mathbb{1}$ and $\mathbb{1}_*$, are 1-types of every model of $T^{\mathbb{Z}}$, as every finite subset of $\mathbb{1}$ or $\mathbb{1}_*$ is satisfied in $\mathbb{Z}$ and hence in every model of $T^{\mathbb{Z}}$. The type $\mathbb{1}$ is realized by $1 \in \mathbb{Z}$, while $\mathbb{1}_*$ is realized by $(0,1)$ in $\mathbb{Z} \oplus \mathbb{Q}$ (that the latter is a model of $T^{\mathbb{Z}}$ was to be proved in Exercise 15.1.4).

For the sequel it is important to keep in mind that, given any realization $1_M$ of $\mathbb{1}$ in a model $\mathcal{M}$ of $T^{\mathbb{Z}}$, the element $1_M + nM$ generates the group $M/nM$ $(0 < n < \omega)$. Otherwise it would be contained in a proper subgroup and would thus have order a proper divisor $k$ of $n$; i. e., $k1_M \in nM$ and $n = km$ for some $m \neq 1$, which would, by torsionfreeness, imply $1_M \in mM$, contradicting $1_M \models \mathbb{1}$.

Our next objective is to verify that, in analogy to Lemma 15.4.2, the types $\mathbb{1}$ and $\mathbb{1}_*$ already determine complete types (over $\emptyset$). By the remarks at the end of §15.2, every $\Delta$-formula in one free variable $x$ is AG$_{\mathrm{tf}}$-equivalent either to $x = x$ or $x = 0$, or else to a $\Delta$-formula of the form $k \mid x$ where $1 < k < \omega$. Any complete 1-type over $\emptyset$ is hence determined by formulas of this kind. In particular, $\mathbb{1}$ and $\mathbb{1}_*$ determine complete types. We summarize.

**Remarks.** Suppose $\mathcal{M}, \mathcal{N} \models T^{\mathbb{Z}}$ and $\Phi \in \mathrm{S}_1^{\mathcal{M}}(\emptyset)$, $\Psi \in \mathrm{S}_1^{\mathcal{N}}(\emptyset)$.
(1) $\Phi = \Psi$ if and only if $\Phi$ and $\Psi$ contain the same formulas of the form $n \mid x$ $(n < \omega)$.
(2) $b \models_{\mathcal{M}} \mathbb{1}$ iff $b \models_{\mathcal{M}} \mathrm{tp}^{\mathbb{Z}}(1)$.
(3) $b \models_{\mathcal{M}} \mathbb{1}_*$ iff $b \models_{\mathcal{M}} \mathrm{tp}^{\mathbb{Z} \oplus \mathbb{Q}}((0,1))$.

'In-between' the extremes $\mathbb{1}$ and $\mathbb{1}_*$ there are continuum many 1-types over $\emptyset$, as we will see in a moment. Once we allow a parameter realizing the type $\mathbb{1}$, we even get as many pairwise inconsistent $\Delta$-1-types.

**Lemma 15.4.3.**
(1) $|\mathrm{S}_1(T^{\mathbb{Z}})| = 2^{\aleph_0}$.
(2) *Suppose* $\mathbb{1}$ *is realized in* $\mathcal{M} \models T^{\mathbb{Z}}$, *by a certain element* $1_M$, *and* $\mathrm{tp}^{\mathcal{M}}(c/M)$ *is a nonalgebraic (1-) type.*
    *If* $\mathcal{N} \succeq \mathcal{M}$ *and* $b \in N \smallsetminus M$ *realizes the type* $\mathrm{tp}_{\Delta^+}^{\mathcal{M}}(c/1_M)$, *then* $b$ *realizes all of* $\mathrm{tp}^{\mathcal{M}}(c/M)$.

*Proof.* Ad (1). Consider, for any set $X \subseteq \mathbb{P}$ of prime numbers, the 1-type $\Phi_X = \{p \mid x : p \in X\} \cup \{p \nmid x : p \in \mathbb{P} \smallsetminus X\}$. (Then $\Phi_\emptyset$ is the type $\mathbb{1}$, while $\mathbb{1}_*$ is $\Phi_{\mathbb{P}} \cup \{x \neq 0\}$.) There are $2^{\aleph_0}$ such sets $X$, and for different $X$ the types $\Phi_X$ are incompatible. Thus they have different completions in

$S_1(T^{\mathbb{Z}})$, hence $|S_1(T^{\mathbb{Z}})| \geq 2^{\aleph_0}$. We have equality here, since, for any theory in a countable language, $|S_1(T^{\mathbb{Z}})| \leq 2^{\aleph_0}$, as there are no more 1-types (over $\emptyset$) than there are sets of formulas, i. e. at most $|\mathfrak{P}(L_{\mathbb{Z}})| = 2^{|L_{\mathbb{Z}}|} = 2^{\aleph_0}$.

(2) is analogous to the proof of Lemma 15.4.2—with the only difference that one can now take $\{i1_M : i < n\}$ as a set of representatives of the partition of $M$ into cosets of $nM$ (for $n > 0$). The formula $n \mid (x - i1_M)$ needs only the parameter $1_M$ then, for $iy$ is an $L_{\mathbb{Z}}$-term. $\qquad\square$

Let us get back to the stability question.

**Theorem 15.4.4.** $T^{\mathbb{Z}}$ *is a superstable, but not small theory.* $T^{\mathbb{Z}}$ *is $\lambda$-stable if and only if $\lambda \geq 2^{\aleph_0}$.*

*Proof.* That $T^{\mathbb{Z}}$ is stable in all these $\lambda$ and thus superstable we have already seen (before Lemma 15.4.1). Further, by Lemma 15.4.3(1), $T^{\mathbb{Z}}$ is neither small nor stable in a cardinal $\lambda < 2^{\aleph_0}$, for $|S_1(T^{\mathbb{Z}})| = |S_1^{\mathcal{M}}(\emptyset)| \leq |S_1^{\mathcal{M}}(M)|$ for any $\mathcal{M} \models T^{\mathbb{Z}}$. $\qquad\square$

**Exercise 15.4.1.** Show that the algebraic type containing $x = 0$ is the only isolated type in $S_1(T^{\mathbb{Z}})$ (another reason that $T^{\mathbb{Z}}$ has no atomic models, cf. Exercise 15.3.1).

**Exercise 15.4.2.** Derive the completeness of $T^{\mathbb{Z}}$ directly from the completeness of $T_1^{\mathbb{Z}}$ (from Exercise 15.2.4).

**Exercise 15.4.3.** Prove that a strongly minimal theory $T$ is stable in every $\lambda \geq |T|$.

**Exercise 15.4.4.** Prove that a theory stable in a regular cardinal $\lambda$ has a saturated model of power $\lambda$. (Here $\lambda$ can well be smaller than $|T|$, as in Theorem 12.1.3.)

## 15.5   Positively saturated models and direct summands

A central role in the description of the models of $T^{\mathbb{Z}}$ play models that always realize certain positive types. One could call these *positively saturated*, however the technical term known from algebra is the following.

A(n abelian) group $\mathcal{A}$ is said to be **algebraically compact** if every $\Delta$-1-type of $\mathcal{A}$ (over $A$) is realized in $\mathcal{A}$.

At the beginning of the previous section we saw that a $\Delta$-type is determined already by *one* instance of each $\Delta$-formula, i. e. by $|\Delta| = \aleph_0$ formulas altogether. This immediately implies

**Remark.** $\aleph_1$-saturated groups are algebraically compact.

We will see in due course that $\aleph_0$-saturation suffices in the case of models of $T^{\mathbb{Z}}$ (Proposition 15.5.3).

A model $\mathcal{N}$ of $T^{\mathbb{Z}}$ is algebraically compact if and only if every $\Delta^+$-1-type of $\mathcal{N}$ is realized in $\mathcal{N}$, for the only $\Delta$-1-types of $\mathcal{N}$ that are not $\Delta^+$-types are, by Lemma 15.4.1 (and the discussion following it), algebraic and hence realized anyway. We will strengthen this in Lemma 15.5.2, but before we state a useful characterization of elementary substructures of models of $T^{\mathbb{Z}}$.

**Remark.** Let $\mathcal{B} \subseteq_{\mathrm{rd}} \mathcal{M} \models \mathrm{AG}$ and $0 < n < \omega$.

If $b \in B$, then $b + nB = B \cap b + nM$ (as $b \in B \cap b + nM$ implies $B \cap b + nM = b + (B \cap nM)$, and purity yields $B \cap nM = nB$.

**Lemma 15.5.1.** *Let* $\mathcal{B} \subseteq_{\mathrm{rd}} \mathcal{M} \models T^{\mathbb{Z}}$.
(1)  $\mathcal{B} \preccurlyeq \mathcal{M}$ *if and only if* $B \cap a + nM \neq \emptyset$ *for every* $n > 0$ *and* $a \in M$ *(if and only if for all* $n > 0$ *and* $a \in M$ *there is* $b \in B$ *such that* $b + nM = B \cap a + nM$ *).*
(2)  *If* $B$ *contains a realization of the type* $\mathbb{1}$ *in* $\mathcal{M}$, *then* $\mathcal{B} \preccurlyeq \mathcal{M}$.

*Proof.* Ad (1). By hypothesis and Lemma 15.3.2, $\mathcal{B} \preccurlyeq \mathcal{M}$ is equivalent to $\mathcal{B} \models T^{\mathbb{Z}}$. If this is the case, $\mathcal{B}$ and $\mathcal{M}$ have the same number of cosets of $nB$, respectively of $nM$ (namely $n$). Then the assertion in parentheses (hence also the other one) follows from the above remark.

If, conversely, every coset of $nM$ contains an element from $B$, then we obtain, for every generating element $a + nM$ of the factor group $M \smallsetminus nM = \{ka + nM : k < n\}$, an element $b \in B$ such that $b - a \in nM$, hence also $kb - ka \in nM$, for all $k < \omega$. This means that different $k < n$ yield different cosets $kb + nB$. As $nM \cap B = nB$, there can't be more such cosets in $B$ than there are cosets of $nM$ in $M$. Hence the elements $kb + nB$ of $B/nB$, where $k < n$, already exhaust all of $B/nB$. But then the latter is generated by $b + nB$, hence cyclic of order $n$. Being a subgroup of $\mathcal{M}$, the group $\mathcal{B}$ is torsionfree and consequently a model of $T^{\mathbb{Z}}$.

(2) is immediate from (1). $\square$

**Lemma 15.5.2.** *A model* $\mathcal{N}$ *of* $T^{\mathbb{Z}}$ *is algebraically compact if and only if* $\mathcal{N}$ *realizes* $\mathbb{1}$, *by an element* $1_N$, *say, and also all* $\Delta^+$-1-types of $\mathcal{N}$ *over this element* $1_N$ *(that is, with the single parameter* $1_N$*).*

*Proof.* Let $\mathcal{N} \models T^{\mathbb{Z}}$ be algebraically compact and $\mathcal{M}$ a joint elementary extension of $\mathcal{N}$ and $\mathbb{Z}$ (which exists by completeness). Then $1 \in \mathbb{Z} \preccurlyeq \mathcal{M}$ realizes $\mathbb{1}$ in $\mathcal{M}$, hence the sentences $\exists x \bigwedge_{p \in \mathbb{P}, p < n} p \mid x - 1$ are true in $\mathcal{M}$ for all $n < \omega$. By the previous lemma, there is, for all $p \in \mathbb{P}$, an element $b_p \in N$ such that $\mathcal{N} \cap 1 + p\mathcal{M} = b_p + p\mathcal{N}$ (and $b_p + p\mathcal{M} = 1 + p\mathcal{M}$). Thus $p \mid x - 1$ is $\mathcal{M}$-equivalent to $p \mid x - b_p$, and the sentences $\exists x \bigwedge_{p \in \mathbb{P}, p < n} p \mid x - b_p$ are true in $\mathcal{M}$ for all $n < \omega$. As their parameters are from $N$, these sentences are also true in $\mathcal{N} \preccurlyeq \mathcal{M}$. Thus $\{p \mid x - b_p : p \in \mathbb{P}\}$ is a $\Delta$-type of $\mathcal{N}$, realized in $\mathcal{N}$ by hypothesis. Any realization $1_N$ of this type, however, realizes also $\mathbb{1}$, for $1_N \in b_p + p\mathcal{N} = \mathcal{N} \cap b_p + p\mathcal{M} = \mathcal{N} \cap 1 + p\mathcal{M}$ implies $\mathcal{N} \models p \nmid 1_N$ for all $p \in \mathbb{P}$ (hence also for all natural numbers $n \neq 1$).

As all $\Delta^+$-1-types of $\mathcal{N}$ over $1_N$ are in particular $\Delta$-1-types, all these are realized by the algebraic compactness of $\mathcal{N}$. This proves one direction of the lemma.

For the converse assume, $1_N \in N$ realizes the type $\mathbb{1}$ in $\mathcal{N}$ and also that all $\Delta^+$-1-types of $\mathcal{N}$ over $1_N$ are realized in $\mathcal{N}$. We mentioned above that for algebraic compactness it suffices to realize all $\Delta^+$-1-types of $\mathcal{N}$. Now, as in the proof of Lemma 15.4.3(2), if $1_N \models \mathbb{1}$, the set $\{i1_N : i < n\}$ is a set of representatives of the partition of $\mathcal{N}$ into cosets of $n\mathcal{N}$ ($n > 0$). Hence every such coset is definable in $\mathcal{N}$ using the single parameter $1_N$. Consequently, every $\Delta^+$-1-type of $\mathcal{N}$ over $N$ is already implied by its $\Delta^+$-1-subtype over $1_N$. □

This lemma provides us with many algebraically compact models of $T^{\mathbb{Z}}$.

**Proposition 15.5.3.**
(1)  *All $\aleph_0$-saturated models of $T^{\mathbb{Z}}$ are algebraically compact.*
(2)  *Every algebraically compact model of $T^{\mathbb{Z}}$ has a power $\geq 2^{\aleph_0}$.*
(3)  *$\mathbb{Z}$ has an algebraically compact elementary extension of power $2^{\aleph_0}$.*
(4)  *Every elementary extension of an algebraically compact model of $T^{\mathbb{Z}}$ is algebraically compact.*

*Proof.* (1) is immediate from the preceding lemma.

Ad (2). In the previous section we saw that there are $2^{\aleph_0}$ pairwise inconsistent $\Delta^+$-1-types over any model of $T^{\mathbb{Z}}$, and in an algebraically compact one they must all be realized.

Ad (3). That $T^{\mathbb{Z}}$ has just *any* algebraically compact model of power $2^{\aleph_0}$ follows already from (1) and (2) and Remark (8) in §12.1. However, we want one that extends $\mathbb{Z}$ elementarily. For this, let $S$ be the set of all 'complete' $\Delta^+$-1-types of $\mathbb{Z}$ over $\mathbb{1}$. ($S$ has power $2^{\aleph_0}$.) Invoking Proposition 11.2.5, choose $\mathcal{M}_S \succcurlyeq \mathbb{Z}$, of power $\leq |\mathbb{Z}| + |L_{\mathbb{Z}}| + |S| = \aleph_0 + \aleph_0 + 2^{\aleph_0} = 2^{\aleph_0}$, realizing all types from $S$. As $\mathbb{1} \in \mathbb{Z} \preccurlyeq \mathcal{M}_S$ realizes $\mathbb{1}$ also in $\mathcal{M}_S$, the assertion follows from the preceding lemma.

Ad (4). If $\mathcal{N} \models T^{\mathbb{Z}}$ is algebraically compact, it realizes, by the above lemma again, the type $\mathbb{1}$ by some $1_N$ and also all $\Delta^+$-1-types of $\mathcal{N}$ over $1_N$. It remains to note that, if $\mathcal{M} \succcurlyeq \mathcal{N}$, all these types are the same, whether considered as types of $\mathcal{N}$ or as types of $\mathcal{M}$. Thus all these types, regarded as types of $\mathcal{M}$ are realized in $\mathcal{N}$, hence also in $\mathcal{M}$. Another application of the preceding lemma yields the algebraic compactness of $\mathcal{M}$. $\qquad\square$

**Recall**, a subgroup $\mathcal{A}$ of a group $\mathcal{B}$ is a **direct summand** of $\mathcal{B}$ iff it has a **complement**, i. e. a subgroup $\mathcal{C}$ of $\mathcal{B}$ such that $\mathcal{B} \cong \mathcal{A} \times \mathcal{C}$ (in the sense of §1.6); where, for abelian groups, one usually writes $\mathcal{B} \cong \mathcal{A} \oplus \mathcal{C}$. This is equivalent to the existence of a subgroup $\mathcal{C}$ such that $\mathcal{A} \cap \mathcal{C} = 0$ and $\mathcal{B} = \mathcal{A} + \mathcal{C}$ (where the latter just means that every element of $\mathcal{B}$ is a sum of an element of $\mathcal{A}$ and an element of $\mathcal{C}$; notice, this is the same as saying $\mathcal{A}$ and $\mathcal{C}$ generate $\mathcal{B}$). An embedding $f : \mathcal{A} \hookrightarrow \mathcal{B}$ is said to **split** (or to *be* **split**) if $f[\mathcal{A}]$ is a direct summand of $\mathcal{B}$.

A group is **injective** iff every embedding of it into another group splits. It is well known that divisible groups are injective. (The converse is true too: every group, in particular, every injective group is, as shown in §6.3 embeddable in a divisible group, and direct summands of divisible groups are obviously divisible.) Utilizing the fact that direct sums of divisible groups are also divisible, one can show that every group contains a greatest divisible subgroup, its so-called **divisible part**. We use $\mathcal{D}_\mathcal{A}$ to denote the divisible part of $\mathcal{A}$.

The existence of a divisible part is particularly easy to verify in the torsionfree case. Namely, if $\mathcal{A}$ is a torsionfree group, $\bigcap_{0<n<\omega} n\mathcal{A}$ is already divisible and (thus the divisible part $\mathcal{D}_\mathcal{A}$, as it contains every divisible subgroup of $\mathcal{A}$). For if $d$ is any element of this subgroup, then, given $0 < n < \omega$, there is $a \in A$ with $d = na$. We show that $a$ must be in this subgroup (which is enough). So let $0 < m < \omega$. By the choice of $d$, there is also an element $b \in A$ such that $d = nmb$. Then torsionfreeness yields $a = mb$, and so $a \in m\mathcal{A}$, as desired.

As mentioned, $\mathcal{D}_\mathcal{A}$ is injective, hence it has a complement $\mathcal{A}_r$, i. e. $\mathcal{A} = \mathcal{A}_r \oplus \mathcal{D}_\mathcal{A}$. By the maximality of $\mathcal{D}_\mathcal{A}$, the group $\mathcal{A}_r$ contains no nontrivial divisible subgroup. Such groups are called **reduced**. As opposed to $\mathcal{D}_\mathcal{A}$, the subgroup $\mathcal{A}_r$ is in general not uniquely determined. But it is so up to isomorphism, as $\mathcal{A} = \mathcal{B} \oplus \mathcal{D}_\mathcal{A}$ implies $\mathcal{B} \cong \mathcal{A}/\mathcal{D}_\mathcal{A} \cong \mathcal{A}_r$. Therefore, allowing a little laxness, we call $\mathcal{A}_r$ a (sic!) **reduced part** of $\mathcal{A}$.

Reducedness of torsionfree groups can be described by omitting a type.

**Lemma 15.5.4.** *A torsionfree group is reduced if and only if it omits the type* $\mathbb{1}_*$.

*Proof.* Suppose $\mathcal{A}$ is a torsionfree group. Every realization of $\mathbb{1}_*$ in $\mathcal{A}$ is obviously contained in $\bigcap_{0<n<\omega} n\mathcal{A} = \mathcal{D}_\mathcal{A}$ and different from 0. Conversely, every nonzero element of $\mathcal{D}_\mathcal{A}$ realizes the type $\mathbb{1}_*$. The assertion follows on noticing that $\mathcal{A}$ is reduced if and only if $\mathcal{D}_\mathcal{A} = 0$. $\qquad\square$

Thus the omitting types theorem provides us with **reduced** models of $T^{\mathbb{Z}}$, i. e. models that are, as a group, reduced (exercise!). But, of course, $\mathbb{Z}$ is easily seen to be one, and the lemma below yields many more concrete examples.

**Remark.** It is not hard to see that every realization of $\mathbb{1}_*$ in a torsionfree group generates an infinite cyclic group all of whose nonzero elements realize this type. Invoking the torsionfreeness again (uniqueness of divisibility!), this can be used to show that, in fact, the group $\mathbb{Q}$ is embedded. As, being divisible, $\mathbb{Q}$ is injective, this argument can be iterated as to show that every divisible torsionfree group is a direct sum of copies of $\mathbb{Q}$ (in other words, it can be regarded as a vector space over $\mathbb{Q}$). Thus these groups are of the form $\mathbb{Q}^{(\kappa)}$ (cf. notation from §11.1; the decomposition theorem cited there is a generalization of this).

**Lemma 15.5.5.** *Let $\mathcal{A} = \mathcal{A}_r \oplus \mathcal{D}_{\mathcal{A}}$ be an arbitrary torsionfree group.*
(1)   $\mathcal{A} \models T^{\mathbb{Z}}$ *iff* $\mathcal{A}_r \models T^{\mathbb{Z}}$ *iff* $\mathcal{A}_r \preccurlyeq \mathcal{A} \models T^{\mathbb{Z}}$.
(2)   $\mathcal{A}$ *is an algebraically compact model of $T^{\mathbb{Z}}$ if and only if so is $\mathcal{A}_r$.*

*Proof.* The divisibility of $\mathcal{D}_{\mathcal{A}}$ yields $\mathcal{A}/n\mathcal{A} \cong \mathcal{A}_r/n\mathcal{A}_r \oplus \mathcal{D}_{\mathcal{A}}/n\mathcal{D}_{\mathcal{A}} \cong \mathcal{A}_r/n\mathcal{A}_r$ for every $n > 0$. Hence we have $\mathcal{A}/n\mathcal{A} \cong \mathbb{Z}_n$ iff $\mathcal{A}_r/n\mathcal{A}_r \cong \mathbb{Z}_n$. As $\mathcal{A}_r$ is a torsionfree group, this yields $\mathcal{A} \models T^{\mathbb{Z}}$ iff $\mathcal{A}_r \models T^{\mathbb{Z}}$.

Being a direct summand of $\mathcal{A}$, the subgroup $\mathcal{A}_r$ is pure in $\mathcal{A}$. If now $\mathcal{A}_r$, hence also $\mathcal{A}$, is a model of $T^{\mathbb{Z}}$, then the inclusion is elementary (by Lemma 15.3.2). This proves (1).

Ad (2). If $\mathcal{A}_r \models T^{\mathbb{Z}}$ is algebraically compact, (1) implies $\mathcal{A}_r \preccurlyeq \mathcal{A} \models T^{\mathbb{Z}}$, and Proposition 15.5.3(4), in turn, that $\mathcal{A}$ is algebraically compact.

For the converse, write every element $a$ of $\mathcal{A} = \mathcal{A}_r \oplus \mathcal{D}_{\mathcal{A}}$, according to this decomposition, as $a = a_r + d_a$. By the algebraic compactness of $\mathcal{A}$, there is such an $a$ realizing $\mathbb{1}$.

Note that $\mathcal{A}_r$ realizes $\mathbb{1}$ too: given $n > 0$, we have $n \mid d_a$, hence $n \nmid a$ implies $n \nmid a_r$; on the other hand, $a_r$ can't be 0, for $n \mid d_a$, but $n \nmid a$.

To be able to apply Lemma 15.5.2, it would be enough to show that $a$ and $a_r$ have the same $\Delta^+$-type over $1_r$. But this is equivalent to saying that, for all $n > 0$, $n \mid (a - i1_r)$ iff $n \mid (a_r - i1_r)$, which is immediate from $n \mid d_a$.                                                                                 $\square$

This leads to a useful decomposition theorem.

**Theorem 15.5.6.**
(1)   *Suppose $f : \mathcal{N} \hookrightarrow_{\mathrm{rd}} \mathcal{M}$, where $\mathcal{N}, \mathcal{M} \models T^{\mathbb{Z}}$. Then, if $\mathcal{N}$ is algebraically compact, then (so is $\mathcal{M}$ and) $f$ is elementary and split.*

(2) Suppose $\mathcal{N} \subseteq_{\mathrm{rd}} \mathcal{M}$, $\mathcal{N}, \mathcal{M} \models T^{\mathbb{Z}}$, and $\mathcal{N}$ is algebraically compact. Then $\mathcal{M}$ is algebraically compact, and $\mathcal{D}_{\mathcal{M}}$ has a direct summand $\mathcal{D}$ with $\mathcal{D}_{\mathcal{M}} = \mathcal{D}_{\mathcal{N}} \oplus \mathcal{D}$ and such that $\mathcal{N}_r \preccurlyeq \mathcal{N} = \mathcal{N}_r \oplus \mathcal{D}_{\mathcal{N}} \preccurlyeq \mathcal{M} = \mathcal{N} \oplus \mathcal{D} = \mathcal{N}_r \oplus \mathcal{D}_{\mathcal{M}}$.
*(Further, $\mathcal{D}$, $\mathcal{D}_{\mathcal{N}}$, and $\mathcal{D}_{\mathcal{M}}$ are $\mathbb{Q}$-vector spaces, whose dimensions satisfy $\dim \mathcal{D}_{\mathcal{M}} = \dim \mathcal{D}_{\mathcal{N}} + \dim \mathcal{D}$.)*

*Proof.* (2) applied to $f[\mathcal{N}] \subseteq_{\mathrm{rd}} \mathcal{M}$ immediately yields (1). Therefore we work in the situation of (2).

$\mathcal{N} \subseteq \mathcal{M}$ implies $\mathcal{D}_{\mathcal{N}} = \bigcap_{0 < n < \omega} n\mathcal{N} \subseteq \bigcap_{0 < n < \omega} n\mathcal{M} = \mathcal{D}_{\mathcal{M}}$ (and, together with purity, even $\mathcal{D}_{\mathcal{N}} = \mathcal{N} \cap \mathcal{D}_{\mathcal{M}}$). Then there must be a decomposition $\mathcal{D}_{\mathcal{M}} = \mathcal{D}_{\mathcal{N}} \oplus \mathcal{D}$. Now the remark before the preceding lemma implies the assertion in parentheses, while the lemma itself yields $\mathcal{N}_r \preccurlyeq \mathcal{N}_r \oplus \mathcal{D}_{\mathcal{N}} \ (= \mathcal{N})$.

By Lemma 15.3.2, $\mathcal{N}$ is elementary in $\mathcal{M}$. Hence Proposition 15.5.3 implies that $\mathcal{M}$ is algebraically compact.

It remains to prove $\mathcal{M} = \mathcal{N}_r \oplus \mathcal{D}_{\mathcal{M}}$ (for then we also have $\mathcal{M} = \mathcal{N}_r \oplus \mathcal{D}_{\mathcal{N}} \oplus \mathcal{D} = \mathcal{N} \oplus \mathcal{D}$). As $\mathcal{N}_r$ is reduced, its intersection with $\mathcal{D}_{\mathcal{M}}$ is trivial. Hence it is enough to prove $\mathcal{M} = \mathcal{N}_r + \mathcal{D}_{\mathcal{M}}$.

For this, pick any element $a$ of $M$. As $\mathcal{N}_r \preccurlyeq \mathcal{M}$, the type $\mathrm{tp}_{\Delta^+}^{\mathcal{M}}(a/N_r)$ is also a type of $\mathcal{N}_r$. By algebraic compactness it must be realized by some $b$ in $\mathcal{N}_r$. Then, for all $n > 0$, the elements $a$ and $b$ lie in the same coset of $n\mathcal{M}$, i. e., $a - b \in n\mathcal{M}$. This means that $a - b \in \mathcal{D}_{\mathcal{M}}$ and hence $a = b + (a - b) \in \mathcal{N}_r + \mathcal{D}_{\mathcal{M}}$, as desired. $\qquad \square$

**Corollary 15.5.7.** *If $\mathcal{N} \preccurlyeq \mathcal{M} \models T^{\mathbb{Z}}$ and $\mathcal{N}$ is algebraically compact, then $\mathcal{N}_r \cong \mathcal{M}_r$.*

*Proof.* The theorem yields $\mathcal{M}_r \oplus \mathcal{D}_{\mathcal{M}} = \mathcal{M} = \mathcal{N}_r \oplus \mathcal{D}_{\mathcal{M}}$, and it remains to apply one of the isomorphism theorems of group theory to infer $\mathcal{M}_r \cong \mathcal{M}/\mathcal{D}_{\mathcal{M}} \cong \mathcal{N}_r$. $\qquad \square$

**Exercise 15.5.1.** Apply the omitting types theorem to obtain a reduced model of $T^{\mathbb{Z}}$.

**Exercise 15.5.2.** Prove that *every* (abelian) group has an algebraically compact elementary extension.

**Exercise 15.5.3.** Let $U$ be a nonprincipal ultrafilter on the set $\mathbb{P}$ of all prime numbers. Prove that $\prod_{p \in \mathbb{P}} \mathbb{Z}_p / U$ is a torsionfree divisible group.

**Exercise 15.5.4.** Show that the factor groups from the introduction to §4.1 are algebraically compact.

## 15.6  Reduced and saturated models

Proposition 15.5.3 provides us with an algebraically compact elementary extension of $\mathbb{Z}$ of power $2^{\aleph_0}$. We use $\widehat{\mathbb{Z}}$ to denote a reduced part of this elementary extension of $\mathbb{Z}$. (The notation stems from the fact that, on the one hand, $\mathbb{Z}$ can be regarded as being contained in $\widehat{\mathbb{Z}}$ (even elementarily), as follows from Corollary 15.6.3 below, and, on the other hand, $\widehat{\mathbb{Z}}$ is, in a certain technical sense, a completion of $\mathbb{Z}$; see §15.8 for references.)

Since $\mathbb{Z}$ is not divisible, none of its elementary extensions is divisible either. Therefore $\widehat{\mathbb{Z}}$ is not 0. Moreover, Lemma 15.5.5 says that $\widehat{\mathbb{Z}}$ is an algebraically compact model of $T^{\mathbb{Z}}$, hence it must have power $2^{\aleph_0}$ (its power is at least this number by Proposition 15.5.3; then it is equal by choice). As we will see shortly, $\widehat{\mathbb{Z}}$ is the only model of $T^{\mathbb{Z}}$ that is at the same time reduced and algebraically compact, and it is in a sense the biggest reduced model of $T^{\mathbb{Z}}$ and the smallest algebraically compact one. This gives it a borderline position among the models of $T^{\mathbb{Z}}$.

**Lemma 15.6.1.** *Suppose* $\mathcal{M} = \mathcal{M}_r \oplus \mathcal{D}_{\mathcal{M}} \models T^{\mathbb{Z}}$ *and* $\mathrm{pr} : \mathcal{M} \to \mathcal{M}_r$ *is the canonical projection induced by this decomposition (cf. §1.6).*

*For every elementary embedding* $f$ *of a reduced model* $\mathcal{N}$ *of* $T^{\mathbb{Z}}$ *into* $\mathcal{M}$ *we have* $\mathrm{pr}\, f : \mathcal{N} \stackrel{\equiv}{\hookrightarrow} \mathcal{M}_r$.

*Proof.* We prove that the restricted map $\mathrm{pr} \restriction f[\mathcal{N}]$ is elementary (then $\mathrm{pr} f$ is elementary by Lemma 8.2.2). In view of Lemma 15.3.2 we need only show that it is pure.

So let $a \in f[\mathcal{N}]$ and $\mathrm{pr}(a)$ be divisible by $n < \omega$ in $\mathcal{M}_r$. We have to show that $a$ is divisible by $n$ (in $f[\mathcal{N}]$, but because of $f[\mathcal{N}] \preccurlyeq \mathcal{M}$ this is the same as being divisible in $\mathcal{M}$). According to the decomposition of $\mathcal{M}$, write $a = b + d$, where $b \in \mathcal{M}_r$ and $d \in \mathcal{D}_{\mathcal{M}}$. Then the divisibility of $d$ in $\mathcal{D}_{\mathcal{M}}$ and that of $b = \mathrm{pr}(a)$ in $\mathcal{M}_r$ implies that of $a$ in $\mathcal{M}$.  $\square$

We are ready to state and prove the aforementioned theorem showing that reduced algebraically compact models resemble both, atomic and saturated models, and in a way also $\mathbb{Z}$ (cf. Theorems 12.1.1 and 12.2.1).

**Theorem 15.6.2.**
(1)  (Universality) *Every reduced model of* $T^{\mathbb{Z}}$ *is elementarily embeddable in* $\widehat{\mathbb{Z}}$.
(2)  (Minimality) *No proper elementary substructure of* $\widehat{\mathbb{Z}}$ *is algebraically compact.*
(3)  (Uniqueness) *All reduced algebraically compact models of* $T^{\mathbb{Z}}$ *are isomorphic to* $\widehat{\mathbb{Z}}$.

(4) (Embeddability) $\widehat{\mathbb{Z}}$ *is elementarily embeddable in every algebraically compact model of* $T^{\mathbb{Z}}$.

*Proof.* Ad (1). Let $\mathcal{M} \models T^{\mathbb{Z}}$ be reduced and $\mathcal{N}$ a joint elementary extension of $\mathcal{M}$ and $\widehat{\mathbb{Z}}$. The above lemma yields an elementary embedding of $\mathcal{M}$ in $\mathcal{N}_r$. Corollary 15.5.7 implies that $\mathcal{N}_r$ is isomorphic to the reduced part of $\widehat{\mathbb{Z}}$, hence to $\widehat{\mathbb{Z}}$ itself. Consequently, $\mathcal{M}$ is elementarily embedded in $\widehat{\mathbb{Z}}$.

Ad (2). Let $\mathcal{N} \preccurlyeq \widehat{\mathbb{Z}}$ be algebraically compact. Theorem 15.5.6 yields a divisible group $\mathcal{D}$ such that $\widehat{\mathbb{Z}} = \mathcal{N} \oplus \mathcal{D}$. Since $\widehat{\mathbb{Z}}$ is reduced, $\mathcal{D}$ must be trivial and $\mathcal{N}$ must be equal to $\widehat{\mathbb{Z}}$.

(3) is immediate from (1) and (2).

Ad (4). Let $\mathcal{N} \models T^{\mathbb{Z}}$ be algebraically compact. Then $\mathcal{N}_r$ is elementarily embedded in $\widehat{\mathbb{Z}}$ by (1). But Lemma 15.5.5 says that $\mathcal{N}_r$ is algebraically compact (and $\mathcal{N}_r \preccurlyeq \mathcal{N}$). Then, by (2), $\mathcal{N}_r$ is isomorphic to $\widehat{\mathbb{Z}}$. Consequently, $\widehat{\mathbb{Z}}$ is elementarily embedded in $\mathcal{N}$. $\square$

(4) says that $\widehat{\mathbb{Z}}$ is a prime structure for the class of algebraically compact models of $T^{\mathbb{Z}}$—even 'elementarily'. For this reason, abusing the language slightly, $\widehat{\mathbb{Z}}$ is often called the **prime algebraically compact model** of $T^{\mathbb{Z}}$.

Invoking e. g. Lemma 15.5.4, the theorem implies

**Corollary 15.6.3.** *The reduced models of* $T^{\mathbb{Z}}$ *are, up to isomorphism, exactly the elementary substructures of* $\widehat{\mathbb{Z}}$. *All these thus have a power* $\leq$ $2^{\aleph_0}$. $\square$

As $\mathbb{Z}$ is reduced, it is (up to isomorphism) an elementary substructure of $\widehat{\mathbb{Z}}$. Referring to the remark before Lemma 15.5.5 about the shape of torsionfree divisible groups we can now give a description of the models of $T^{\mathbb{Z}}$.

**Corollary 15.6.4.** *The models of* $T^{\mathbb{Z}}$ *are, up to isomorphism, exactly the groups of the form* $\mathcal{M} \oplus \mathbb{Q}^{(\kappa)}$, *where* $\mathcal{M} \preccurlyeq \widehat{\mathbb{Z}}$ *and* $\kappa \in \mathbf{Cn}$. $\square$

For obvious reasons we call $\kappa$ the $\mathbb{Q}$-**dimension** of the model $\mathcal{N} = \mathcal{M} \oplus \mathbb{Q}^{(\kappa)}$ (provided $\mathcal{M} \preccurlyeq \widehat{\mathbb{Z}}$), in symbols $\mathbb{Q}$-dim $\mathcal{N} = \kappa$.

**Remarks.**

(1) In the above, $\mathcal{M}$ is a reduced part, and $\mathbb{Q}^{(\kappa)}$ is the divisible part of $\mathcal{N}$. Since both are determined up to isomorphism, any model $\mathcal{N}$ of $T^{\mathbb{Z}}$ is (up to isomorphism) uniquely determined by the isomorphism type of its reduced part $\mathcal{N}_r$ and its $\mathbb{Q}$-dimension $\mathbb{Q}$-dim $\mathcal{N}$.

(2) (On the cardinality of models of $T^{\mathbb{Z}}$.)
If $\mathcal{N} \models T^{\mathbb{Z}}$, then $|\mathcal{N}| = \max\{|\mathcal{N}_r|, \mathbb{Q}\text{-dim}\,\mathcal{N}\}$ (cf. Exercise 7.6.2).

**Corollary 15.6.5.** *The elementary diagram* $\mathrm{Th}(\widehat{\mathbb{Z}}, \widehat{\mathbb{Z}})$ *of* $\widehat{\mathbb{Z}}$ *is* $\lambda$-*categorical for all* $\lambda > 2^{\aleph_0}$.

*Proof.* The (reducts of the) models of this elementary diagram are the models of $T^{\mathbb{Z}}$ in which $\widehat{\mathbb{Z}}$ is elementarily embeddable (in the extended language with constants for all (images of) elements of $\widehat{\mathbb{Z}}$). Hence they are the models whose reduced part is (isomorphic to) all of $\widehat{\mathbb{Z}}$. These are precisely the groups of the form $\widehat{\mathbb{Z}} \oplus \mathbb{Q}^{(\lambda)}$, where $\lambda$ runs over all cardinals. They are uniquely determined by their $\mathbb{Q}$-dimension $\lambda$ and have, by the preceding remark, power $\lambda$, provided $\lambda > 2^{\aleph_0}$. But in every such cardinality there is, up to isomorphism, exactly one such group. $\square$

Following Shelah, complete theories with a model whose elementary diagram is categorical in all bigger[1] powers are called **unidimensional**. We will return to these in the exercises as well as in the last section.

Our next objective is to describe the saturated models of $T^{\mathbb{Z}}$. There is no countable one, as $T^{\mathbb{Z}}$ is not small (see Theorem 12.1.3; this can also be derived from Proposition 15.5.3(1) and (2)).

**Lemma 15.6.6.** *Let* $\lambda \in \mathbf{Cn}$ *be uncountable and* $\mathcal{M} \models T^{\mathbb{Z}}$.
*$\mathcal{M}$ is* $\lambda$-*saturated if and only if* $\mathcal{M}$ *realizes the type* $\mathbb{1}$ *and, in all models* $\mathcal{N} \preccurlyeq \mathcal{M}$ *with* $|N| < \lambda$ *that also realize* $\mathbb{1}$, *all* $\Delta^+$-$1$-*types over* $N$ *(or just over a realization of* $\mathbb{1}$ *in* $\mathcal{N}$) *are realized by an element of* $M \smallsetminus N$ *in* $\mathcal{M}$.

*Proof.* $\Longrightarrow$. Already $\aleph_0$-saturation ensures that $\mathbb{1}$ is realized. Let now $\mathcal{N} \preccurlyeq \mathcal{M}$ contain such a realization $1_N$ (which, clearly, realizes $\mathbb{1}$ also in $\mathcal{M}$). We have used repeatedly (e. g. in the second half of the proof of Lemma 15.5.2) that the $\Delta^+$-$1$-types over $N$ are determined by those over $1_N$. Thus it is enough to realize these latter types in $M \smallsetminus N$. Let $\Phi$ be such a $\Delta^+$-$1$-type over $1_N$, and consider the type $\Psi = \Phi \cup \{x \neq b : b \in N\}$.

We claim, the latter is a type of $\mathcal{N}$ (hence also of $\mathcal{M} \succcurlyeq \mathcal{N}$). Every finite conjunction of formulas from $\Phi$ (which corresponds to a finite conjunction of D-formulas) defines, as we saw in §15.2, an infinite set in $\mathcal{N}$. Hence every such conjunction is consistent with finitely many of the inequations $x \neq b$ ($b \in N$) occurring in $\Psi$. Thus $\Psi$ is a type of $\mathcal{M}$ (cf. Proposition 11.2.1(ii)). The set of parameters occurring in $\Psi$ is $N$ (remember, $1_N \in N$). Since this has power $< \lambda$, the $\lambda$-saturation of $\mathcal{M}$ ensures that $\Psi$ is realized in $\mathcal{M}$. It was built into $\Psi$ that then $\Phi$ is realized in $M \smallsetminus N$.

$\Longleftarrow$. In order to check $\lambda$-saturation it suffices, by Remark (4) from §12.1, to consider sets of parameters of power $< \lambda$ that are themselves elementary

---
[1](bigger than the power of that diagram)

substructures (remember, $|L_{\mathbb{Z}}| = \aleph_0 < \lambda$). Hence we need only realize the complete non-algebraic 1-types over every given $\mathcal{N} \preccurlyeq \mathcal{M}$ with $|N| < \lambda$ (the algebraic ones are realized anyway). Lemma 15.4.3(2) (with the roles of $\mathcal{M}$ and $\mathcal{N}$ interchanged) says that for this it would be enough to realize in $M \smallsetminus N$ every $\Delta^+$-1-type over any realization of $\mathbb{1}$—which is guaranteed by the hypothesis, provided $\mathcal{N}$ contains such a realization. Of course, it need not to begin with, but, as $\mathcal{M}$ is assumed to contain such a realization, $\mathbb{1}_M$ say, we can simply adjoin this to $N$ (and replace $\mathcal{N}$ by a slightly bigger elementary substructure of $\mathcal{M}$ of the same power which contains $\mathbb{1}_M$). Since this adjunction can only *add* types, we can assume without loss of generality that $\mathcal{N}$ is of the kind as in the hypothesis, and we are done. $\qquad\square$

**Proposition 15.6.7.** $\mathcal{M} \models T^{\mathbb{Z}}$ *is saturated if and only if* $\mathcal{M}_r \cong \widehat{\mathbb{Z}}$ *and* $\mathbb{Q}\text{-dim}\,\mathcal{M} = |M|$.

*Proof.* If the model $\mathcal{M}$ is saturated, it is algebraically compact (e. g. by Proposition 15.5.3(1)). Then, by Lemma 15.5.5, $\mathcal{M}_r$ is a (reduced!) algebraically compact model, hence by the above uniqueness theorem isomorphic to $\widehat{\mathbb{Z}}$. If now the $\mathbb{Q}$-dimension of $\mathcal{M}$ is $\kappa$, then $|\mathcal{D}_\mathcal{M}| = |\mathbb{Q}^{(\kappa)}| = \kappa + \aleph_0$. We claim, $\kappa = |\mathcal{M}|$. If not, every 1-type over $\mathcal{D}_\mathcal{M}$ would have to be realized in $\mathcal{M}$; in particular, the type $\mathbb{1}_* \cup \{x \neq d : d \in \mathcal{D}_\mathcal{M}\}$ would. But this is merely impossible (cf. Lemma 15.5.4 and the remark thereafter). This proves the claim, and so also one direction of the proposition (from left to right).

For the other direction, consider $\mathcal{M} = \widehat{\mathbb{Z}} \oplus \mathbb{Q}^{(\kappa)}$ with $\kappa = |M|$. Consider $\mathcal{N} \preccurlyeq \mathcal{M}$ of power less than $\kappa$. By the above Remark (2), we have $\kappa \geq 2^{\aleph_0}$ and $\mathbb{Q}\text{-dim}\,\mathcal{N} < \kappa$. Hence $\mathcal{D}_\mathcal{M} \smallsetminus \mathcal{D}_\mathcal{N} \neq \emptyset$ for all such $\mathcal{N}$. It is these $\mathcal{N}$ that we have to consider for the criterion of the preceding lemma. So pick $d \in \mathcal{D}_\mathcal{M} \smallsetminus \mathcal{D}_\mathcal{N}$. Being an elementary extension of $\widehat{\mathbb{Z}}$, the model $\mathcal{M}$ is algebraically compact and so realizes all $\Delta^+$-1-types over $N$. As $d$ is divisible by all $n > 0$, given any $a \in M$, the elements $a$ and $a + d$ have the same $\Delta^+$-1-type over $N$ in $\mathcal{M}$. On the other hand, $a + d$ lies outside $N$, whenever $a$ lies inside. Hence, in any case all the $\Delta^+$-1-types over $N$ are realized in $M \smallsetminus N$, as required in order to apply the lemma. $\qquad\square$

**Corollary 15.6.8.** *The saturated models of* $T^{\mathbb{Z}}$ *are (up to isomorphism) exactly the groups* $\widehat{\mathbb{Z}} \oplus \mathbb{Q}^{(\kappa)}$, *where* $\kappa \geq 2^{\aleph_0}$. $\qquad\square$

**Exercise 15.6.1.** Find a direct proof of the fact that every reduced model of $T^{\mathbb{Z}}$ has power $\leq 2^{\aleph_0}$.

**Exercise 15.6.2.** As reduced parts of models are unique only up to isomorphism, in Lemma 15.6.1 one cannot expect $f[\mathcal{N}]$ to be *equal* to $\mathcal{M}_r$. Find a counterexample.

**Exercise 15.6.3.** Show that every $2^{\aleph_0}$-saturated model of $T^{\mathbb{Z}}$ is saturated.

As (according to Remark (5) of §12.1) saturated models of the same power are isomorphic, a complete (!) theory with the property that all $2^{\aleph_0}$-saturated models are saturated has, in fact, the property that all $2^{\aleph_0}$-saturated models of the same power are isomorphic—a property that, in the case of $T^{\mathbb{Z}}$, could be read off directly from the description of its saturated models. The next exercise deals with this property in the far more general context of arbitrary unidimensional theories as defined in fine print after Corollary 15.6.5.

**Exercise 15.6.4.** Prove that, for every unidimensional theory, there is a cardinal $\kappa$ such that all $\kappa$-saturated models of the same power are isomorphic.

**Exercise 15.6.5.** Prove that a theory categorical in all powers bigger than a fixed cardinal must be unidimensional.

## 15.7   The spectrum

Recall the spectrum function from §13.3, assigning the number $I(\lambda, T)$ of nonisomorphic models of power $\lambda$ to every cardinal $\lambda$ (and every theory $T$). The problem of finding this function for a given theory $T$ is known as the **spectrum problem** for $T$.

**Remarks.**
(1)   $T$ is $\lambda$-categorical if and only if $I(\lambda, T) = 1$.
(2)   $I(\lambda, T) > 0$ for all $L$-theories $T$ having infinite models and all $\lambda \geq |T|$.
(3)   A complete theory $T$ has an infinite model if and only if $I(n, T) = 0$ for all $n < \omega$.

Our last task in the analysis of the theory $T^{\mathbb{Z}}$ is the spectrum problem, that is, to describe its spectrum function. We do not want to do this in full detail—simply because the missing details require some finer analysis of the elementary submodels of $\widehat{\mathbb{Z}}$, which is rather algebraic than model-theoretic (see the next section). Nevertheless, we will pin down this function to quite an extent. But first, some very general properties of the spectrum function, some general bounds for just any theory.

**Lemma 15.7.1.** *Let $\lambda \in \mathbf{Cn}$.*
(1)   *There are at most $\max\{|L|, 2^\lambda\}$ nonisomorphic $L$-structures of power $\lambda$.*
(2)   *If $\lambda \geq |L|$, there are at most $2^\lambda$ nonisomorphic $L$-structures of power $\lambda$.*

*Proof.* Every interpretation of an $n$-place relation (as well as of every $n-1$-place function) in a structure $\mathcal{M}$ corresponds to a subset of $M^n$. If $|M| = \lambda$ is infinite, then $|M^n| = \lambda^n = \lambda$, hence there are at most $2^\lambda$ such subsets. For each of the (at most) $|L|$ non-logical symbols of $L$ there are then at most $2^\lambda$ possibilities of interpretation on any universe $M$ of power $\lambda$. Thus there are at most $|L| \cdot 2^\lambda$ possibilities of making $M$ an $L$-structure. This implies (1) for infinite $\lambda$ and hence also (2).

If $|M| = \lambda$ is finite, $M^n$ has only finitely many subsets, hence $\aleph_0$ is an, albeit crude, bound for the number of possibilities of interpretation of each non-logical symbol of $L$ on $M$. Altogether, this makes $|L| \cdot \aleph_0 = |L|$ ways of turning $M$ into an $L$-structure. From this, (1) follows also for finite $\lambda$. $\quad\square$

Clearly, these are also bounds for $I(\lambda, T)$ for any $L$-theory $T$. We leave as an exercise to show that these are best possible (for infinite $\lambda$, even if $T$ is assumed to be complete).

Let's get back to our example and consider $I(\lambda, T^{\mathbb{Z}})$ for some infinite cardinal $\lambda$. By Corollary 15.6.4 and the cardinality considerations following it, the models of power $\lambda$ are (up to isomorphism) exactly the groups $\mathcal{M} \oplus \mathbb{Q}^{(\kappa)}$ for which $\mathcal{M} \preccurlyeq \widehat{\mathbb{Z}}$ and $\kappa \leq \lambda$, where $\kappa = \lambda$ in case $|\mathcal{M}| < \lambda$. For every given reduced model $\mathcal{M}$, there is exactly one such model in case $|\mathcal{M}| < \lambda$, namely $\mathcal{M} \oplus \mathbb{Q}^{(\lambda)}$, while in the case $|\mathcal{M}| = \lambda$ there are as many such models as there are cardinals $\leq \lambda$, namely one model $\mathcal{M} \oplus \mathbb{Q}^{(\kappa)}$ for every $\kappa \leq \lambda$.

This reduces the spectrum problem for $T^{\mathbb{Z}}$ to the question of how many reduced models this theory has. Even though we haven't classified the reduced models, we are able to say something substantial about the spectrum function of this theory.

**Theorem 15.7.2.**
(1) $I(\aleph_0, T^{\mathbb{Z}}) = 2^{\aleph_0}$.
(2) If $\aleph_0 \leq \lambda \leq 2^{\aleph_0}$, then $2^{\aleph_0} \leq I(\lambda, T^{\mathbb{Z}}) \leq 2^\lambda$.
(3) $2^{\aleph_0} \leq I(\lambda, T^{\mathbb{Z}}) \leq 2^{2^{\aleph_0}}$ for all $\lambda \geq \aleph_0$.

*Proof.* (1) follows from the fact that there are $2^{\aleph_0}$ distinct $\Delta^+$-1-types of $\mathbb{Z}$ over 1, each of which has to be realized in some countable model (and every countable model can realize only countably many types). This also yields the lower bounds in (2) and (3), while the upper bound in (2) follows from the above lemma.

If $\lambda > 2^{\aleph_0}$, the $\mathbb{Q}$-dimension of a model of power $\lambda$ is confined to $\lambda$, and $I(\lambda, T^{\mathbb{Z}})$ is equal to the number of reduced models of power $\lambda$. But, each of them being a subset of $\widehat{\mathbb{Z}}$, there can't be more than $2^{2^{\aleph_0}}$ of them.

This proves (3) in the case that $\lambda > 2^{\aleph_0}$. The remaining case, however, falls under (2), for $\lambda \leq 2^{\aleph_0}$ implies $2^{\lambda} \leq 2^{2^{\aleph_0}}$.                                      □

It can be shown that these upper bounds are assumed in both cases, i. e. that $I(\lambda, T^{\mathbb{Z}}) = 2^{\min(\lambda, 2^{\aleph_0})}$ for all $\lambda \geq \aleph_0$ (see the next section for references). This then completely solves the spectrum problem for $T^{\mathbb{Z}}$.

Above, we came across counting the number of cardinals less than or equal to a given cardinal $\lambda$. In case, $\lambda$ is finite, this number obviously is $\lambda + 1$. For infinite $\lambda$, alephs are useful again: suppose $\lambda = \aleph_\alpha$; then there are easily seen to be $\aleph_0 + |\alpha| + 1 = \aleph_0 + |\alpha|$ cardinals $\leq \lambda$.

**Exercise 15.7.1.** Given an infinite cardinal $\kappa$, find a language $L$ of power $\kappa$ and an $L$-theory $T$ such that $I(2, T) = \kappa$.

**Exercise 15.7.2.** Find a countable complete theory $T$ such that $I(\lambda, T) = 2^{\lambda}$ for all $\lambda \geq \aleph_0$.

Let $T^*$ denote the elementary diagram $\mathrm{Th}(\widehat{\mathbb{Z}}, \widehat{\mathbb{Z}})$ of $\widehat{\mathbb{Z}}$. Since Corollary 15.6.5 we know that $T^*$ is categorical in all $\lambda > 2^{\aleph_0}$, i. e. $I(\lambda, T^*) = 1$ for all $\lambda > 2^{\aleph_0}$ (and clearly $I(\lambda, T^*) = 0$ for all $\lambda < 2^{\aleph_0}$, for $|\widehat{\mathbb{Z}}| = 2^{\aleph_0}$). The question arises of what the number $I(2^{\aleph_0}, T^*)$ is. In order to find an arithmetical expression for this, write $2^{\aleph_0}$ as $\aleph_\alpha$ for an appropriate $\alpha \in \mathbf{On}$ (we don't know which though, cf. §7.6).

**Exercise 15.7.3.** Prove $I(2^{\aleph_0}, T^*) = \aleph_0 + |\alpha|$, provided $2^{\aleph_0} = \aleph_\alpha$.

**Exercise 15.7.4.** Let $\mathcal{K}$ be an infinite division ring (and $T_{\mathcal{K}}^{\infty}$ the complete theory of all infinite $\mathcal{K}$-vector spaces, cf. Exercise 8.4.2). Determine $I(|\mathcal{K}|, T_{\mathcal{K}}^{\infty})$.

## 15.8   A sort of epilogue

We have arrived at the end of our treatise. The example $T^{\mathbb{Z}}$ that has occupied us for so long not only serves as an illustration but also is instructive with respect to further topics to study. We use the opportunity to draw the readers attention to some of those, while more general hints are to be found in the appendix.

As mentioned before, we have not completely solved the spectrum problem for $T^{\mathbb{Z}}$. The gap we have left consists of the computation of the number of elementary substructures of $\widehat{\mathbb{Z}}$ in the corresponding cardinalities. This would require a more concrete description of $\widehat{\mathbb{Z}}$, for which

[Nadel, M. and Stavi, J. : On models of the elementary theory of $(\mathbb{Z}, +, 1)$, J. Symbolic Logic **55** (1990) 1 – 20]

can be consulted. (The authors deal only with the expansion $T_1^{\mathbb{Z}}$ of $T^{\mathbb{Z}}$, which we considered in Exercise 15.2.4. It turns out, however, that this is enough.) The treatment is to a large extent elementary. Besides, other interesting properties of $T^{\mathbb{Z}}$—or rather $T_1^{\mathbb{Z}}$—are investigated.

That the group $\widehat{\mathbb{Z}}$ can be regarded as a certain *completion* of $\mathbb{Z}$ (whence the notation) and that it decomposes into a direct product of the groups of so-called *p-adic integers*, where $p$ runs over the primes, can be found in

[Fuchs, L. : **Infinite Abelian Groups I**, Academic Press, N. Y. 1970]

or the original

[Kaplansky, I. : **Infinite Abelian Groups**, University of Michigan Press, Ann Arbor [2]1956],

which also treat the general theory of algebraically closed groups.

In Theorem 15.5.6(1) we showed that pure embeddings of algebraically compact models of $T^{\mathbb{Z}}$ split. This is true for all algebraically compact groups and, in fact, equivalent to algebraic compactness (even in the context of arbitrary structures, see below). Algebraically compact groups are therefore often called **pure-injective** (*injective* for *pure* embeddings). More about this connection can be found in the above two sources. The $\widetilde{\Delta}$-elimination for $T^{\mathbb{Z}}$ (and hence also the connection between purity and elementariness as described in Lemma 15.3.2) can be extended to all (abelian) groups (even simultaneously for all AG, if some extra sentences are added to $\widetilde{\Delta}$), as was shown in

[Szmielew, W. : Elementary properties of Abelian groups, Fundamenta Math. **41** (1955) 203 – 271].

This then implies, using the same argument as Remark (2) in §15.4, that every (abelian[2]) group is stable. [Hint to the model-theoretic jargon: a structure is said to be **stable** if its complete theory is; similarly for other properties of theories rather than structures.] We leave it as an exercise to find a group that is not superstable.

Mutatis mutandis, what we said about (abelian) groups applies as well to *modules* over arbitrary rings with 1, where $\Delta$-formulas have to be replaced by positive primitive formulas in general. In particular, modules admit elimination up to positive primitive formulas, see

[Baur, W. : Elimination of quantifiers for modules, Israel J. Mathematics **25** (1976) 64 – 70].

More recent proofs of this are more conceptional and easier to read and yield more or less directly also the (stronger) $\widetilde{\Delta}$-elimination for (abelian)

---

[2]The situation for nonabelian groups is completely different: no infinite power of a nonabelian group is stable, see Exercise 9.1.15 in Hodges' *Model Theory*.

groups, see

[Prest, M. : **Model Theory and Modules**, London Mathematical Society Lecture Notes Series 130, Cambridge University Press, Cambridge 1988, 380 pp.],

[Ziegler, M. : Model theory of modules, Ann. Pure and Applied Logic **26** (1984) 149 – 213],

or

[Hodges, W. : **Model Theory**, Encyclopedia of mathematics and its applications 42, Cambridge University Press, Cambridge 1993, 772 pp.].

(The latter contains, among many other things, a discussion—in the context of arbitrary structures—of the connection of algebraic compactness with other saturation properties like *atomic compactness.*) Also the introductory chapter of

[Baldwin, J. T. : **Fundamentals of Stability Theory**, Perspectives in Mathematical Logic, Springer, N. Y. 1988]

has a proof of the elimination result for arbitrary modules. The book by Mike Prest covers the whole circle of topics considered so far and contains an abundance of model- and stability-theoretic classification results for modules. The work by Martin Ziegler (which is more or less covered by Prest's book, but inspiring reading in its own right) provides a solution to the spectrum problem for *all* infinite modules over countable rings. As part of this, Ziegler proved Theorem 15.6.2(4) about the existence of a(n elementarily) prime structure for the class of all algebraically compact models of $T^{\mathbb{Z}}$ in much more generality, including all superstable complete theories of modules (over *arbitrary* rings).

A great deal of our analysis of $T^{\mathbb{Z}}$ applies to arbitrary unidimensional theories (cf. fine print after Corollary 15.6.5)—at least to those of so-called $U$- (or *Lascar*) *rank 1*. This is part of stability or classification theory as developed by Saharon Shelah in

[Shelah, S. : **Classification Theory (and the Number of Non-Isomorphic Models)**, Studies in Logic and the Foundations of Mathematics 92, North-Holland, Amsterdam ²1990, 705 pp.].

The case of unidimensional *modules* is much easier, presupposes, however, some familiarity with the theory of algebraically closed modules, cf. Chapter 7 of Prest's book.

Shelah's monograph is hard to read and unsuitable for an introductory reading (neither is Baldwin's book cited above). Some more suitable reading is mentioned in the appendix. It would include the classification of those countable unidimensional theories that are at the same time $\aleph_0$-stable (one also says $\omega$-*stable*): for these are exactly the $\aleph_1$-categorical countable theories, which Morley's theorem is about, a theorem that was a starting point for stability theory, cf. §8.5.

Let us conclude with a final remark about the spectrum problem in order to avoid a common misconception.

First, some historical remark about the spectrum problem. In its original version from

[Scholz, H. : Ein ungelöstes Problem der symbolischen Logik, J. Symbolic Logic **17** (1952) 160]

it was not formulated for the spectrum function, but rather, as it were, its support, i. e. the set $\{\lambda \in \mathbf{Cn} : \mathrm{I}(\lambda, T) > 0\}$. And this only for finitely axiomatized theories $T$, that is—and this is how Heinrich Scholz formulated it—for single sentences. In view of Löwenheim-Skolem this problem is trivial for infinite $\lambda$ (a priori, only for $\lambda \geq |L|$, but single sentences, of course, can always be formulated in a countable fragment of the language). More concretely, the above set contains either every infinite cardinal or none. Therefore one confines oneself to the finite case and considers the set $\{n < \omega : \mathrm{I}(n, T) > 0\}$ and calls *this* the **spectrum** of $T$, in symbols $\mathrm{Spec}(T)$. The description of such spectra, even of single sentences, is an extremely hard problem and to date very far from being solved, see also

[Börger, E. : **Berechenbarkeit, Komplexität, Logik**, Vieweg, Braunschweig 1985],

[Fagin, R. : Finite-model theory – a personal perspective, Theoretical Computer Science **116** (1993) 3 – 31],

the *Finite Model Theory* by Ebbinghaus and Flum, and Hodges' *Model Theory*. (This is touched upon in some of the exercises below.)

In the infinite, where Scholz' original problem is trivial, the next question arising naturally is that for the *values* of the spectrum function—and this, for arbitrary theories—that is, what *we* called spectrum problem.

This problem is programmatic for Shelah's theory, and one might think that, searching for certain cardinal valued functions, this theory is rather set-theoretically oriented. Far off the mark! One is interested most of all in results that are independent of the properties of the given universe of set theory, and what one needs for this, is *model*-theoretic insight, insight into the structure of things that does not change when passing from one set-theoretic universe to another.

The best example is Vaught's theorem about the impossibility of the spectral value 2 at the argument $\aleph_0$. In itself it appears to be no more than a curiosity, and it does not seem to have yielded many corollaries or applications. Its proof, however, has had momentous structural consequences.

**Exercise 15.8.1.** Show that the group $\mathbb{Z}^{(\omega)}$ is not superstable.

We write $\mathrm{Spec}(\varphi)$ instead of $\mathrm{Spec}(\{\varphi\})$ (the **spectrum** of $\varphi$). A subset $X$ of $\mathbb{N}_{>0}$ is called a **spectrum** if there exists a sentence $\varphi$ such that $X$ is the spectrum of $\varphi$.

**Exercise 15.8.2.** Verify that, for any sentence $\varphi$, at least one of the sets, $\mathrm{Spec}(\varphi)$ and $\mathrm{Spec}(\neg\varphi)$, has to be cofinite.

**Exercise 15.8.3.** Show that the following subsets of $\mathbb{N}_{>0}$ are spectra.
(1)  All finite and all cofinite subsets,
(2)  $X \cup Y$ and $X \cap Y$, provided $X$ and $Y$ are spectra,
(3)  $\{nm + k : n > 0\}$, for all $m > 0$ and $k \geq 0$,
(4)  $\{n^2 : n \in \mathbb{N}_{>0}\}$,
(5)  $\mathbb{N}_{>0} \smallsetminus \mathbb{P}$,
(6)  $\mathbb{P}$.

**Exercise 15.8.4.** Is every subset of $\mathbb{N}_{>0}$ a spectrum?

# Hints to selected exercises

## Chapter 2

2.2.3   One way of doing this is to first show that in a term the number of left and right parentheses must coincide, while a proper initial segment of such a term has more left than right ones.

2.3.2   Similarly as Exercise 2.2.3.

2.6.1   See the rule adopted at the end of the definition of substitution.

## Chapter 3

3.1.2   On the complexity of formulas.

3.3.3   Prove (4) by induction on the complexity of $\psi$, where only the conjunction step requires a little trick of renaming variable.

3.3.5   Verify $\varphi \sim (\varphi \wedge y = y)$ for arbitrary formulas $\varphi$ and variables $y$.

3.4.1   Inductively on the complexity of formulas.

3.6.1   Show that the deductive closure contains all $L$-sentences. Consider also $\forall x(x \neq x)$.

3.6.2   Note that in the empty structure all formulas beginning with an existential quantifier are false, while all formulas beginning with a universal quantifier are true.
You could also use the next exercise to find a solution to this problem.

## Chapter 4

4.1.4   Prove that if $F = \mathrm{F}(B)$, then $\mathcal{N}^I/F \cong \mathcal{N}^B$.

4.3.3   If $\mathbf{K}'$ is the complement of $\mathbf{K}$, $\mathbf{K} = \operatorname{Mod}\Sigma$, and $\mathbf{K}' = \operatorname{Mod}\Sigma'$, consider $\Sigma \cup \Sigma'$.

## Chapter 5

5.2.2   Löwenheim-Skolem.

5.2.4   Given a collection of groups, $\{\mathcal{G}_i : i \in I\}$, and a filter, $F$, on $I$, show that $\prod_F \mathcal{G}_i =_{\mathrm{def}} \{g \in \prod_{i \in I} \mathcal{G}_i : I \smallsetminus \operatorname{supp} g \in F\}$ is a normal subgroup of $\prod_{i \in I} \mathcal{G}_i$, where in analogy with §4.1, $\operatorname{supp} g = \{i \in I : g(i) \neq 1\}$. (If the $\mathcal{G}_i$ are abelian and additively written, and if further $F$ is the Fréchet filter on $\mathbb{N}$, then $\prod_F \mathcal{G}_i$ is nothing more than the direct sum $\bigoplus_{i \in I} \mathcal{G}_i$.)

5.3.4   $\forall x(x^2 = 0 \to x = 0)$ suffices!

5.3.5   Cf. Exercise 5.2.4.

5.3.6   Given a proper ideal, $X$, of the direct product of the division rings $\mathcal{K}_i$ $(i \in I)$, show that $U_X =_{\mathrm{def}} \{I \smallsetminus \operatorname{supp} r : r \in X\}$ is a filter on $I$.

5.4.1   Only one axiom needs adjustment.

5.4.2   One has to express that, for every structure $\mathcal{M}$, the definable set $K(\mathcal{M})$ is a division ring and the definable set $V(\mathcal{M})$ is a left $K(\mathcal{M})$-vector space.

5.5.3   Consider irreflexivity,

5.6.6   That every filter is a theory rests on the finiteness theorem.

5.7.1   $\bigcap_{\varphi \in \Sigma} \langle \varphi \rangle = \emptyset$ if and only if $\Sigma$ is contradictory.

## Chapter 6

6.1.6   Use the preceeding exercise as well as Exercise 2.3.1.

6.1.8   Map also the parameters.

6.2.1   Use the remark after Lemma 6.2.2 and the Galois correspondence of Exercise 3.4.4.

6.2.5   Use Exercise 6.2.1.

6.2.6   Use the previous exercise for the first part (this involves showing that $\mathrm{Th}_\forall \mathcal{N} \cup T$ is consistent whenever $\mathcal{N} \models T_\exists$) and proceed like in Exercise 6.2.1 for the second.

6.3.5   Use Prop. 5.2.1 and consider e. g. the Prüfer group $\mathbb{Z}_{p^\infty}$ of Ch. 11.

6.4.2   Consider pairs of elements.

6.5.1   Consider the (hereditary and elementary) properties $\Sigma_1 = (x \cdot y = y \cdot x)\}$.

## Chapter 7

7.2.1   Note that the elements of the same rank form an antichain, i. e. a set of pairwise incomparable elements.

7.2.3   Use the second part of the preceeding exercise and the fundamental theorem of abelian groups mentioned after Henkin's criterion (Proposition 6.3.1). Patch the orders of the direct summands together lexicographically.

7.4.2   Cf. Exercise 7.5.3 below.

7.5.2   Use Lemma 7.5.3(4).

7.5.3   Cf. Exercise 7.4.2.

7.6.1   Concerning (8), notice that every function from $(\kappa^\lambda)^\mu$ is uniquely determined by one from $\kappa^{\lambda \times \mu}$, and vice versa. Concerning (11) and (13), note that although $^0X$ contains $\emptyset = \emptyset \times X$, the set $^X\emptyset$ is empty in case $X \neq \emptyset$.

7.6.2   Every vector has finite support, cf. §4.1. Find estimates first for the number of finite supports and then for the number of vectors with a fixed support. Then apply Lemma 7.6.6.

7.6.3   Using Exercise 7.5.2, show that, given a set $X \subseteq \mathbf{On}$, the supremum of $\{\aleph_\alpha : \alpha \in X\}$ in $\mathbf{Cn}$ is $\aleph_{\bigcup X}$.

## Chapter 8

8.1.2   Proposition 6.1.3 and Löwenheim-Skolem.

8.2.1   Appropriately define the $\Delta$-diagram of $\mathcal{M}$.

8.3.2    Cf. §15.1 below.

8.3.4    Cf. Exercise 6.4.4.

8.3.5    Consider the ordering $\omega$ and prove that every automorphism of an elementary extension of $\omega$ fixes every natural number.

8.4.2    Use Löwenheim-Skolem 8.4.2, Lemma 8.2.2(2), and the property that the isomorphism type of a vector space is given by its dimension (hence, in cardinalities greater than the language, simply by its cardinality, cf. Exercise 7.6.2).

8.4.4    First classify the models of $T = \mathrm{Th}\,\mathcal{M}$, then find elementary extensions of both, $\mathcal{M}$ and $\mathcal{N}'$, of a fixed uncountable cardinality. Then note that they are isomorphic.

8.4.5    With the notation from Exercise 8.4.4 set $T = \mathrm{Th}\,\mathcal{N}$. Using Exercise 6.2.6, show that $\mathcal{M} \models T_{\exists}$. Finally note that $\mathcal{M}$ does not embed $\mathcal{N}$. (The same works with $<$ replaced by the successor function.)

8.5.1    Cf. Exercise 7.3.2.

8.5.2    For $\lambda \geq \mathbb{R}$ consider two models, one consisting of $\lambda$ linearly ordered copies of $\rho$ (the order of the reals), and the other consisting of $\lambda$ linearly ordered copies of $\eta$ (the order of the rationals). For smaller uncountable $\lambda$ replace $\rho$ by an elementary substructure of power $\lambda$.

8.5.4    Cf. Exercise 8.1.3.

8.5.5    Look at the hint to that exercise.

# Chapter 9

9.1.1    Translate the proof given, using inconsistency and finiteness.

9.2.4    Use Löwenheim-Skolem and the idea of proof of Corollary 9.2.3.

9.2.5    Instead of $\exists^*$ consider the set of all formulas of the form $\exists x\,\psi$, where $\psi$ is a finite conjunction of formulas from $\Delta$ and negations of formulas from $\Delta$.

9.2.6    Cf. example at end of §5.7.

9.2.7    Use that every subspace is a direct summand and apply Löwenheim-Skolem upward (or just Exercise 9.2.3).

9.2.8    First show that all terms are equivalent to terms of the form $g(x)$ and all atomic formulas are equivalent to term equations of the form $x = g(y)$, where $g \in G$ (and possibly $x$ and $y$ denote the same variable).

9.4.2    Consider the axiomatization of characteristic 0.

9.4.3    Use that adjoining finitely many algebraic elements makes a finite algebraic extensions, which is itself a finite structure.

9.5.2    Find $\sigma^*(\bar{y})$ as before (but possibly with parameters from $\mathcal{K}_0$) such that, for all $\mathcal{K} \models \mathrm{ACF}$ extending $\mathcal{K}_0$, $\sigma^*(\bar{y})$ is a criterion for the solvability of $\sigma(\bar{x}, \bar{y})$. On the level of the formal language,

$$\mathrm{ACF} \cup \mathrm{D}(\mathcal{K}_0) \models \forall \bar{y}(\exists \bar{x}\sigma(\bar{x}, \bar{y}) \leftrightarrow \sigma^*(\bar{y})).$$

9.5.4    Consider the polynomials $f_0, \ldots, f_{m-1}, 1 - g \cdot z$ in $\mathcal{K}[\bar{x}, z]$.

9.5.5   Interpret, in a slightly more general sense than in §6.4*, the group $GL_2(\mathcal{K})$
        in $\mathcal{K}^4$ by regarding the elements of $GL_2(\mathcal{K})$ as quadruples in $\mathcal{K}$, and find
        $(0, 1; +, -, \cdot)$-formulas defining the 'right' quadruples, the group multipli-
        cation, and conjugation. Do all this universally for all fields $\mathcal{K}$.

9.6.1   The embeddings of $\mathcal{M}_0$ and $\mathcal{M}_1$ into a joint elementary extension does not,
        in general, respect any given embedding of $\mathcal{M}_0$ in $\mathcal{M}_1$, cf. proof of 8.4.4.

9.6.6   Use Corollary 9.2.3.

9.6.5   Cf. 9.3.2

9.6.8   Consider proper subgroups of $\mathbb{Z}$.

## Chapter 10

10.1.2  Use transfinite induction over $\alpha$, where the limit case works like for $\omega$,
        while the successor case $\alpha = \beta + 1$ splits into two trivial cases, according
        to whether $\beta$ is a successor or a limit ordinal.

10.2.4  Use Lemma 10.2.2 (as in the proof of Theorem 10.2.5) and Lemma 10.2.3.

10.2.5  List the existential sentences in $L_0(M)$ as $\{\varphi_i : i < \alpha\}$ and construct an as-
        cending continuous chain of structures $\mathcal{M}_i$ $(i < \alpha)$, where $\mathcal{M}_i \subseteq \mathcal{M}_{i+1} \models T$
        and $\mathcal{M}_{i+1} \models \varphi_i$, provided such an $\mathcal{M}_{i+1}$ exists (and $\mathcal{M}_{i+1} = \mathcal{M}_i$ other-
        wise).

10.2.6  Use Löwenheim-Skolem in every step of the transfinite construction.

10.3.1  Analogously to (1) one has to show that $\mathrm{Th}(\mathcal{M}, M) \cup \mathrm{Th}_-(\mathcal{N}, N)$ is con-
        sistent.

10.3.3  Notice, the axiom $0 \neq 1$ is the only nonpositive axiom of TF.

## Chapter 11

11.1.3  Analogous to Exercise 9.2.7. Use that all models are divisible, hence, as
        groups, injective, i. e., they split off as a direct summand wherever they are
        embedded, cf. fine print in §15.5.

11.2.1  Cf. proof of Lemma 12.1.2.

11.3.2  Consider Dedekind cuts.

11.3.3  Note: $\rho_{B,A}(\mathrm{tp}(\bar{a}/B)) = \mathrm{tp}(\bar{a}/B) \cap L_n(A)$, and to be continuous means for
        this map that, for every formula $\varphi \in \mathrm{tp}(\bar{a}/A)$, there is $\psi \in \mathrm{tp}(\bar{a}/B)$ such
        that the restriction onto $A$ of every type in $S_n^{\mathcal{M}}(B)$ containing $\psi$ contains
        $\varphi$.

11.4.1  Finite sets have only finitely many subsets.

11.4.2  Use topology.

11.4.3  Look at $T_=^\infty$.

11.4.4  Look at $T_=^\infty$.

11.4.7  For the direction from right to left use Exercise 11.4.1 to obtain not only
        an algebraic formula $\varphi(\bar{y})$ in $\mathrm{tp}(\bar{b})$, but one that isolates this type. Note,
        then the algebraicity of $\psi(\bar{x}, \bar{c})$ follows already from $\mathcal{M} \models \varphi(\bar{c})$.

11.5.4 Modify the proof from Exercise 12.2.3 taking into account an(y) enumeration of $\mathrm{acl}_{\mathcal{M}}(\mathrm{dom}\, f) \smallsetminus \mathrm{dom}\, f$ that is 'constructible'.

## Chapter 12

12.1.3 Given a model $\mathcal{M}$ of $T$, construct $\mathcal{M}_\omega$ as in Lemma 12.1.2, where the application of Proposition 11.2.5 is replaced by that of Exercise 10.2.5. For the cardinality bound see Exercise 10.2.6.

12.1.5 For uniqueness use Exercise 11.5.4.

12.1.7 By induction on $n$. Replace every formula $\varphi(x_0, \ldots, x_n)$ in an $n + 1$-type over $A$ by $\exists x_n\, \varphi(x_0, \ldots, x_n)$ and show that the so-obtained set is an $n$-type over $A$.

12.1.8 Use that $\aleph_1$ is regular (Exercise 7.6.4).

12.1.10 Make use of the preceding exercise.

12.2.1 The embedding argument from the theorem now goes through (independently of the cardinalities).

12.2.2 Transitivity of isolation.

12.2.3 Cf. proof of Lemma 11.5.2(2).

12.2.5 The standard model $(\omega; +, \cdot)$ of number theory is atomic (even purely algebraic), since every element therein forms, as a singleton, a definable set (without using parameters), i. e., there are formulas $\varphi_n(x)$ such that for all $\mathcal{M} \equiv (\omega; +, \cdot)$ we have $\mathcal{M} \models \varphi_n(a)$ iff $a = n$. Consequently $(\omega; +, \cdot)$ is an elementarily prime model of its own complete theory, i. e. of true arithmetic.

12.2.8 Use the previous exercise and the uniqueness part of the Theorem.

## Chapter 13

13.1.2 Look at the example in §9.2 (p.135).

13.1.3 See G. Fuhrken : Bemerkung zu einer Arbeit E. Engelers. Zeitschrift f. Math. Logik u. Grundlagen d. Mathematik, 8(1962) 277–79.

13.1.4 The only non-algebraic 1-type over $\emptyset$ can't be isolated.

13.1.5 First expand the language by a binary relation $<$ and add appropriate axioms about it to PA. Then omit an appropriate type over $M$.

13.2.2 Cf. Exercise 12.2.8 and Theorem 13.2.1.

13.2.3 Use Exercise 12.2.5.

13.3.3 Cf. Exercise 13.2.2 and Theorem 13.3.1.

13.3.4 Consider $T_{\cong}^\infty$ or the theory of infinite vector spaces over a finite field $\mathcal{K}$.

13.4.1 Argue as in Exercise 9.6.1 noticing that the embedding of $\mathcal{M}_1$ in $\mathcal{M}_2$ will be respected, as there is only one such embedding. Similarly for the embedding into $\mathcal{M}_3$.

13.4.2 See the suggestions before the example $T_3$.

13.4.3 Cf. Exercise 11.2.3.

13.4.4 Cf. Exercise 11.2.2.

## Chapter 14

14.1.1  Use quantifier elimination (as to be shown in Exercise 11.1.3).

14.1.2  Use Exercise 9.2.7.

14.1.3  Use Exercise 12.1.10.

14.1.4  Show first that for every formula $\psi(x, \bar{y})$ without parameters there is a natural number $k_\psi$ such that for all models $\mathcal{M}$ of $T$ the following implication holds: $|\varphi(\mathcal{M}) \cap \psi(\mathcal{M}, \bar{a})| > k_\psi \implies \varphi(\mathcal{M}) \cap \psi(\mathcal{M}, \bar{a})$ is infinite.

14.1.6  Consider the ordering of the natural numbers.

14.1.7  Every finite $aR$ is 0.

14.1.8  Use the quantifier elimination from Exercise 9.2.8 for the second equivalence and notice that the last condition is preserved under elementary equivalence.

14.1.9  (1) All cosets of a given subgroup have the same power.

14.1.10 (4) Note that, by Fermat's little theorem (saying that if $n$ is a natural number not divisible by $p$, then $n^{p-1} = 1 (\mathrm{mod}\ p)$) or, equivalently, by the fact that the multiplicative group of the prime field $\mathbb{F}_p$ is cyclic of order $p - 1$, the set of roots of the polynomial $x^p - x$ is precisely $\mathbb{F}_p$.

(5) Finite-dimensional vector spaces over finite fields are clearly finite. Field extensions obtained by adjoining finitely many algebraic elements are finite, cf. v. d. Waerden's *Algebra*, §41.

(6) The multiplicative group of a finite field is finite, hence all of its elements have finite order.

(7) Every element in $\mathcal{K}_0$ is algebraic over $\mathbb{F}_p$.

(8) Every element algebraic over $\mathcal{K}$ is already algebraic over a finitely many elements, hence over a finite subfield of $\mathcal{K}$.

14.1.11 (1) Note that $x = x$ is also minimal in the additive group, $\mathcal{K}^+$, of $\mathcal{K}$, and that the map given by $x \mapsto x^p - x$ is a homomorphism of this group.

(3) Note that $x = x$ is also minimal in the multiplicative group, $\mathcal{K}^\times$, of $\mathcal{K}$, and that the map given by $x \mapsto x^n$ is a homomorphism of this group.

(4) Choose a minimal $n > 0$ such that this is not the case and notice that then, by the fact, $\mathcal{K}$ contains all $n$th roots of unity.

14.1.12 (2) Moreover, the $\varphi^n(a)$ are pairwise distinct.

(3) Automorphism preserve sets definable without parameters.

14.1.13 (1) $\mathcal{K}_0$ is locally finite.

14.2.1  Show that $\mathcal{M}_0$ is constructible, cf. end of §12.2, and apply Exercise 12.2.1 instead of Theorem 12.2.1.

14.3.4  Use the exchange lemma for (ii) $\implies$ (i).

14.3.7  Consider a directed graph such that each vertex has infinitely many incoming, but only one outgoing arrow.

14.3.8  Choose a low basis of $C$ and, using homogeneity (Theorem 12.1.1(3)), successively move it away by some automorphism over what has been moved already in such a way that it becomes nonalgebraic over what has been moved already *and* $B$. Finally make use of a certain exchange phenomenon. Another version of this result appears as (2.16) in Cameron (1990).

14.3.9   Use compactness to reduce the problem to finite sets.

14.4.2   Use Lemmas 14.4.1 and 14.4.2, Proposition 14.5.1, and Exercise 14.3.1 (and recall Lemma 7.6.4).

14.4.3   Use Lemma 14.4.2 or Theorem 14.4.3.

14.5.1   Consider separately vector spaces over finite fields and vector spaces over infinite fields.

14.5.4   A model as mentioned must have infinite dimension and is thus $\omega$-saturated.

14.6.2   For the nontrivial direction, successively build a partial elementary map from *all of* $\mathcal{M}$ onto $\mathcal{M}$.

14.6.3   Relativize the entire chapter to strongly minimal formulas. Then apply the resulting dimension theory for $D$ and Lemma 14.4.2.

14.7.2   (1)   Denote by $\operatorname{supp} g$ or $\operatorname{supp} a$ the set of indeces of the corresponding sequence where the entry is 1. Find a condition in these terms for $a$ to be in $\operatorname{fix} g$ and use this to show that the set of all elements of $A$ of length at most $n$ is definable in $\mathcal{A}$ without parameters.

         (2)   Consider the type saying that $x$ is in no fixed set (and notice that this is the same as saying that all $g(x)$ are different, where $g$ runs over the elements of $G$).

         (3)   Use the previous assertion.

         (4)   Use the quantifier elimination from Exercise 9.2.8.

## Chapter 15

15.1.4   Use one of Noether's isomorphism theorems and that $\mathcal{A} = \mathcal{B} \oplus \mathcal{C}$ implies $\mathcal{C} \cong \mathcal{A}/\mathcal{B}$.

15.2.2   The essential step is to directly find a quantifier-free formula that is $\mathbb{Z}_n$-equivalent to the formula $\exists x \, (\bigwedge_{i<k} m_i x = y_i \wedge \bigwedge_{k \leq i < j} m_i x \neq y_i)$. The rest is as before (using Lemma 9.2.1, of course).

15.2.3   Use the previous exercise.

15.3.1   Use Exercise 12.2.7 (and Theorem 12.2.1).

15.4.2   Every model of $T^{\mathbb{Z}}$ has an elementary extension that can be expanded to a model of $T_1^{\mathbb{Z}}$.

15.4.3   Use Proposition 14.1.3.

15.4.4   Cf. proof of Lemma 12.1.2.

15.5.1   Cf. Lemma 15.5.4.

15.5.2   Cf. Exercises 8.4.3 and 12.1.6.

15.5.3   See Malcev (1973), §8.2.

15.6.1   Cf. end of proof of Theorem 15.5.6.

15.6.2   Show e. g. that the homomorphism from $\mathbb{Z}$ to $\mathbb{Z} \oplus \mathbb{Q}$ given by $n \mapsto (n, n)$ is pure, and hence elementary.

15.6.4   Generalizing Theorem 12.1.1(1), prove that $\kappa$-saturated structures are $\kappa$-**universal** in the sense that every elementarily equivalent structure of a power $\leq \kappa$ can be elementarily embedded.

15.7.1 Consider $\kappa$ constant symbols (and $T = \emptyset^\vDash$).

15.7.3 Find a relationship between the cardinality of the ordinal $\alpha + 1$ and the number of infinite cardinals $\leq \aleph_\alpha$.

15.7.4 Cf. previous exercise.

15.8.1 Consider a descending chain of definable subgroups with infinite (!) factors. This allows to find models where the factors have any prescribed cardinality $\kappa$. Then, over only $\kappa$ parameters (for all the coset representatives of all the (countably many) subgroups, one can produce $\kappa^{\aleph_0}$ types.

15.8.3 (1) Use formulas of the form $\exists^{=n} x \, (x = x)$.

(2) Use the disjunction, resp. the conjunction, of the formulas given by the condition on $X$ and $Y$, choosing the signatures disjoint.

(3). Let $\approx$ be a binary relation symbol and $\varphi_\approx$ the formula saying that $\approx$ is an equivalence relation. For $k > 0$ use the formula $\varphi_\approx \wedge \exists^{=1} x (\exists^{=k} y \, x \approx y \wedge \forall z (z \not\approx x \to \exists^{=m} y \, z \approx y))$. Delete the "$x$"-part if $k = 0$.

(4). Consider the language with a unary predicate $P$ and a binary function symbol $f$, and a formula $\varphi$ expressing that $f$ is a bijection between $P(\mathcal{M})^2$ and $M$, i. e. $\forall x \exists yz \, (Py \wedge Pz \wedge f(y,z) = x) \wedge \forall uvxy \, (Px \wedge Py \wedge Pu \wedge Pz \wedge f(u,v) = f(x,y) \to u = v \wedge x = y)$.

(5). Expand the group language by a unary predicate $U$ and take a formula from this language such that every model has power 1 or is a group with a nontrivial proper subgroup.

Another solution is to use two binary relation symbols, $R$ and $S$, together with a unary function symbol $f$. Set $\varphi_0 = \varphi_R \wedge \varphi_S \wedge \forall xy \, ((xRy \wedge xSy \to x = y) \wedge \exists z \, (xRz \wedge zSy))$, $\varphi_1 = \forall xy \, ((f(x) = f(y) \to x = y) \wedge (xRy \to f(x)Sf(y)))$, $\varphi_2 = \exists xy \, (x \neq y \wedge xRy) \wedge \exists xy \, (x \neq y \wedge xSy)$. Use the conjunction of these three formulas.

(6). Expand the ring language by a binary relation symbol $<$. Let $\varphi$ be the conjunction of the field axioms, the linear order axioms for $<$ and the formula $\forall x \, (x + 1 \neq 0 \to x < x + 1)$. Then the only finite models of $\varphi$ are the finite prime fields. For, if the finite field $\mathcal{K}$ is a model of $\varphi$ of characteristic $p$, but not a prime field, then there exists an element $a$ such that $0 \notin \{a, a + 1, \ldots, a + (p - 1)\}$. But then we have $a < a + 1 < \ldots < a + p = a$, contradicting the fact that $<$ is a linear order.

15.8.4 How many are there? Only countably many!

# Solutions for selected exercises

## Chapter 5

5.3.7    $a \leq b$ iff $\mathbb{R} \models \varphi(a,b)$, where $\varphi(x,y)$ is the formula $\exists z \, y = x + z^2$.

5.6.1    (The countable case.) Let $b_0, b_1, \ldots$ be an enumeration of $\mathcal{B}$. Given a filter $F_0$, we inductively define extending filters $F_{n+1}$ as follows. If $b_n \cdot a = 0$ for some $a \in F_n$, then set $F_{n+1} = F_n \cup \{\overline{b_n}\}$. Otherwise, set $F_{n+1} = F_n \cup \{b_n\}$. The union of the $F_n$ is the desired ultrafilter.

5.5.1    Orderings of power smaller than 2 could arise.

5.6.6    It is routine to check that a theory forms a filter in the Lindenbaum-Tarski algebra. All we need for the converse is the finiteness theorem in order to derive that every consequence of the filter is a consequence of finitely many elements of the filter. For then, by filter axiom (ii), every consequence of the filter is a consequence of a single element, hence, by filter axiom (iii), itself an element of the filter. Consequently, a filter is deductively closed. So, if it were contradictory, it would contain (the equivalence class of) a contradiction. But the class of any contradiction is $\perp / \sim$, i. e. the 0 of the Lindenbaum-Tarski algebra, which is contained in no filter, by axiom (i).

5.7.2    Without using that $S_L$ is the Stone space of $\mathcal{B}_L$, we know that its elements are of the form $\mathrm{Th}\mathcal{M}$. Then the consistency of a set of sentences, $\Sigma$, i. e. the containment of $\Sigma$ in some $\mathrm{Th}\mathcal{M}$, is equivalent to $\bigcap_{\varphi \in \Sigma} \langle \varphi \rangle = \emptyset$ in $S_L$, that is to $\bigcup_{\varphi \in \Sigma} \langle \neg\varphi \rangle = S_L$. Now finiteness is immediate from compactness.

## Chapter 7

7.3.1    See Cameron (1990).

## Chapter 9

9.2.8    Application of the theorem. Let $\exists x \psi(x, \bar{c})$ be a simply primitive formula with parameters $\bar{c}$ from a joint substructure $\mathcal{C}$ of two models of our theory. If some term equation $x = g(c_i)$ occurs positively in $\psi(x, \bar{c})$, where $c_i$ is an entry of $\bar{c}$, then only $g(c_i)$ satisfies this formula, which is already in $\mathcal{C}$, and we are done. If, on the other hand, all term equations occur negated in $\psi(x, \bar{c})$, then all this formula says is that $x$ is not in a certain finite subset of the substructure of $\mathcal{C}$ generated by $\bar{c}$. Then we find solutions of it in *any* infinite structure extending it, in particular in all such models of our theory.

9.4.3    $\mathrm{ACF}_p$ is not finitely axiomatizable.

9.5.1    By quantifier elimination $\exists \bar{x} \, \sigma(\bar{x}, \bar{y})$ is ACF-equivalent to a quantifier-free formula $\sigma^*(\bar{y})$, which is of the desired form by Lemma 5.3.3(1).

9.5.2    Write $\sigma(\bar{x}, \bar{y})$ as $\sigma'(\bar{x}, \bar{y}, \bar{a})$, where $\bar{a}$ contains all the parameters from $\mathcal{K}_0$, and then apply the previous exercise to $\sigma'(\bar{x}, \bar{y}, \bar{z})$.

9.5.3　Let $\sigma$ have a solution in some $\mathcal{K}' \supseteq \mathcal{K}$. We have to show it has one in every algebraically closed $\mathcal{K}^* \supseteq \mathcal{K}$. Write $\sigma$ as $\sigma(\bar{x}, \bar{c})$, where $\bar{c}$ is from $\mathcal{K}$. Choose $\sigma^*(\bar{y})$ as in Exercise 9.5.1. Choose further an algebraically closed field $\mathcal{K}'' \supseteq \mathcal{K}'$. As $\sigma$ has a solution in $\mathcal{K}' \supseteq \mathcal{K}$, it has one in $\mathcal{K}''$ too (for $\exists \bar{x}\, \sigma(\bar{x}, \bar{y})$ is an $\exists$-formula, cf. Corollary 6.2.6). Then $\sigma^*(\bar{c})$ holds in $\mathcal{K}''$, and, by the substructure completeness of ACF (9.4.2 and 9.2.2), we have $(\mathcal{K}^*, \mathcal{K}) \equiv (\mathcal{K}'', \mathcal{K})$. But $\bar{c}$ is in $\mathcal{K}$, hence $\mathcal{K}'' \models \sigma^*(\bar{c})$ implies $\mathcal{K}^* \models \sigma^*(\bar{c})$, and $\sigma$ has a solution in $\mathcal{K}^*$.

## Chapter 11

11.3.4　See any source in stability theory.

11.3.5　Assume, $\sigma$ and $\sigma'$ are such continuous sections of $\rho_{B,N}$ differing on a certain type $\Phi \in S_n^{\mathcal{M}}(N)$, i. e., $\sigma(\Phi) \neq \sigma'(\Phi)$. Then there is a formula $\varphi \in \sigma(\Phi)$ such that $\neg\varphi \in \sigma'(\Phi)$. By continuity, there must be $\psi, \psi' \in \Phi$ such that $\psi \in \Psi \in S_n^{\mathcal{M}}(N)$ implies $\varphi \in \sigma(\Psi)$, and $\psi' \in \Psi \in S_n^{\mathcal{M}}(N)$ implies $\neg\varphi \in \sigma'(\Psi)$. Then $\psi \wedge \psi'$ is consistent. Pick $\bar{a} \in \psi(\mathcal{N}) \cap \psi'(\mathcal{N})$. Then $\bar{x} = \bar{a} \in \mathrm{tp}(\bar{a}/N) \subseteq \sigma(\mathrm{tp}(\bar{a}/N))$ and $\bar{x} = \bar{a} \in \mathrm{tp}(\bar{a}/N) \subseteq \sigma'(\mathrm{tp}(\bar{a}/N))$, which yields the contradiction that $\bar{a}$ satisfies both, $\varphi$ and $\neg\varphi$ in $\mathcal{M}$.

11.3.6　Let $\varphi \in \sigma(\Phi)$. By continuity of $\sigma$, there must be a formula $\psi \in L_n(N)$ such that $\psi \in \Psi \in S_n^{\mathcal{M}}(N)$ implies $\varphi \in \sigma(\Psi)$. Pick $\bar{a} \in \Psi(N)$. Then $\psi \in \mathrm{tp}(\bar{a}/N)$, hence $\varphi \in \sigma(\mathrm{tp}(\bar{a}/N))$. But $\bar{x} = \bar{a} \in \mathrm{tp}(\bar{a}/N) \subseteq \sigma(\mathrm{tp}(\bar{a}/N))$, so the latter type contains the formula $\bar{x} = \bar{a} \wedge \varphi$. Hence this formula is consistent, i. e. satisfied in $\mathcal{M}$. But the only tuple satisfying it is $\bar{a}$, which is in $\mathcal{N}$.

11.3.7　For the first assertion, use Exercise 11.3.4 above and the fact that continuous maps are closed. The second assertion then follows, since, being a map, a section has a well-defined image.

11.4.1　Consider an algebraic $n$-type $\Phi$ over a subset $A$. Choose an algebraic formula $\varphi$ in the type. If this does not yet isolate $\Phi$, there is a formula $\psi$ in $L_n(A)$ such that both, $\varphi \wedge \psi$ and $\varphi \wedge \neg\psi$, are consistent with $\Phi$. Then both of them are algebraic with *less* solutions than $\varphi$. Replace $\varphi$ by one of them and continue with the same argument. Since this process decreases the number of solutions of the formulas in question, we must arrive, after finitely many steps, at a formula $\varphi'$ in $L_n(A)$ consistent with $\Phi$ and such that there is no formula $\psi'$ in $L_n(A)$ such that both, $\varphi' \wedge \psi'$ and $\varphi' \wedge \neg\psi'$, are consistent with $\Phi$. This means that $\varphi'$ isolates $\Phi$.

11.5.4　Let $\mathrm{acl}\, A \setminus A = \{a_i : i < \kappa\}$ and $\mathrm{acl}\, B \setminus B = \{b_j : j < \lambda\}$. Let further $f$ be an elementary bijection between $A$ and $B$.

We have the following inductive back-and-forth argument.

Starting with $a_{i_0} = a_0$, choose a formula $\eta(x, \bar{a})$ over $A$ which implies the type $p = tp(a_0/A)$ (this is possible by Exercise 11.4.1). Since $f$ is elementary, also $\eta(x, f(\bar{a}))$ is consistent and implies a complete type over $B$, namely $f(p)$ (just check, using the inverse map $f^{-1}$, that it must be an atom over $B$!). But for the same reason this formula is algebraic, too.

Hence it has a solution in $\operatorname{acl} B$. Again by the same argument, this solution cannot be in $B$ (in fact, each of them is in $B \smallsetminus \operatorname{acl} B$). Therefore there is some $j_0 < \lambda$ such that $b_{j_0}$ satisfies $\eta(x, f(\bar{a}))$, hence realizes $f(p)$. So $f$ extends to a bijection $f_1$ between $Aa_{i_0}$ and $Bb_{j_0}$. It is easily checked that, since we worked with principal types, this bijection is again elementary. Proceeding like this we can successively embed $\operatorname{acl} A$ into $\operatorname{acl} B$ by an elementary map. However, we want the map to be surjective. Therefore we switch sides after every step: Let $j_1$ be the smallest index $< \lambda$ such that $b_{j_1} \notin Bb_{j_0}$, and choose $a_{i_1}$ as before with $Bb_{j_0}$ playing the role of $A$ and $Aa_{i_0}$ playing that of $B$ (and $(f^{-1})$ playing that of $f$). We find a corresponding $a_{i_1}$ in $\operatorname{acl} A$, and so forth.

At limit stages we simply take unions of what we have constructed so far. The back-and-forth guarantees that we exhaust both $\operatorname{acl} A$ and $\operatorname{acl} B$ in this process. Consequently, at the limit stage $i = \kappa$, say (or $j = \lambda$, depending on which is apparently bigger) we have an elementary map from $\operatorname{acl} A$ to $\operatorname{acl} B$.

11.5.7 Being deductive closure, the one on $L_0$ is finitary. To see that the one on the classes of $L$-structures need not be, consider $\mathbf{K}$, the class of all finite sets (in $L_=$); then $\operatorname{Mod} \operatorname{Th} \mathbf{K}$ consists of *all* nonempty sets, however, given a finite subclass $\mathbf{K}_0$ of $\mathbf{K}$, the theory $\operatorname{Th} \mathbf{K}_0$ can have no infinite model.

# Chapter 14

14.1.7 [Podewski, K.-P. : Minimale Ringe, Math.-Physik. Semesterberichte **22** (1975) 193 – 197]

14.1.8 The dimension is 0, as every element is algebraic over the empty set. A bases of the model $\mathbb{Z}_{p^\infty} \oplus \mathbb{Q}^{(\beta)}$ are precisely the bases of the vector space $\mathbb{Q}^{(\beta)}$.

14.1.10 (1) v. d. Waerden, §37.

(4) If $b \in \mathcal{K}$ is a solution of $x^p - x - a$ and $c$ is another one, then $c^p - c = b^p - b$, hence, by the first part of the exercise, $(c - b)$ is a root of the polynomial $x^p - x$, hence (by the corresponding hint) in $\mathbb{F}_p \subseteq \mathcal{K}$, whence $c\mathcal{K}$.

14.1.11 (4) Let $b \in \mathcal{K}$ and let $c$ be a root of $x^n - b$ outside $\mathcal{K}$. Pick any root, $a$, of the latter in $\mathcal{K}$. Then $a^n = c^n$ implies that $a^{-1}c$ is an $n$th root of unity, hence in $\mathcal{K}$ (by what was said in the corresponding hint). Then so is $c$, contradiction.

14.1.12 (1) They are roots of $x^{p^n} - x$.

(3) If there were $a \in \operatorname{acl}_{\mathcal{K}} \emptyset \smallsetminus \mathcal{K}_0$, there would be a finite subset of $\mathcal{K}$ definable without parameters (in $\mathcal{K}$). As every $\varphi^n$ is an automorphism, this finite set would have to contain the infinitely many images $\varphi^n(a)$, a contradiction.

14.1.13 (2) By Exercise 14.1.11 (4), every root of unity is in $\mathcal{K}$, hence, by its definition, also in $\mathcal{K}_0$. So the assertion follows from the preceding part of the exercise.

(3) Being algebraically closed, $\mathcal{K}_0$ is infinite. Now it remains to see that

Lemma 14.1.1 applies also to elementary substructures of $\mathcal{K}$ (in which $x = x$ is minimal).

## Chapter 15

15.1.2   All subgroups of cyclic groups are cyclic. Every subgroup of a finite cyclic group is uniquely determined by its order. The map assigning the order to any subgroup of a given cyclic group of order $n$ is a bijection between the set of all subgroups of the given group and the set of all divisors of $n$. The automorphisms of a cyclic group of order $m$ are the maps of multiplication by a number $< m$ prime to $m$. Every automorphism of a subgroup of a cyclic group of prime power order can be extended to an automorphism of the latter. Every finite cyclic group is a direct sum of (finitely many) cyclic groups of pairwise prime prime power orders.

15.5.4   See Corollary 42.2 in Fuchs' book (Appendix F).

15.8.2   (Proposition 2.2.12 in *Finite Model Theory* by Ebbinghaus and Flum) Choose a (finite) signature $\sigma$ with $\varphi \in L(\sigma)_0$ and a constant symbol $c \in \mathbf{C}$. Let $\Phi$ be the set

$$\{\exists^{>n} x\, x = x \,:\, n \in \mathbf{N}\} \cup \{\forall \bar{x}\, R\bar{x} \,:\, R \in \mathbf{R}\} \cup$$
$$\{\forall \bar{x}\, f(\bar{x}) = c \,:\, f \in \mathbf{F}\} \cup \{d = d' \,:\, d, d' \in \mathbf{C}\}.$$

It is easy to see, that the deductive closure of $\Phi$ is (totally-categorical and hence) a complete theory. Therefore, $\Phi \models \varphi$ or $\Phi \models \neg\varphi$. The compactness theorem yields a finite subset $\Phi_0$ of $\Phi$ such that $\Phi_0 \models \varphi$ or $\Phi_0 \models \neg\varphi$. Hence, $\mathrm{Spec}(\varphi)$ or $\mathrm{Spec}(\neg\varphi)$ is cofinite.

# Bibliography and hints for further reading

A few hints to certain special topics were already given in the text, most of all in the last section. Some more general advice on further model-theoretic reading follows.

Those who intend to systematically study model theory should try and read Bruno Poizat's *Cours de théorie des modèles* (the White Album).[3] Those who intend to rather get an overview of the variety of model-theoretic methods (also for higher logics) should consult Wilfrid Hodges' comprehensive *Model Theory* (or just his *shorter model theory*). All three books are very well written—from a point of view of contemporary research—and make a most pleasant reading.

These and a few more texts suitable for further reading are listed in Section A. Most of them treat Morley's theorem, whose importance was pointed out in §8.5. Exceptions are Hodges' 'game-theoretic' text and his *Model Theory*, where the proof is part of the exercises (his *shorter model theory*, however, has a chapter giving a complete proof). The third edition of the classic by Chen Chung Chang and Jerome Keisler contains, among many other things, a section about nonstandard analysis, a theory developed by Abraham Robinson that raises Leibnitz' infinitesimals into the realm of exact mathematics.[4]

The monographs cited in Section B may be too demanding. But it should be mentioned at least that Anand Pillay's *Geometric Stability Theory* reflects much of the more recent direction in model theory and, in its introduction, gives a beautiful account of its formation and development, from Boris Zilber to Ehud Hrushovski.

The titles about particular classes of structures, cited in Section C, are for the most part—at least model-theoretically speaking—quite accessible. The only exception here is Poizat's (second) book (which, in spite of its adult group-theoretic contents, rather belongs into the circle of monographs mentioned in Section B).

The collections of papers listed in Section B may help the interested reader to get a fresher impression of current research. They contain articles of a wide variety (of both, topics and level of presentation). To name but one, Poizat's *An introduction to algebraically closed fields and varieties* (written in French!), from the volume edited by Ali Nesin and Anand Pillay (pp. 41–67), surveys interesting connections between model theory and algebraic geometry.

One almost unforgivable omission in this book is that we have not discussed *real* closed fields, a class of (ordered) fields in many ways complementary to and of the same historic significance as that of *algebraically* closed fields. The reason for this omission is that, because of their significance for stability theory, I preferred to include—as a natural and powerful generalization of algebraically closed fields— a presentation of strongly minimal structures. Interestingly enough, there is an analogous generalization in the case of ordered structures. Namely one considers

---

[3]Unfortunately Poizat's book, as well as his second one cited in Section C, is published privately and, to date, available only from the author (Université Lyon, France). An English translation, however, is in preparation.

[4]A chapter, written by Alexander Prestel, on nonstandard analysis (and a little bit of everything) is contained in the book on numbers cited in G).

BIBLIOGRAPHY

ordered structures (in a language containing among other things a binary relation symbol $<$) that are minimal with respect to the sets that are definable in its models. Since we have the ordering around, many more sets than just the finite and cofinite ones are now definable, for instance, all intervals are definable too, and hence all finite unions of these are. Regarding single points as (degenerate closed) intervals, one defines, more precisely, a theory containing the theory of linear orderings (in a language containing at least a binary relation symbol $<$) to be **o-minimal**,[5] if in all of its models the sets definable by 1-place formulas (with parameters) are unions of finitely many intervals. The theory of dense linear orderings is an example of an o-minimal theory, as is easily seen from quantifier elimination (Theorem 9.3.2). The most prominent example though is the theory of real closed fields. Analogously to the case of strongly minimal theories and algebraic geometry, there is an interesting connection between o-minimal theories and real algebraic geometry and analysis. Lou van den Dries' recent book cited in C is an excellent source to learn more about this (while real closed fields are dealt with in almost all of the model-theoretic literature).

For a briefer complementary reading some model-theoretic chapters of the logic texts listed in D might prove appropriate. On the other hand, these texts may be consulted about general logical topics. To mention but one, Manin's ambitious monograph contains a proof of the independence of the continuum hypothesis, as well as the solution of Hilbert's 10th problem about Diophantine equations, a proof of Gödel's incompleteness theorem, and a chapter related to word problems in groups.

Finally, a summary of the bulk of the classical part of mathematical logic can be found in

[**Handbook of Mathematical Logic**, ed. by J. Barwise, Studies in Logic and the Foundations of Mathematics 90, North-Holland, Amsterdam 1977, 1165 pp.]

(one third of which is dedicated to model theory).

The numbers in brackets at the end of some of the selected entries to follow refer to the corresponding pages in the text.

## A  Texts

Buechler, S. : **Essential Stability Theory**, Perspectives in Mathematical Logic, Springer, Berlin 1996, 385 pp.

Chang, C. C. and Keisler, H. J. : **Model Theory**, Studies in Logic and the Foundations of Mathematics 73, North-Holland, Amsterdam [3]1990, 650 pp. [159]

Hodges, W. : **Building Models by Games**, London Mathematical Society Student Texts 2, Cambridge University Press, Cambridge 1985, 311 pp. [281]

Hodges, W. : **Model Theory**, Encyclopedia of mathematics and its applications 42, Cambridge University Press, Cambridge 1993, 772 pp.[38, 44, 84, 90, 131, 159, 204, 265, 266, 267, 281]

---

[5]The letter o stands for 'ordered.'

Hodges, W. : **a shorter model theory**, Cambridge University Press, Cambridge 1997, 310 pp. [124, 204, 281]

Lascar, D. : **Stability in model theory**, Pitman Monographs and Surveys 36, Longman Science & Technical, Harlow 1987, 193 pp. (Translated from the French.)

Pillay, A. : **An introduction to stability theory**, Oxford Logic Guides 8, Clarendon Press, Oxford 1983, 146 pp.

Poizat, B. : **Cours de théorie des modèles**, Nur al-Mantiq wal-Ma'rifah n°1, Villeurbanne 1985, 584 pp. (English translation **A Course in Model Theory**, Springer, to appear 2000.) [281]

Sacks, G. E. : **Saturated Model Theory**, Benjamin, Reading 1972, 335 pp. [195]

# B   Newer monographs and collections of papers

Baldwin, J. T. : **Fundamentals of Stability Theory**, Perspectives in Mathematical Logic, Springer, N. Y. 1988, 447 pp. [266]

Droste, M. and Göbel, R. (eds.) : **Advances in Algebra and Model Theory**, Selected surveys presented at conferences in Essen 1994 and Dresden 1995, Algebra, Logic and Applications 9, Gordon & Breach, N. Y. 1997, 499 pp.

Hart, B. T., Lachlan, A. H., and Valeriote, M. A. (eds.) : **Algebraic Model Theory**, Proceedings of the NATO Advanced Study Institute in Toronto 1996, NATO ASI Series C: Mathematical and Physical Sciences 496, Kluwer Academic Publishers, Dordrecht 1997. 277 pp.

Haskell, D., Pillay, A., and Steinhorn, C. I. (eds.) : **Model Theory, Algebra and Geometry**, Proceedings of the Model Theory Semester at MSRI, Berkeley 1998, Cambridge University Press, Cambridge, to appear 2000.

Nesin, A. H. and Pillay, A. (eds.) : **The model theory of groups**, Notre Dame Mathematical Lectures 11, Univ. of Notre Dame Press, Notre Dame 1989, 209 pp. [281]

Pillay, A. : **Geometric Stability Theory**, Oxford Logic Guides 32, Clarendon Press, Oxford 1996, 361 pp. [235, 237, 281]

Shelah, S. : **Classification Theory (and the Number of Non-Isomorphic Models)**, Studies in Logic and the Foundations of Mathematics 92, North-Holland, Amsterdam [2]1990, 705 pp. [266]

Zilber, B. I. : **Uncountably categorical theories**, Translations of mathematical monographs 117, Providence, RI: American Mathematical Society, 1993, 122 pp. (Translated from the Russian.)

## C   Model theory of particular structures

Borovik, A. and Nesin, A. : **Groups of Finite Morley Rank**, Oxford Logic Guides 26, Clarendon Press, Oxford 1994, 409 pp.

Cherlin, C. : **Model-Theoretic Algebra**, Lecture Notes in Mathematics 521, Springer, Berlin 1976, 232 pp. (Valued fields, rings, pure-injective abelian groups and modules, $\aleph_1$-categorical fields.)

van den Dries, L. : **Tame Topology and O-minimal Structures**, London Mathematical Society Lecture Notes Series 248, Cambridge University Press, Cambridge 1998, 180 pp. [282]

Ebbinghaus, H.-D. and Flum, J. : **Finite Model Theory**, Perspectives in Mathematical Logic, Springer, Berlin $^2$1999, 360 pp. [267]

Ершов, Ю. Л. ( Ershov, Y. L. [read: 'Yershov']) : Проблемы разрешимости и конструктивные модели (In Russian, **Decidability problems and constructive models**), Nauka, Moskva 1980, 416 pp. (Distributive lattices, algebraically compact abelian groups, valued fields.)

Hurd, A. e. and Loeb, P. A. : **An Introduction to Nonstandard Real Analysis**, Pure and Applied Mathematics, Academic Press, Orlando 1985, 232 pp.

Jensen, C. U. and Lenzing, H. : **Model theoretic algebra (With particular emphasis on fields, rings and modules)**, Algebra, Logic and Applications 2, Gordon & Breach, N. Y. 1989, 443 pp.

Kaye, R. : **Models of Peano Arithmetic**, Oxford Logic Guides 15, Clarendon Press, Oxford 1991, 292 pp.

Poizat, B. : **Groupes stables**, Nur al-Mantiq wal-Ma'rifah n°2, Villeurbanne 1987, 215 pp. [281]

Poizat, B. : **Les petits cailloux (une approche modèle-théoretique de l'Algorithmie)**, Nur al-Mantiq wal-Ma'rifah n°3, Aléas Editeur, Lyon 1995, 217 pp.

Prest, M. : **Model Theory and Modules**, London Mathematical Society Lecture Notes Series 130, Cambridge University Press, Cambridge 1988, 380 pp. [266]

Пунинский, Г. е. и Туганбаев, А. А. (Puninski, G. e. and Tuganbaev A. A.): Кольца и модули (In Russian, **Rings and modules**),[6] Soyuz, Moskva 1998, 420 pp. (One model-theoretic chapter on modules over serial rings.)

Rosenstein, J. G. : **Linear Orderings**, Academic Press, N. Y. 1982, 487 pp. [104]

Wagner, F. : **Stable Groups**, London Mathematical Society Lecture Notes Series 240, Cambridge University Press, Cambridge 1997, 309 pp.

---

[6]English translation of one half of the book—entitled **Serial rings** written by Puninski—in preparation.

# D Logic

Barnes, D. W. and Mack, J. M. : **An Algebraic Introduction to Mathematical Logic**, Graduate Texts in Mathematics 22, Springer, N. Y. 1975, 121 pp.

Bell, J. L. and Machover, M. : **A course in mathematical logic**, North-Holland $^3$1993, 599 pp.

Bridge, J. : **Beginning model theory : the completeness theorem and some consequences**, Oxford Logic Guides 1, Clarendon Press, Oxford 1977, 143 pp.

Cori, R. and Lascar, D. : **Logique mathématique I, II**, Masson, Paris $^2$1994, 385+347 pp. (English translation, Oxford University Press, to appear 2000.)

Ebbinghaus, H.-D., Flum, J., and Thomas, W. : **Mathematical Logic**, Undergraduate Texts in Mathematics, Springer, N. Y. 1984, 216 pp. (Translated from the German.)

Enderton, H. B. : **A mathematical introduction to logic**, Academic Press, N. Y. $^{14}$1997, 295 pp.

Goldstern, M. and Judah, H. : **The Incompleteness Phenomenon**, A New Course in Mathematical Logic, A K Peters, Natick Massachusetts 1998, 247 pp.

Keisler, H. J. et al : **Mathematical Logic and Computability**, 1998, 484 pp.

Machover, M. : **Set theory, logic and their limitations**, Cambridge University Press, Cambridge 1996, 288 pp.

Manin, Yu. I. : **A Course in Mathematical Logic**, Graduate Texts in Mathematics 53, Springer, N. Y. 1977, 286 pp. [282]

Monk, J. D. : **Mathematical Logic**, Graduate Texts in Mathematics 37, Springer, N. Y. 1976, 531 pp.

# E Set theory and topology

Ciesielski, K. : **Set Theory for the Working Mathematician**, London Mathematical Society Student Texts 39, Cambridge University Press, Cambridge 1997, 236 pp.

Devlin, K. J. : **The joy of sets: fundamentals of contemporary set theory**, Undergraduate Texts in Mathematics, Springer, N. Y. $^2$1993, 192 pp.

Enderton, H. B. : **Elements of set theory**, Academic Press, N. Y. 1996, 279 pp.

Halmos, P. R. : **Naive set theory**, Springer, N. Y. 1974, 104 pp.

Jänich, K. : **Topology**, Undergraduate Texts in Mathematics, Springer, N. Y. 1984, 192 pp. (Translated from the German.)

Just, W. and Weese, M. : **Discovering modern set theory**, Graduate studies in mathematics, American Mathematical Society, Providence 1996.

Levy, A. : **Basic Set Theory**, Perspectives in Mathematical Logic, Springer, Berlin 1979, 391 pp.

Machover, M. : **Set theory, logic and their limitations**, Cambridge University Press, Cambridge 1996, 288 pp.

## F   Algebra

Cohn, P. M. : **Algebra**, Vol. 2, Wiley, Chichester 1977.

Fuchs, L. : **Infinite Abelian Groups I**, Academic Press, N. Y. 1970. [265]

Halmos, P. R. : **Lectures on Boolean Algebras**, Springer, N. Y. 1974, 147 pp.

Kaplansky, I. : **Infinite Abelian Groups**, University of Michigan Press, Ann Arbor $^2$1956. [265]

Kargapolov, M. I. and Merzljakov, Ju. I. : **Fundamentals of the Theory of Groups**, Graduate Texts in Mathematics 62, Springer, N. Y. 1979. (Translated from the Russian.) [84]

Sikorski, R. : **Boolean Algebras**, Ergebnisse der Mathematik und ihrer Grenzgebiete 25, Springer, Berlin $^3$1969, 237 pp.

van der Waerden, B. L. : **Algebra**: based in part on lectures by E. Artin and E. Noether, Springer, N. Y. 1991. (Translated from the German.) [274, 279]

## G   A little bit of everything

Ebbinghaus, H. D., Hermes, H., Hirzebruch, F., Koecher, M., Mainzer, K., Neukirch, J., Prestel, A., and Remmert, R. : **Numbers**, Graduate Texts in Mathematics 123, Springer, N. Y. 1990. (Translated from the German.) [281]

## H   Further literature cited in the text

Ahlbrandt, G. and Ziegler, M. : Quasi-finitely axiomatizable totally categorical theories, Annals of Pure and Applied Logic **30** (1986) 63 – 82. [234]

Bachmann, H. : **Transfinite Zahlen**, Ergebnisse der Mathematik und ihrer Grenzgebiete 1, Springer, Berlin 1955, 204 pp. [101]

Baldwin, J. T., Blass, A. R., Glass, A. M. W., and Kueker, D. W. : A 'natural' theory without a prime model, algebra universalis **3** (1973) 152 – 155. [246]

Baldwin, J. T. and Lachlan, A. H. : On strongly minimal sets, J. Symbolic Logic **36** (1971) 79 – 96. [229]

J. Barwise, J. (ed.) : **Handbook of Mathematical Logic**, Studies in Logic and the Foundations of Mathematics 90, North-Holland, Amsterdam 1977, 1165 pp. [282]

Baur, W. : Elimination of quantifiers for modules, Israel J. Mathematics **25** (1976) 64 – 70. [265]

Börger, E. : **Berechenbarkeit, Komplexität, Logik**, Vieweg, Braunschweig 1985. [267]

Cameron, P. J. : **Oligomorphic Permutation Groups**, London Mathematical Society Lecture Notes Series 152, Cambridge University Press, Cambridge 1990, 160 pp. [274, 277]

Cantor, G. : Beiträge zur Begründung der transfiniten Mengenlehre I, Mathematische Annalen **46** (1895) 481 – 512. [89]

Cantor, G. : Beiträge zur Begründung der transfiniten Mengenlehre II, Mathematische Annalen **49** (1897) 207 – 246. [101]

Chang, C. C. : On unions of chains of models, Proceedings Am. Math. Soc. **10** (1959) 120 – 127. [155]

Engeler, E. : A characterization of theories with isomorphic denumerable models, Notices Am. Math. Soc. **6** (1959) 161. [201]

Fagin, R. : Finite-model theory – a personal perspective, Theoretical Computer Science **116** (1993) 3 – 31. [267]

Fuhrken, G. : Bemerkung zu einer Arbeit E. Engelers, Zeitschrift f. Math. Logik u. Grundlagen d. Mathematik **8** (1962) 277–79. [273]

Gödel, K. : Die Vollständigkeit der Axiome des logischen Funktionenkalküls, Monatshefte f. Math. u. Phys. **37** (1930) 349 – 360. [46]

Gordon, K. E. : **The Well-Tempered Sentence**, Ticknor & Fields, N. Y. 1983, 96 pp. [287]

Grzegorczyk, A., Mostowski, A., and Ryll-Nardzewski, C. : Definability of sets in models of axiomatic theories, Bulletin Acad. Polon. Sci. Sér. Sci. Math. Astronom. Phys. **9** (1961) 163 – 167. [196]

Henkin, L. : Some interconnections between modern algebra and mathematical logic, Transactions of the American Mathematical Society **74** (1953) 410 – 427. [78]

Hrushovski, E. : Totally categorical structures, Transactions of the American Mathematical Society **313** 131 – 159. [235]

Huntington, E. V. : The continuum as a type of order: an exposition of the modern theory, Annals of Math., **6** (1904/05) §45. [90]

Keisler, H. J. : Some applications of infinitely long formulas, J. Symbolic Logic **30** (1965) 339 – 349. [159]

Laskowski, M. C. : Uncountable theories categorical in a higher power, J. Symbolic Logic **53** (1988) 512 – 530. [230]

Laskowski, M. C., Pillay, A., and Rothmaler, Ph. : Tiny models of categorical theories, Archive for Math. Logic **31** (1992) 385 – 396. [237]

Löwenheim, L. : Über Möglichkeiten im Relativkalkül, Mathematische Annalen **76** (1915) 447 – 470. [49, 120]

Łoś, J. : On the categoricity in power of elementary deductive systems and some related problems, Colloqium Math. **3** (1954) 58 – 62. [123]

Łoś, J. : Quelques remarques, théorèmes et problèmes sur les classes définissables d'algèbres, in **Mathematical interpretations of formal systems** (L. e. J. Brouwer et al., ed.), North-Holland, Amsterdam, 1955, pp. 98 – 113. [44]

Łoś, J. : On extending of models I, Fundamenta Math. **42** (1955) 38 – 54. [73]

Łoś, J. and Suszko, R. : On the extending of models IV: Infinite sums of models, Fundamenta Math. **44** (1957) 52 – 60. [154]

Lyndon, R. C. : Properties preserved under homomorphism, Pacific J. Math. **9** (1959) 143 – 154. [157]

Macintyre, A. : On $\omega_1$-categorical theories of fields, Fundamenta Math. **71** (1971) 1 – 25. [212]

Makowsky, J. A. : Kategorizität und endliche Axiomatisierbarkeit, Diplomarbeit, ETH Zürich, 1971, mimeographed. [234]

Makowsky, J. A. : On some conjectures connected with complete sentences, Fundamenta Math. **81** (1974) 193 – 202. [234]

Malcev, A. I. : Untersuchungen aus dem Gebiet der mathematischen Logik, Rec. Math. N.S. (Mat. Sbornik) **1** (1936) 323 – 336, (In German, English translation in Malcev (1971).) [46]

Мальцев, А. И. (Malcev, A. I.): Об одном общем методе получения локальных теорем теории групп, Uchenye Zapiski Ivanov. Ped. Inst. 1, **1** (1941) 3 – 9. (English translation as *A general method for obtaining local theorems in group theory* in Malcev (1971) 15–21.) [67]

Malcev, A. I. : **The metamathematics of algebraic systems**, Collected Papers: 1936 - 1967, Studies in logic and the foundations of mathematics 66, North-Holland, Amsterdam 1971, 494 pp. [46, 67]

Malcev, A. I. : **Algebraic systems**, Grundlehren der mathematischen Wissenschaften 192, Springer, Berlin 1973, 317 pp. (Translated from the Russian.) [7]

Marsh, W. E. : On $\omega_1$-categorical and not $\omega$-categorical theories, Dissertation, Dartmouth College 1966, unpublished. [207]

Morley, M. : Categoricity in power, Transactions of the American Mathematical Society **114** (1965) 514 – 538. [123]

Morley, M. and Vaught, R. L. : Homogeneous universal models, Math. Scandinavica **11** (1962) 37 – 57. [186]

D. Mumford, **Algebraic Geometry I (Complex Projective Varieties)**, Grundlehren der Mathematischen Wissenschaften 221, Springer, Berlin 1976. [145]

Nadel, M. and Stavi, J. : On models of the elementary theory of $(\mathbb{Z}, +, 1)$, J. Symbolic Logic **55** (1990) 1 – 20. [264]

v. Neumann, J. : Zur Einführung der transfiniten Zahlen, Acta Litt. Scient. Univ. Szeged., Sectio scient. math. **1** (1923) 199 – 208. [101]

v. Neumann, J. : Die Axiomatisierung der Mengenlehre, Math. Zeitschrift **27** (1928) 669 – 752. [108]

Podewski, K.-P. : Minimale Ringe, Math.-Physik. Semesterberichte **22** (1975) 193 – 197. [279]

Прутков, К. (Prutkov, K.) : Полное собрание сочинений, St. Peterburg ²1885 [xi]

Reineke, J. : Minimale Gruppen, Z. Math. Logik Grundlagen Math. **21** (1975) 357 – 359. [213]

Robinson, A. : **On the metamathematics of algebra**, Studies in Logic and the Foundations of Mathematics, North-Holland, Amsterdam, 1951. [55]

Robinson, A. : **Complete Theories**, Studies in Logic and the Foundations of Mathematics, North-Holland, Amsterdam, 1956. [125, 146]

Robinson, A. : **Introduction to Model Theory and to the Metamathematics of Algebra**, Studies in Logic and the Foundations of Mathematics 66, North-Holland, Amsterdam ²1965, 284 pp. [142]

Ryll-Nardzewski, C. : On the categoricity in power $\aleph_0$, Bulletin Acad. Polon. Sci. Sér. Sci. Math. Astronom. Phys. **7** (1959) 545 – 548. [201]

Scholz, H. : Ein ungelöstes Problem der symbolischen Logik, J. Symbolic Logic **17** (1952) 160. [267]

Skolem, Th. : Logisch-kombinatorische Untersuchungen über die Erfüllbarkeit oder Beweisbarkeit mathematischer Sätze nebst einem Theorem über dichte Mengen, Skrifter, Videnskabsakademie i Kristiania I. Mat.-Nat. Kl. no. 4 (1920) 1 – 36. [49, 120]

Skolem, Th. : Einige Bemerkungen zur axiomatischen Begründung der Mengenlehre, in **Proc. 5th Scand. Math. Congress**, Helsinki 1922, 217 – 232. [120]

Skolem, Th. : Über die Nichtcharakterisierbarkeit der Zahlenreihe mittels endlich oder abzählbar unendlich vieler Aussagen mit ausschließlich Zahlenvariablen, Fundamenta Math. **23** (1934) 150 – 161. [50]

Skolem, Th. : **Selected works in logic**, p. 366, Universitetsforlaget, Oslo, 1970, Bem. 3. [50]

Steinitz, E. : Algebraische Theorie der Körper, J. f. Reine u. Angewandte Math. **137** (1910) 167 – 309, Reprint, R. Baer, H. Hesse (eds.), Chelsea, N. Y. 1950, 176 pp. [225]

Stone, M. H. : The representation theorem for boolean algebra, Transactions of the American Mathematical Society **40** (1936) 37 – 111. [62]

Svenonius, L.: $\aleph_0$-categoricity in first-order predicate calculus, Theoria (Lund) **25** (1959) 82 – 94. [201]

Szmielew, W. : Elementary properties of Abelian groups, Fundamenta Math. **41** (1955) 203 – 271. [265]

Szpilrajn, E. : Sur l'extension de l'ordre partiel, Fundamenta Math. **16** (1930) 386 – 389. [88]

Tarski, A. : Über einige fundamentale Begriffe der Mathematik, Comptes Rendus Séances Soc. Sci. Lettres Varsovie Cl. III **23** (1930) 22 – 29. [29, 36]

Tarski, A. : Une contribution à la théorie de la mesure, Fundamenta Math. **15** (1930) 42 – 50. [46]

Tarski, A. : Der Wahrheitsbegriff in den formalisierten Sprachen, Studia Philosoph. **1** (1935) 261 – 405. [21]

Tarski, A. : Contributions to the theory of models I, II, Koninkl. Ned. Akad. Wetensch. Proc. Ser. A **57** (1954) 572 – 588. [73]

Tarski, A. : **Collected Papers**, eds. S. R. Givant, R. N. McKenzie, Birkhäuser, Basel 1986 (Vol. 1: 1921-34, 659 pp., Vol. 2: 1935-44, 699 pp., Vol. 3: 1945-57, 682 pp., Vol. 4: 1958-79, 757 pp.)

Tarski, A. and Vaught, R. : Arithmetical extensions of relational systems, Compositio Math. **13** (1957) 81 – 102. [115, 151]

P. Terentius Afer (Terence), **Drei Komödien**, Reclam, Leipzig 1973. (Translated from Latin.) [xiii]

Vaught, R. : Applications of the Löwenheim-Skolem-Tarski theorem to problems of completeness and decidability, Koninkl. Ned. Akad. Wettensch. Proc. Ser. A **57** (1954) 467 – 472. [123]

Vaught, R. : Denumerable models of complete theories, **Infinitistic Methods Pergamon**, London and Panstwowe Wydawnictwo Naukowe, Warszawa 1961, 303 – 321. [190, 192, 196]

Wagner, F. : Minimal fields, J. Symbolic Logic, to appear. [212]

Ziegler, M. : Model theory of modules, Ann. Pure and Applied Logic **26** (1984) 149 – 213. [266]

Зильбер, Б. И. (Zilber, B. I.) : О решении проблемы конечной аксиоматизируемости для теорий, категоричных во всех бесконечных мощностях, in Исследования по теоретическому программированию, Kazakh. Gos. University, Alma-Ata 1981, pp. 69 – 74. (English translation as *On the solution of the problem of finite axiomatizability for theories categorical in all infinite powers* in American Mathematical Society Translations (2) **135** (1987) 13 – 17.) [235]

# Symbols

## Generalities, sets, filters, etc.

## Structures and operators thereon

## Particular structures

## Signatures and languages

## Relations between structures

## Types

# Index

**Other titles in the Algebra, Logic and Applications series**

Printed in the United States
by Baker & Taylor Publisher Services